Lecture Notes in Economics and Mathematical Systems

D0066623

Founding Editors:

M. Beckmann
H. P. Künzi

Editorial Board:

H. Albach, M. Beckmann, P. Dhrymes, G. Feichtinger, W. Hildenbrand
W. Krelle, H. P. Künzi, K. Ritter, U. Schittko, R. Selten

Managing Editors:

Prof. Dr. G. Fandel
Fachbereich Wirtschaftswissenschaften
Fernuniversität Hagen
Feithstr. 140/AVZ II, W-5800 Hagen 1, FRG

Prof. Dr. W. Trockel
Institut für Mathematische Wirtschaftsforschung (IMW)
Universität Bielefeld
Universitätsstr. 25, W-4800 Bielefeld 1, FRG

3) In Table 4.9 (Page 262), p_{PF} should be:

"about \$92" = (92, 91, 94) in State 1; and

"about \$90" = (90, 89.5, 91.5) in state 2.

4) In Page 263, we should have:

$E_1 = \ldots + (0, -.2173, .0109)S_F - \ldots$ (Line 16); and

$E_2 = \ldots + (.0217, .0054, .0272)S_F \ldots$ (Line 18).

5) In the equation of Pages 264 and 265, we should have:

$R_1 - 0.2D_1 - 0.025S_F \geq 0$ (Line 4); $\qquad R_2 - 0.2D_2 - 0.025S_F \geq 0$ (Line 5);

$D_1 \leq 1010$ (Line 12); $\qquad\qquad D_2 \leq 1130$ (Line 12);

$R_1 + STS_1 + LTS_1 + L - D_1 - E_1 = 0$ (Line 20);

$R_2 + STS_2 + LTS_2 + L - D_2 - E_2 = 0$ (Line 21);

$E_1^L \ldots + 0.01086S_F \ldots$ (Line 24); $\qquad E_2^L \ldots - 0.13586S_F \ldots$ (Line 26);

$E_1^U \ldots - 0.005434S_F \ldots$ (Line 28); $\qquad E_2^U \ldots - 0.024456S_F \ldots$ (Line 30); and

$E^U = 0.5E_1^U + 0.5E_2^U$ (Line 32).

6) The corrected solution tables, Table 4.10 in Page 264 and Table 4.11 in Page 266, are (Figure 4.20 should also be redrawn):

Table 4.10   Solutions of the banking hedging decision problem (α = 0.7)

β	PIS		NIS		Compromise solution				Γ
					E_1		E_2		
	E^{L*}	E^{U*}	E^L	$E^{U.}$	E_1^L	E_1^U	E_2^L	E_2^U	
1.0	117.42	117.42	95.89	95.89	117.329	117.329	120.772	120.772	1.000
0.9	116.44	117.95	95.20	96.31	116.384	117.822	120.020	121.239	1.000
0.8	115.54	118.48	94.50	96.73	115.440	118.315	119.271	121.706	1.000
0.7	115.17	119.01	93.81	97.15	114.495	118.808	118.524	122.172	1.000
0.6	114.80	119.54	93.12	97.58	113.554	119.300	117.779	122.637	1.000
0.5	114.43	120.06	92.43	98.00	112.613	119.792	117.036	123.101	1.000
0.4	114.08	120.64	91.59	98.42	112.513	119.579	115.590	121.645	0.999
0.3	113.74	121.23	90.46	98.84	112.175	120.105	114.608	121.684	0.985
0.2	113.40	121.00	89.22	99.26	112.187	120.139	113.649	120.994	0.980
0.1	113.05	122.43	88.03	99.69	110.829	120.817	113.381	122.326	0.962
0.0	112.71	122.96	86.84	100.05	110.203	121.151	112.781	122.612	0.953

Table 4.11   The optimal balance sheet and decisions (α = 0.7 and β = 0.5)

The corrected ending values of State 1 and State 2 are:

State 1: R=208.56, STS=0.00, LTS=147.23, L=774.00, D=1010.00, E=119.79, P_{STS}=0.00, P_{LTS}=0.00, P_L=424.00, S_F= 62.50, S_{STS}=50.00, S_{LTS}=115.27

State 2: R=232.56, STS=0.00, LTS=246.54, L=774.00, D=1130.00, E=123.10, P_{STS}=0.00, P_{LTS}=0.00, P_L=424.00, S_F=262.50, S_{STS}=50.00, S_{LTS}=15.96

Errata

Lecture Notes in Economics and Mathematical Systems, Vol. 394
Young-Jou Lai, Ching-Lai Hwang, Fuzzy Mathematical Programming
ISBN 3-540-56098-X

© Springer-Verlag Berlin Heidelberg 1992

CORRECTIONS

1) The equation in Page 253 should include the following modifications:

$$R - 0.2D - 0.025S_F \geq 0 \quad \text{(Line 4)};$$

$$LTS - P_{LTS} + S_{LTS} = 262.5 \quad \text{(Line 8); and}$$

$$R + STS + LTS + L - D - E = 0 \quad \text{(Line 10)}.$$

2) The corrected solution tables, Table 4.7 in Page 253 and Table 4.8 in Page 254, are (Figure 4.18 should also be redrawn):

Table 4.7 Solutions of the bank hedging decision problem ($\alpha = 0.7$)

β	PIS z^{L*}	z^{U*}	NIS z^{L-}	z^{U-}	Compromise solution z^L	z^U	Γ
1.0	117.43	117.43	99.62	99.62	117.43	117.43	1.000
0.9	116.89	117.86	98.51	99.89	116.77	117.74	0.993
0.8	116.49	118.42	97.41	100.16	116.13	118.07	0.981
0.7	116.09	118.99	96.31	100.43	115.48	118.41	0.969
0.6	115.68	119.56	95.20	100.71	114.81	118.76	0.958
0.5	115.27	120.13	94.10	100.98	114.15	119.11	0.947
0.4	114.87	120.70	93.00	101.25	113.48	119.46	0.936
0.3	114.46	121.27	91.90	101.52	112.80	119.81	0.926
0.2	114.05	121.84	90.79	101.79	112.12	120.18	0.917
0.1	113.64	122.41	89.69	102.07	111.43	120.54	0.908
0.0	113.24	122.99	88.59	102.34	110.75	120.90	0.899

Table 4.8 The optimal balance sheet and decisions ($\alpha = 0.7$)

β	1.0	0.9	0.8	0.7	0.6	0.5	0.4	0.3	0.2	0.1	0.0
Balance sheet											
R	197.20	198.48	198.48	198.53	198.60	198.64	198.67	198.71	198.75	198.79	198.82
STS	0.00	0.00	0.00	0.00	0.00	0.00	0.00	0.00	0.00	0.00	0.00
LTS	132.23	131.26	131.59	131.88	132.16	132.47	132.79	133.10	133.43	133.75	134.08
L	774.00	774.00	774.00	774.00	774.00	774.00	774.00	774.00	774.00	774.00	774.00
D	986.00	986.00	986.00	986.00	986.00	986.00	986.00	986.00	986.00	986.00	986.00
E	117.43	117.74	118.07	118.41	118.76	119.11	119.46	119.82	120.18	120.54	120.90
Decision											
1st stage											
P_{STS}	0.00	0.00	0.00	0.00	0.00	0.00	0.00	0.00	0.00	0.00	0.00
P_{LTS}	132.23	131.26	131.59	131.88	132.17	132.47	132.79	133.10	133.43	133.75	134.08
P_L	424.00	424.00	424.00	424.40	424.00	424.00	424.00	424.00	424.00	424.00	424.00
S_F	0.00	51.28	51.31	53.31	55.81	57.76	58.96	60.56	62.03	63.40	64.70
2nd stage											
S_{STS}	50.00	50.00	50.00	50.00	50.00	50.00	50.00	50.00	50.00	50.00	50.00
S_{LTS}	262.50	262.50	262.50	262.50	262.50	262.50	262.50	262.50	262.50	262.50	262.50

Young-Jou Lai   Ching-Lai Hwang

Fuzzy Mathematical Programming

Methods and Applications

Springer-Verlag

Berlin Heidelberg New York
London Paris Tokyo
Hong Kong Barcelona
Budapest

Authors

Prof. Young-Jou Lai
Prof. Ching-Lai Hwang
Department of Industrial Engineering
Kansas State University
Manhattan, KS 66506, USA

ISBN 3-540-56098-X Springer-Verlag Berlin Heidelberg New York
ISBN 0-387-56098-X Springer-Verlag New York Berlin Heidelberg

Typesetting: Camera ready by author/editor
42/3140-543210 - Printed on acid-free paper

PREFACE

This monograph is intended for an advanced undergraduate or graduate course as well as for researchers who want a compilation of developments in the rapidly growing fields of operations research, management science, economics, banking/finance, engineering and computer science. This is a sequel of our previous works entitled *Multiple Objective Decision Making - Methods and Applications*, *Multiple Attribute Decision Making - Methods and Applications*, *Group Decision Making Under Multiple Criteria - Methods and Applications*, and *Fuzzy Multiple Attribute Decision Making - Methods and Applications* (respectively, Nos. 164, 186, 281 and 375 of the Lecture Notes).

The literature on methods and applications of fuzzy mathematical programming (FMP) has been systematically and thoroughly reviewed and classified. This state-of-the-art survey provides readers with a capsule look into the existing methods, and their characteristics and applicability to analysis of FMP problems.

In the last 25 years, the fuzzy set theory has been applied in many disciplines such as operations research, management science, control theory, artificial intelligence/expert system, etc. The growth of applications of the fuzzy set theory has been accumulating. This study presents systematically the state-of-the-art of fuzzy mathematical programming in both techniques and applications. In addition to fuzzy mathematical programming, we also present all existing possibilistic mathematical programming problems, methods and applications, which are among the latest important topics.

We will present our material in two volumes: the current volume covers single-objective fuzzy and possibilistic mathematical programming; and the second includes fuzzy multiple objective decision-making and possibilistic multiple objective decision-making problems.

To realize practical fuzzy modelling, we present important real-world problems including production/manufacturing, transportation, assignment, game, environmental management, water resources management, nutrition management, personnel management, accounting, banking/finance, marketing management, trade balance, and agricultural economics. It should be noted that all concepts and methods are developed for solving practical problems.

For this volume, we first systematically and thoroughly define and classify all existing fuzzy and possibilistic (single-objective) mathematical programming problems and solution techniques in Chapter 1.

In Chapter 2, basic fuzzy set theories, membership functions, fuzzy decisions, operators and fuzzy arithmetic are introduced. We present them with simple numerical examples in an easy-to-read and easy-to-follow manner.

Basic concepts, computational procedures and characteristics of all existing fuzzy mathematical programming are concisely presented in Chapter 3. The computational procedure of each approach is illustrated by solving a practical problem or a simple numerical example. Six problems and eleven approaches of fuzzy linear programming are discussed. Fuzzy non-linear programming and fuzzy integer programming are also discussed in this chapter. We also present our expert decision-making support system IFLP which improves flexibility and robustness of linear programming techniques by considering and solving all possibilities of a specific domain of linear programming problems.

The practical applications discussed in this chapter include production/manufacturing, transportation, environmental management and agriculture economics problems under fuzzy environments.

Since the 1980s, the possibility theory has become more and more important in the fields of operations research, management science, expert systems, etc. In Chapter 4, we concisely present primary concepts, computational procedures and characteristics of all existing possibilistic mathematical programming. The computational procedure of each approach is illustrated by solving a practical problem or a simple numerical example. Seven problems and fourteen approaches of possibilistic linear programming are discussed.

The practical applications discussed in this chapter include project investment, game, banking and finance problems under imprecise environments.

In Chapter 5, brief discussions of probability theory versus fuzzy set theory, and stochastic versus fuzzy mathematical programming are given. Concluding remarks and future studies are also presented.

An updated bibliographical listing of 62 books, monographs or conference proceedings, and about 300 selected papers, reports or theses is presented in the end of this study.

We would like to thank all the researchers in this field, especially all scholars listed in Bibliography. The scope of this study could not have been completed without many excellent researchers carrying out and publishing their research. We would like to especially thank Professor L. A. Zadeh for his creation of fuzzy set theory; Professors H.-J. Zimmermann, C.

V. Negoita, H. Tanaka, K. Asai, S. Lee and P.L. Yu for their momentous contributions in operations research under fuzziness; Professors D. Dubois and H. Prade for their material contributions in fuzzy set theory and possibility theory; and Professors A. Kaufmann and M. M. Gupta for their important contributions in fuzzy arithmetic.

Finally, we thank our wives for their patience while we devoted the many hours it takes to see a project of this nature to completion. We also thank Ms. Jochun Wu and Ms. M. Rankin for editing this monograph.

This study was partly supported by HTX International, Inc.

Y. J. Lai
Kansas State University
Manhattan, Kansas

C. L. Hwang
Kansas State University
Manhattan, Kansas

TABLE OF CONTENTS

PREFACE   V

1 INTRODUCTION   1

    1.1 Objectives of This Study   1

    1.2 Fuzzy Mathematical Programming Problems   2

    1.3 Classification of Fuzzy Mathematical Programming   5

    1.4 Applications of Fuzzy Mathematical Programming   8

    1.5 Literature Survey   10

2 FUZZY SET THEORY   14

    2.1 Fuzzy Sets   14

    2.2 Fuzzy Set Theory   19

        2.2.1 Basic Terminology and Definition   20

            2.2.1.1 Definition of Fuzzy Sets   20

            2.2.1.2 Support   21

            2.2.1.3 α-level Set   21

            2.2.1.4 Normality   22

            2.2.1.5 Convexity and Concavity   22

            2.2.1.6 Extension Principle   23

            2.2.1.7 Compatibility of Extension Principle
                     with α-cuts   23

            2.2.1.8 Relation   24

            2.2.1.9 Decomposability   24

            2.2.1.10 Decomposition Theorem   25

            2.2.1.11 Probability of Fuzzy Events   25

            2.2.1.12 Conditional Fuzzy Sets   27

        2.2.2 Basic Operations   27

            2.2.2.1 Inclusion   27

            2.2.2.2 Equality   27

            2.2.2.3 Complementation   28

2.2.2.4   Intersection   28

2.2.2.5   Union   29

2.2.2.6   Algebraic Product   29

2.2.2.7   Algebraic Sum   29

2.2.2.8   Difference   29

2.3   Membership Functions   30

2.3.1   A Survey of Functional Forms   31

2.3.2   Examples to Generate Membership Functions   34

2.3.2.1   Distance Approach   35

2.3.2.2   True-Valued Approach   40

2.3.2.3   Payoff Function   45

2.3.2.4   Other Examples   48

2.4   Fuzzy Decision and Operators   48

2.4.1   Fuzzy Decision   49

2.4.2   Max-Min Operator   51

2.4.3   Compensatory Operators   53

2.4.3.1   Numerical Example for Operators   56

2.5   Fuzzy Arithmetic   58

2.5.1   Addition of Fuzzy Numbers   58

2.5.2   Subtraction of Fuzzy Numbers   61

2.5.3   Multiplication of Fuzzy Numbers   63

2.5.4   Division of Fuzzy Numbers   65

2.5.5   Triangular and Trapezoid Fuzzy Numbers   66

2.6   Fuzzy Ranking   72

3   FUZZY   MATHEMATICAL PROGRAMMING   74

3.1   Fuzzy Linear Programming Models   75

3.1.1   Linear Programming Problem with Fuzzy Resources   75

3.1.1.1   Verdegay's Approach   79

3.1.1.1a Example 1: The Knox Production-Mix

Selection Problem   80

3.1.1.1b Example 2: A Transportation Problem   85

3.1.1.2 Werners's Approach   87

3.1.1.2a Example 1: The Knox Production-Mix
         Selection Problem   89

3.1.1.2b Example 2: An Air Pollution Regulation
         Problem   90

3.1.2 Linear Programming Problem with Fuzzy Resources and
       Objective   95

3.1.2.1 Zimmermann's Approach   95

3.1.2.1a Example 1: The Knox Production-Mix
         Selection Problem   97

3.1.2.1b Example 2: A Regional Resource Allocation
         Problem   98

3.1.2.1c Example 3: A Fuzzy Resource Allocation
         Problem   101

3.1.2.2 Chanas's Approach   104

3.1.2.2a Example 1: An Optimal System Design
         Problem   107

3.1.2.2b Example 2: An Aggregate Production
         Planning Problem   110

3.1.3 Linear Programming Problem with Fuzzy Parameters
       in the Objective Function   119

3.1.4 Linear Programming with All Fuzzy Coefficients   121

3.1.4.1 Example: A Production Scheduling Problem   123

3.2 Interactive Fuzzy Linear Programming   127

3.2.1 Introduction   127

3.2.2 Discussion of Zimmermann's, Werners's
       Chanas's and Verdegay's Approaches   128

3.2.3 Interactive Fuzzy Linear Programming - I   130

3.2.3.1 Problem Setting   130

3.2.3.2 The Algorithm of IFLP-I   136

3.2.3.3 Example: The Knox Production-Mix

Selection Problem   139

3.2.4   Interactive Fuzzy Linear Programming - II   146

3.2.4.1 The Algorithm of IFLP-II   151

3.3 Some Extensions of Fuzzy Linear Programming Problems   155

3.3.1   Membership Functions   155

3.3.1.1 Example: A Truck Fleet Problem   161

3.3.2   Operators   168

3.3.3   Sensitivity Analysis and Dual Theory   170

3.3.4   Fuzzy Non-Linear Programming   170

3.3.4.1 Example: A Fuzzy Machining Economics
         Problem   171

3.3.5   Fuzzy Integer Programming   174

3.3.5.1 Fuzzy 0-1 Linear Programming   176

3.3.5.1a Example: A Fuzzy Location Problem   180

4 POSSIBILISTIC PROGRAMMING   187

4.1 Possibilistic Linear Programming Models   189

4.1.1   Linear Programming with Imprecise Resources and
        Technological Coefficients   189

4.1.1.1 Ramik and Rimanek's Approach   189

4.1.1.1a Example: A Profit Apportionment Problem 191

4.1.1.2 Tanaka, Ichihashi and Asai's Approach   196

4.1.1.3 Dubois's Approach   198

4.1.2   Linear Programming with Imprecise Objective
        Coefficients   203

4.1.2.1 Lai and Hwang's Approach   203

4.1.2.1a Example: A Winston-Salem Development
         Management Problem   208

4.1.2.2 Rommelfanger, Hanuscheck and Wolf's
         Approach   213

4.1.2.3 Delgado, Verdegay and Vila's Approach   217

4.1.3   Linear Programming with Imprecise Objective and

Technological Coefficients   223

  4.1.4 Linear Programming with Imprecise Coefficients 225

    4.1.4.1 Lai and Hwang's Approach   225

    4.1.4.2 Buckley's Approach   227

    4.1.4.2a Example: A Feed Mix (Diet) Problem 230

    4.1.4.3 Negi's Approach   232

    4.1.4.4 Fuller's Approach   236

  4.1.5 Other Problems   237

 4.2 Some Extensions of Possibilistic Linear Programming 238

  4.2.1 Linear Programming with Imprecise Coefficients and
      Fuzzy Inequalities   238

    4.2.1a Example: A Fuzzy Matrix Game Problem 241

  4.2.2 Linear Programming with Imprecise Objective
      Coefficients and Fuzzy Resources   243

    4.2.2a Example: A Bank Hedging Decision Problem 247

  4.2.3 Stochastic Possibilistic Linear Programming 256

    4.2.3a Example: A Bank Hedging Decision Problem 260

5 CONCLUDING REMARKS   268

 5.1 Probability Theory versus Fuzzy Set Theory   268

 5.2 Stochastic versus Possibilistic Programming   270

 5.3 Future Research   275

 5.4 Introduction of the Following Volume   276

 5.5 Fuzzy Multiple Attribute Decision Making   277

BIBLIOGRAPHY   279

 Books, Monographs and Conference Proceedings   279

 Journal Articles, Technical Reports and Theses   283

1 INTRODUCTION

At the turn of the century, reducing complex real-world systems into precise mathematical models was the main trend in science and engineering. In the middle of this century, Operations Research (OR) began to be applied to real-world decision-making problems and thus became one of the most important fields in science and engineering. Unfortunately, real-world situations are often not so deterministic. Thus precise mathematical models are not enough to tackle all practical problems.

To deal with imprecision/uncertainty, the concepts and techniques of probability theory are usually employed. In the 1960s, meanings of the probability theory has been reconsidered and criticized when modelling practical problems, especially in artificial intelligence. Around the same time as the development of chaos theory to handle non-linear dynamic systems in physics and mathematics, fuzzy set theory was developed in 1965 by L.A. Zadeh. Since then, fuzzy set theory has been applied to the fields of operations research, management science, artificial intelligence/expert system, control theory, statistics and many other fields.

Fuzzy set theory is a theory of graded concepts (a matter of degree), but not a theory of chance. Therefore, figures and numerical tables are considered paramount in the study of fuzzy set theory. They, unlike confusing mathematical slang, difficult functions, etc., provide the best ways to communicate with outsiders. This is an important concept in the new generation of operations research.

In operations research, fuzzy set theory has been applied to techniques of linear and non-linear programming, dynamic programming, queueing theory, multiple criteria decision-making, group decision-making and so on. In this monograph, we provide readers with a capsule look into all existing fuzzy linear and nonlinear mathematical programming problems, methods, and applications.

1.1 Objectives of This Study

This study is an introduction to the applications of fuzzy set theory toward Fuzzy Mathematical Programming. It gives a state-of-the-art survey of all existing problems, methods and practical applications concerning fuzzy mathematical programming. Many diversified methods and applications are thoroughly and critically reviewed and systematically classified. Some basic concepts and terminologies are defined so that we can explain the literature in a consistent manner. This study thus offers its readers a capsule look into the existing methods, and their characteristics and applicability to analysis of fuzzy mathematical programming problems. It is the first time these problems, methods and applications have been presented together with both deep and broad discussions.

Practical applications are never over-emphasized. All concepts and techniques are developed for solving real-world problems. Thus the most important issue is to involve real applications with real decision makers. Unfortunately, these real applications have very rarely been reported in the literature. To fill the gap between researchers and real decision makers for solving real applications, discussions of practical problems are considered as important as concepts and techniques in this study. Practical problems discussed here include production/manufacturing, transportation, assignment, game, environmental management, banking/finance and agricultural economics. We hope that through the discussions of these practical problems, real decision makers will obtain more confidence in the usefulness of fuzzy modelling.

1.2 Fuzzy Mathematical Programming Problems

While L. V. Kantorovich, a Soviet mathematician and economist, formulated and solved linear programming dealing with the organization and planning of production in 1939, G. B. Dantzig, M. Wood and their associates of the U. S. Department of the Air Force developed the general problem of linear programming named by A. W. Tucker. In order to apply mathematical and related techniques to military programming and planning problems, Dantzig proposed "that interrelations between activities of a large organization be viewed as a linear programming type model and the optimizing program determined by minimizing a linear objective function." Conceptually, some concepts originated from Leotief's input-output model. Since that time, linear programming has become an important tool of modern theoretical and applied mathematics.

Linear programming is concerned with the efficient allocation of limited resources to activities with the objective of meeting a desired goal such as maximizing profit or minimizing cost. The distinct characteristic of linear programming models is that the interrelations between activities are linear relationships which are the satisfactions of the proportionality and additivity requirements. Symbolically, the general linear programming problem may be stated as:

$$\text{maximize} \quad z = f(c,x) = cx$$
$$\text{subject to} \quad g(A,x) = Ax \leq b \text{ and } x \geq 0 \tag{1.1}$$

where c is the vector of profit coefficients of the objective function and b is the vector of total resources available. x is the vector of decision variables (or alternatives), and A is the matrix of technical coefficients. These input data (of c, b and A) are usually fuzzy/imprecise because of incomplete or non-obtainable information. For instance, available labor hours and available material (b) may be "around 1220" hours and "about 1550" units, respectively. Similarly, the unit profit (c) of new products may be expected to be "about $2.3" per unit and the estimates of technological coefficients (A) may be "around 3" units per labor hour.

To formulate these fuzzy/imprecise numbers, we can use membership functions or possibility distribution (depending on specific problems). The functional forms of membership functions and possibility distributions are depicted in Figures 1.1 and 1.2, respectively. With these fuzzy/imprecise input data, Equation (1.1) is then called fuzzy/possibilistic (linear) programming. In this study, the grade of a membership function indicates a subjective degree of satisfaction within given tolerances. On the other hand, the grade of possibility indicates the subjective or objective degree of the occurrence of an event. It is important to realize this distinction while modelling fuzziness/imprecision in mathematical programming problems.

In Figure 1.1, we illustrate four cases of the preference-based membership functions of fuzzy resources. When the constraints are $Ax \leq \tilde{b}$, the rational preference-based membership functions can be assumed to be non-increasing. Similarly, non-decreasing functions can be assumed for $Ax \geq \tilde{b}$. For equality constraints, triangular or trapezoid functions might be appropriate. For the maximization (or minimization) problem, the preference-based membership functions of $\tilde{c}$ can be assumed to be non-decreasing as $\tilde{b}$ in $Ax \geq \tilde{b}$ or (or non-increasing as $\tilde{b}$ in $Ax \leq \tilde{b}$). As to the preference-based membership functions of $\tilde{A}$, they may be either non-increasing for $\tilde{A}x \leq b$, or non-decreasing for $\tilde{A}x \geq b$. Sometimes, triangular or trapezoid functions might be adopted. For possibilistic (linear) programming, the possibility distributions of $\tilde{A}$, $\tilde{b}$ and/or $\tilde{c}$ are often assumed to be triangular or trapezoid functions as shown in Figure 1.2.

It is noted that non-linear functions, such as piece-wise linear, exponential, power, hyperbolic,etc., may be adopted as indicated in Figures 1.1 and 1.2.

When any of f(c,x) and/or g(A,x) is a non-linear function, Equation (1.1) becomes a non-linear programming problem. If x is restricted to be an integer, then Equation (1.1) will become an integer programming problem. Both cases with fuzzy/imprecise input data as shown in Figures 1.1 and 1.2, Equation (1.1) becomes fuzzy (possibilistic) non-linear and integer programming, respectively.

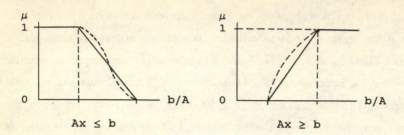

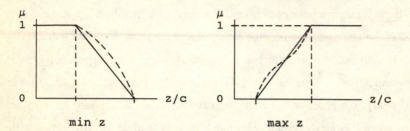

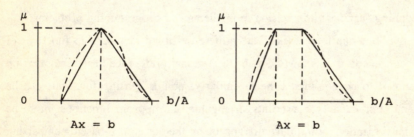

Figure 1.1 Preference-based membership functions of fuzzy b, fuzzy A, fuzzy c or fuzzy z

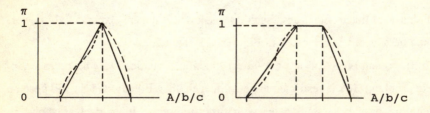

Figure 1.2 Possibility distributions of imprecise A, imprecise b or imprecise c

1.3 Classification of Fuzzy Mathematical Programming

The most recent literature surveys of fuzzy mathematical programming are Zimmermann [304, 306, 1985; BM64, 1987], Slowinski [227, 1986], Wierzchon [264, 1987], Leung [BM38, 1988] and Luhandjula [161, 1989]. Zimmermann first classified fuzzy mathematical programming into symmetrical and nonsymmetrical models. Leung classified fuzzy mathematical programming problems into the following four categories: (1) a precise objective and fuzzy constraints; (2) a fuzzy objective and precise constraints; (3) a fuzzy objective and fuzzy constraints; (4) robust programming (which is considered as one of the possibilistic mathematical programming problems in this study). Luhandjula categorized fuzzy mathematical programming into flexible programming, mathematical programming with fuzzy parameters and fuzzy stochastic programming. Fuzzy parameters or so-called fuzzy numbers are essentially characterized by possibility distributions. For methodologies, he further classified flexible programming methods into symmetrical and asymmetrical approaches as Zimmermann did. For mathematical programming problems with fuzzy parameters, Luhandjula grouped them into two major classes: problems with a deterministic objective function and problems with a fuzzy objective function. Fuzzy stochastic programming concerns the parameters involving both fuzzy and stochastic natures.

In this study, we systematically classify all possible problems and existing approaches after an extensive and thorough literature survey. We distinguish fuzzy linear programming problems from possibilistic linear programming problems. Fuzzy linear programming problems will associate fuzzy input data which should be modelled by subjective preference-based membership functions. On the other hand, possibilistic linear programming problems will associate with imprecise data which should be modelled by possibility distributions. The possibility distributions are an analogue of the probability distributions and can be either subjective or objective. Furthermore, based on the possible combinations of crisp and/or imprecise coefficients (input data), we also sub-classify fuzzy linear programming and possibilistic linear programming problems as shown in Figures 1.3 and 1.4. Obviously, these classifications are problem-oriented.

Based on possible combinations of the fuzziness of A, b, c and/or z, we classify fuzzy (linear) programming into the following five major problems: (1) fuzzy b; (2) fuzzy b and fuzzy z; (3) fuzzy c; (4) fuzzy A, fuzzy b and fuzzy c, fuzzy A and fuzzy b, fuzzy A and fuzzy c, or fuzzy A, fuzzy b and fuzzy c; (5) fuzzy A and fuzzy z, or fuzzy A, fuzzy b and fuzzy z. Moreover, we include the sixth problem of "an expert decision-making support system" which

max/min z = cx	Fuzzy z	Fuzzy c
s.t. Ax $\{\leq,=,\geq\}$ b	Fuzzy b or $\{\leq,=,\geq\}$	Fuzzy A

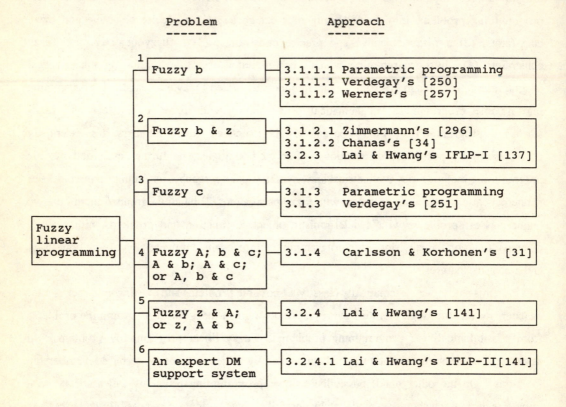

Figure 1.3 A taxonomy of fuzzy linear programming (Lai and Hwang [141])

max/min z = cx		Imprecise c
s.t. Ax {≤,=,≥} b	Imprecise b	Imprecise A

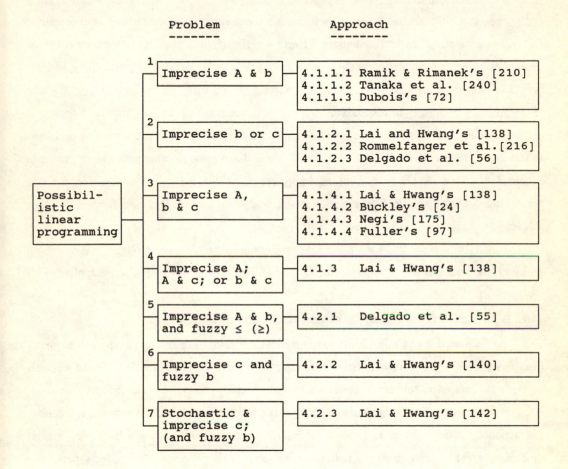

```
                    Problem              Approach
                    -------              --------

              1
              ┌─ Imprecise A & b ──── 4.1.1.1 Ramik & Rimanek's [210]
              │                       4.1.1.2 Tanaka et al. [240]
              │                       4.1.1.3 Dubois's [72]
              │
              2
              ├─ Imprecise b or c ─── 4.1.2.1 Lai and Hwang's [138]
              │                       4.1.2.2 Rommelfanger et al.[216]
              │                       4.1.2.3 Delgado et al. [56]
              3
              ├─ Imprecise A, ─────── 4.1.4.1 Lai & Hwang's [138]
              │  b & c                4.1.4.2 Buckley's [24]
Possibil-     │                       4.1.4.3 Negi's [175]
istic   ──────┤                       4.1.4.4 Fuller's [97]
linear        4
programming   ├─ Imprecise A; ─────── 4.1.3    Lai & Hwang's [138]
              │  A & c; or b & c
              5
              ├─ Imprecise A & b, ─── 4.2.1    Delgado et al. [55]
              │  and fuzzy ≤ (≥)
              6
              ├─ Imprecise c and ─── 4.2.2    Lai & Hwang's [140]
              │  fuzzy b
              7
              └─ Stochastic & ─────── 4.2.3    Lai & Hwang's [142]
                 imprecise c;
                 (and fuzzy b)
```

Figure 1.4 A taxonomy of possibilistic linear programming (Lai [139])

systematically integrates all possible problems listed above. Approaches to solve each fuzzy (linear) programming problem are listed.

Similarly, we group possibilistic (linear) programming problems into the following four major problems: (1) <u>imprecise A and imprecise b</u>; (2) <u>imprecise b</u>, or <u>imprecise c</u>; (3) <u>imprecise A, imprecise b and imprecise c</u>; (4) <u>imprecise A</u>, <u>imprecise A and imprecise c</u>, or <u>imprecise b and imprecise c</u>. Occasionally, some coefficients have preference-based membership functions and others have possibilistic distributions. Here, we classify this problem into the category of possibilistic linear programming as shown in Figure 1.4. The fifth and sixth possibilistic (linear) programming problems are (5) <u>imprecise A, imprecise b and fuzzy inequalities ($\leq$ or $\geq$)</u>, and (6) <u>imprecise c and fuzzy b</u>, respectively. When there are coefficients involving both fuzziness and randomness simultaneously, we have a stochastic possibilistic (linear) programming problem. In this case, we also classify it into the seventh possibilistic programming problem as shown in Figure 1.4. That is: (7) <u>stochastic and imprecise c, and/or fuzzy b</u>. The reason for this arrangement is for ease of our presentation in this study. Of course, this classification is rather subjective. Approaches to solve each problem are listed.

Finally, it is noted that for a non-linear programming problem or an integer programming problem, the concepts and approaches as shown Figures 1.3 and 1.4 may also be applied.

1.4 Applications of Fuzzy Mathematical Programming

As indicated by White [262], mathematical programming has been applied to many disciplines such as advertising, assignment, bacteria identification, blending, blood control, budgeting, computer selection, credit, diamond dressing, diet, disease control, drilling, employment, engineering design, finance, income tax, information services, investment, location, maintenance, managerial compensation, manpower, marketing, materials handling, media, medicare, mergers, metal cutting, networks, organizational design, perishable products, personnel assignment, pollution, production, promotions, projection selection, portfolio analysis, purchasing, quality control, racial balance, reliability, replacement, research and development, safety, salaries, sales, security scheduling, stock control, tendering, timetabling, transportation and visitor quotas. However, in these real-world problems, the input data or parameters are often fuzzy/imprecise because of incomplete or non-obtainable information. Traditional mathematical programming techniques, obviously, cannot solve all fuzzy programming problems.

The drive to have practical models for management systems under fuzziness has been the impetus towards the development of many excellent methods for fuzzy mathematical

Table 1.1 List of practical problems

Problem	Fuzziness/ Imprecision	Approach	Section
I. PRODUCTION/MANUFACTURING:			
Production-Mix Selection	Fuzzy Resources	Verdegay's Werners's	3.1.1.1a 3.1.1.2a
	Fuzzy Resources & Fuzzy Objective	Zimmermann's Lai & Hwang's	3.1.2.1a 3.2.3.3
Optimal System Design	Fuzzy Resources & Fuzzy Objective	Chanas's	3.1.2.2a
Aggregate Production Planning	Fuzzy Resources & Fuzzy Objective	Chanas's	3.1.2.2b
Production Scheduling	Fuzzy Coefficients	Carlsson & Korhonen's	3.1.4.1
Machining Economics	Fuzzy Resources & Fuzzy Objective	Zimmermann's	3.3.4.1
II. TRANSPORTATION:			
Transportation	Fuzzy Resources	Verdegay's	3.1.1.1b
Truck Fleet	Fuzzy Resources	Nakamura's	3.3.1.1
III. ASSIGNMENT:			
Network Location	Fuzzy Resources	Zimmermann & Pollatschek's	3.3.5.1a
IV. GAME:			
Matrix Game	Impr. Coeffs. & Fuzzy Inequalities	Campos's	4.2.1a
V. ENVIRONMENTAL MANAGEMENT:			
Air Pollution Regulation	Fuzzy Resources	Werners's	3.1.1.2b
VI. BANKING/FINANCE:			
Profit Apportionment	Impr. Resources & Impr. Tech. Coeffs.	Ramik & Rimanek's	4.1.1.1a
Project Investment	Impr. Coeffs.	Lai & Hwang's	4.1.2.1
Bank Hedging Decision	Impr. Objective Coeffs. & Fuzzy Resources	Lai & Hwang's	4.2.2a
	Impr. & Stochastic Objective Coeffs., & Fuzzy Resources	Lai & Hwang's	4.2.3a
VII. AGRICULTURE ECONOMICS:			
Regional Resource Allocation	Fuzzy Resources & Fuzzy Objective	Zimmermann's	3.1.2.1b
Feed Mix	Impr. Coeffs.	Buckley's	4.1.4.2a

Note: Impr. --> Imprecise; Coeffs. --> Coefficients.

programming problems. However, few methods have been yet tested with real-world problems by real decision makers. To fill the gap between researchers and real decision makers for solving real applications, discussions of practical problems are considered as important as concepts and techniques in this study.

The practical fuzzy/possibilistic problems discussed in this monograph include the following seven fields: (I) production/manufacturing, (II) transportation, (III) assignment, (IV) game, (V) environmental management, (VI) banking/finance and (VII) agricultural economics (as shown in Table 1.1).

In Table 1.1, the first column indicates the types of problems studied in this volume. The assumed fuzziness and/or imprecision are listed in the second column and approaches to solve each problem are listed in the third column. Finally, the sections where these applications are discussed are indicated in the last column.

It is our belief that this concise presentation of practical problems is necessary for teaching and research in order to realize the meaning of fuzziness while modelling real-world problems. After all, all concepts and methods are developed for solving real-world problems. We expect that this study will persuade more students, researcher and real decision makers/practitioners to discover, investigate and solve all possible real problems.

1.5 Literature Survey

A thorough and systematical literature survey has been carried out. Books, monographs and proceedings, related to the concerned problems, are listed in Table 1.2. Among 300 journal articles, technical reports and theses, the most related are listed in Table 1.3. Both tables provide pertinent readings for each category of problems.

Because of the broad interdisciplinary character of this field, the literature is diversified in many journals as shown in Table 1.4. Journals in which related articles on fuzzy approaches appear frequently are indicated in the lists.

This monograph will provide readers and researchers with a capsule look into the existing methods, their characteristics and applicability to analyzing fuzzy mathematical programming. Although we have tried to give a reasonably complete survey, those papers not included were either inadvertently overlooked or considered not to bear directly on the topics in this survey. We apologize to both the readers and the researchers if we have omitted any relevant papers.

Before going into the actual review of fuzzy and possibilistic mathematical programming in Chapters 3 and 4, we first review necessary backgrounds of fuzzy set theory in Chapter 2.

Table 1.2 Classification of books, monographs and proceedings

Class	Reference
Fuzzy Set Theory and Its Operations	BM7, BM8, BM29, BM31, BM32, BM35, BM36, BM46, BM47, BM55, BM56, BM57, BM63
Fuzzy Arithmetic	BM31, BM32
Fuzzy Statistics	BM27, BM36, BM41
Fuzzy Mathematical Programming	BM2, BM4, BM27, BM30, BM32, BM40, BM47, BM49, BM60, BM62, BM63, BM64
Fuzzy Multiple Objective Decision Making	BM1, BM26, BM27, BM38, BM45, BM64
Fuzzy Multiple Attribute Decision Making	BM3, BM64
Fuzzy Modelling in Industrial Engineering	BM10, BM22, BM23, BM27, BM30, BM32, BM55, BM56, BM59, BM62
Fuzzy Modelling in Other Decision-making Problems	BM1, BM2, BM6, BM14, BM15, BM16, BM22, BM23, BM25, BM27, BM38, BM34, BM37, BM39, BM43, BM44, BM49, BM50, BM51, BM52, BM53, BM55, BM56, BM60, BM62, BM63, BM64

Table 1.3   Classification of journal articles, technical reports and theses

Class	Reference
FUZZY LINEAR PROGRAMMING (FLP)	
Fuzzy Resources	12, 43, 77, 122, 139, 176-181, 185, 222, 231, 236, 250-252, 257-259, 266, 306
Fuzzy Resources & Objectives	12, 34, 35, 37, 43, 79, 87-89, 91, 105, 120, 122, 128, 139, 143, 145, 148, 149, 151-155, 174, 176-181, 192-197, 202, 220, 255, 259, 263, 296-306
Fuzzy Objective Coefficients	12, 139, 251, 252
Fuzzy Coefficients	12, 31, 139
Sensitivity Analysis	112, 199, 215, 233, 242
Interactive FLP - An Expert Decision-Making Support System	12, 137, 139, 141, 309
FUZZY NON-LINEAR PROGRAMMING	12, 84, 139, 248
FUZZY INTEGER PROGRAMMING	12, 83, 84, 139, 303
POSSIBILISTIC LINEAR PROGRAMMING (PLP)	
Imprecise Resources & Technological Coefficients	20, 64, 66, 69, 72, 86, 121, 139, 144, 175, 200, 201, 205, 209-212, 225-228, 232, 240, 243
Imprecise Objective Coefficients	52, 53, 56, 138, 139, 159
Imprecise Objective & Technological Coefficients	138, 139
Imprecise Coefficients	20, 24-27, 43, 97-99, 121, 138, 139, 151, 157-161, 175, 192-197, 216, 217, 221, 225-228, 238-244, 264
Imprecise Coefficients & Fuzzy Inequalities	29, 30, 55, 56, 139
Imprecise Objective Coefficients & Fuzzy Resources	140, 217
STOCHASTIC POSSIBILISTIC LINEAR PROGRAMMING	131, 142, 156, 264a

Table 1.4 List of journals

---

1.   Automata
2.   Colloquia Mathematica Societatis Janos Bolyai
3.   Computers and Mathematics with Applications
4.   Computers and Operations Research
5. * Control and Cybernetics

6.   Decision Sciences
7.   Decision Support Systems
8.   Economic Computer and Economic Cybernetic Studies
      and Researches
9.   Environment and Planning A
10. * European Journal of Operational Research

11. * Fuzzy Sets and Systems
12.   Geographical Analysis
13.   Human Systems Management
14. * IEEE Transactions on Systems, Man, and Cybernetics
15.   Information and Control

16.   Information Science
17.   International Journal of General System
18.   International Journal of Intelligent System
19.   International Journal of Man-Machine Studies
20. * International Journal of Production Research
21.   International Journal of System Science

22.   Journal of Cybernetics
23.   Journal of Experimental Psychology: General
24.   Journal of Experimental Psychology: Human Perception
      and Performance
25.   Journal of Mathematical Analysis and Its Applications
26.   Journal of Operational Research Society

27.   Kybernetika
28.   Kybernetes
29.   Large Scale System
30.   Literary and Linguistic Computing
31.   Machine Intelligence

32.   Management Sciences
33.   Mathematical Analysis and Applications
34.   Mathematical Modelling
35.   Naval Research Logistics
36.   ORSA Journal on Computing
37.   Tekhnicheskaya Kibernetika

---

* indicates that articles of fuzzy mathematical programming and fuzzy
multiple objective decision-making appear frequently in this journal.

2 FUZZY SET THEORY

The term "fuzzy" was proposed by Zadeh in 1962 [273]. In 1965, Zadeh formally published the famous paper "Fuzzy Sets"[274]. The fuzzy set theory is developed to improve the oversimplified model, thereby developing a more robust and flexible model in order to solve real-world complex systems involving human aspects. Furthermore, it helps the decision maker not only to consider the existing alternatives under given constraints (optimize a given system), but also to develop new alternatives (design a system).

The fuzzy set theory has been applied in many fields, such as operations research, management science, control theory, artificial intelligence/expert system, human behavior, etc. In this study we will concentrate on fuzzy mathematical programming and fuzzy multiple objective decision-making. To do so, we will first introduce the required knowledge of the fuzzy set theory and fuzzy mathematics in this chapter.

2.1 Fuzzy Sets

Because of incomplete knowledge and information, precise mathematics are not sufficient to model a complex system. Traditionally, the probability theory is the prevailing approach to handle this incompleteness/uncertainty. One of the fundamentals in the probability theory is the law of the excluded middle ($p(A \cup A^c) = 1$) and contradiction ($p(A \cap A^c) = 0$). For instance, a fruit is either an apple or not an apple, an animal (normally speaking) is either male or female. For this case, the probability is certainly a good approach to represent some knowledge or information whose boundaries can be clearly defined. Throwing coins into the air, you can guess that either heads or tails will be up. Rotating a dice, the result will be 6, 5, 4, 3, 2, or 1, but never be 4.5 or 2.1. Of course, there are a lot of problems satisfying the laws of the excluded middle and contradiction. But, intuitively and commonsensically, this is not true in other problems. An evidence favoring a particular hypothesis to some degree does not simultaneously disconfirm it to any degree, because it may not give any support to the contrary. For example, a man may be <u>smart</u>, <u>not smart</u>, or <u>a little smart</u>. A color can be <u>red</u>, <u>not red</u>, or <u>reddish</u>. Therefore, it is difficult to define the sets of smart men and red colors with sharp/crisp boundaries. Similarly, how can we define and classify "<u>good</u> credit," "<u>beautiful</u> women," "<u>good</u> taste," "<u>reasonable</u> price," "<u>good</u> personality" and so on? Obviously, the probability theory cannot model all the possible problems of incompleteness. The fuzzy set theory is developed to define and solve these problems without sharp boundaries. That is: fuzzy set theory considers the partial relationship/membership.

To clearly distinguish the fuzzy sets from classical (crisp) sets, let us first consider the following examples.

Example 1. In the classical set theory, an element may either "belong to" set A or "not belong to" set A in the given universe. For example, suppose there is a rather large target and shooters always hit inside the target (see Figure 2.1). A circle is located in the center of the target. If a shooter hits inside the circle, region A, he is given the title "good shooter." Otherwise, he is called a "poor shooter." Shooter a^1 shoots inside region A, so he is a good shooter. On the other hand, shooter a_2, who shoots far away from region A, is a poor shooter. The classical set theory with binary relationship shows some problems. For example, if there are three shooters, a^3, a^4 and a^5, who hit the target within close range of one another (see Figure 2.1), yet only a^3 is within the target, then shooter a^3 is a good shooter, and shooters a^4 and a^5 are poor shooters. This is obviously unreasonable.

As a matter of fact, these three people should obtain some similar designation, at least up to a certain degree. Therefore, a measure of the degree to which the shooter belongs to the set "good shooters" should be developed in order to discern how good the shooter is. The set composed of good shooters is actually fuzzy because there is no crisp boundary. It is rational to consider the distance from the boundary of the region A as a measure for indicating the degree to which shooter a^i belongs to the set of "good shooters." In Figure 2.1, a^1 and a^3 are absolutely good shooters. On the other hand, a^4 and a^5 are not absolutely good shooters or absolutely poor shooters. They are, to some degree, good shooters. By giving a numerical measure which is assumed linearly proportional to the distance, d, of each shooter from region A, one can say that shooter a^4 has 0.8 degree of membership in the set of good shooters versus 0.2 for shooter a^5. $\mu(a^4) = 0.8$ and $\mu(a^5) = 0.2$. Of course, the numerical measure, μ, can be any number. A normalized measure which is in [0, 1] is always adapted. In the above example, the preference concept is implied while assigning the numerical measures for each shooter to represent the degree of "A SHOOTER BELONGING TO THE SET OF GOOD SHOOTERS." The shorter the hit spot is from the center of the target, the larger the grade assigned (small d is preferred to large d). The grade values are actually preference values. The function μ (called membership function in fuzzy set theory), constituted by these grades as shown in Figure 2.1, is then a preference function. Similarly, in practice, the decision maker may feel that: "<u>Around</u> $20,000

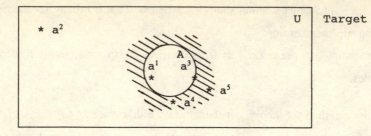

Assigned grades indicating how good the shooters are	Assumed distances from center of the target

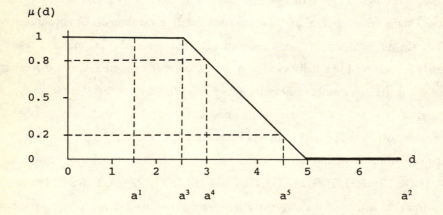

Figure 2.1 Graphic explanation for the fuzzy set of GOOD SHOOTER

profit is acceptable;" "The budget should be <u>around</u> $9,000;" "Dividends should not be <u>higher</u> than 6%;" "Overtime should be <u>less</u> than 5% of the regular man-hours;" and so on. Obviously, these linguistic statements cannot be described by probability. The fuzzy set theory, on the other hand, gives us a way to handle such linguistic situations. Meanwhile, the preference concept is often assumed in building the membership functions of the above linguistic statements.

In order to be able to more clearly understand the fuzzy sets concept, let us consider one more example.

Example 2. Consider a universe composed of 4 male students with the same height of 5'9", Hans, John, George and Young. That is, U = { Hans, John, George, Young }. The weights for the 4 male students are given as follows:

$$\text{Hans : 164 pounds} \qquad \text{John : 190 pounds}$$
$$\text{George : 180 pounds} \qquad \text{Young : 160 pounds.}$$

Now, let consider the linguistic proposition "<u>fat</u> male students." The students who belong to "<u>fat</u> male students" then constitute a fuzzy set, A. Is Hans ϵ A, John ϵ A, George ϵ A or Young ϵ A?

One plots the weights on a real line (see Figure 2.2) in order to present the relative differences. According to common sense, a male student weighing more than 185 pounds at 5'9" in height is absolutely considered a fat male student (actually the term "fat" depends on culture, race, etc., which are beyond our research). On the other hand, a male student weighing less than 160 pounds is not fat at all. Therefore, it is obvious that John is absolutely fat and Young is absolutely not fat. How about Hans and George? Actually Hans approaches Young with respect to weight and George is close to John. It is true that the heavier the male student is, the greater the degree to which he belongs to fuzzy set A. Thus, a degree-scaled line (see Figure 2.2) can be drawn corresponding to the previous weight-scaled line in order to represent the degree of membership indicating that a male student belongs to A. The scale on the degree-scaled line is linearly proportional to the weight-scaled line when the weight belongs to the interval [160, 185]. As a result, the following grades of membership are available:

$$\text{degree(Hans } \epsilon \text{ A)} = \mu(x=H) = 0.2$$
$$\text{degree(John } \epsilon \text{ A)} = \mu(x=J) = 1$$
$$\text{degree(George } \epsilon \text{ A)} = \mu(x=G) = 0.8$$

$$\text{degree(Young } \epsilon \text{ A)} = \mu(x=Y) = 0$$

where $\mu(\cdot)$ (detailed definition is given in the next section) is the membership function of the fuzzy subset A of the set X. Like the previous example, the preference concept is used to elicit the membership function $\mu(\cdot)$.

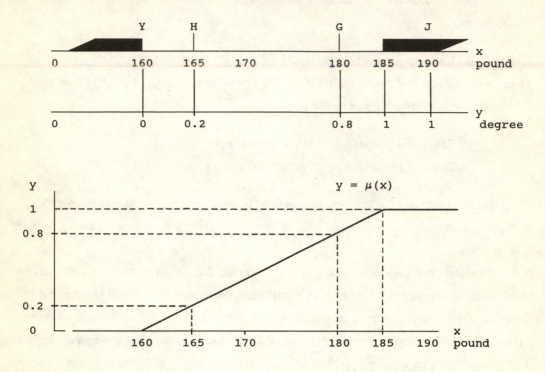

Figure 2.2 Derived degree of membership for the fuzzy set A

Example 3. Consider the statement of "the interest rate will be around 8% to 8.2%." The most possible interest rate is between 8% and 8.2%. The least possible interest rate is 6.5% (optimistic value) and 10% (pessimistic value). It is not likely that the interest rate will be less than 6.5% or larger than 10% in the opinion of the decision maker. Thus, we give 8% - 8.2% interest rate a 1 possibility, and 6.5% or less and 10% or more a 0 possibility. Between them, let us assume the grades are as shown in Figure 2.3. Here, we did not use the preference concept to grade various interest rates, but used the possibility concept: most possible, least possible, or in between. Therefore, we call this membership function a possibility function or possibility distribution, denoted by $\pi(\cdot)$.

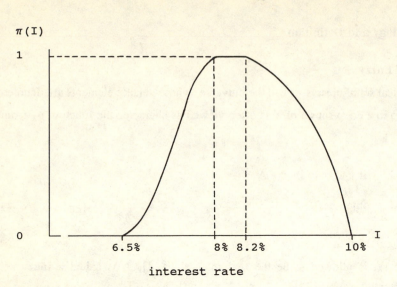

Figure 2.3 The membership (possibility) function of the interest rate

2.2 Fuzzy Set Theory

Consider a crisp set A = {x | x = 2y, y is nature number(N)}. Obviously, A is a set of all even nature numbers. Thus, any nature number y ε N is either belonging to A if it is even, or not belonging to A if it is odd. The crisp set A can then be represented as:

A = {(1, 0), (2, 1), (3, 0), (4, 1), ...}

where the second number of the ordered pairs, 0 or 1, is the measure of membership. 0 indicates that the first number of the ordered pairs is not even, and 1 indicates it is even. By using this notation, one can apply it to Example 2 and get the representation of the fuzzy set A as:

A = {(Hans, 0.2), (John, 1), (George, 0.8), (Young, 0)}

where the second number is no longer just 0 and 1, but is [0, 1]. This is the main characteristic that makes the fuzzy sets different from the classical (crisp) sets. Now, it is time to consider the following terminologies and definitions.

2.2.1 Basic Terminology and Definition

2.2.1.1 Definition of Fuzzy Sets

Let X be a classical set of objects, called the universe, whose generic elements are denoted by x. The membership in a crisp subset of X is often viewed as characteristic function μ_A from X to $\{0, 1\}$ such that:

$$\mu_A(x) = 1 \quad \text{if and only if } x \in A$$

$$= 0 \quad \text{otherwise} \tag{2.1}$$

where $\{0, 1\}$ is called a valuation set.

If the valuation set is allowed to be the real interval $[0, 1]$, A is called a fuzzy set proposed by Zadeh [274]. $\mu_A(x)$ is the degree of membership of x in A. The closer the value of $\mu_A(x)$ is to 1, the more x belongs to A. Therefore, A is completely characterized by the set of ordered pairs:

$$A = \{(x, \mu_A(x)) \mid x \in X\}. \tag{2.2}$$

It is worth noting that the characteristic function can be either a membership function as shown in Examples 2 and 3, or a possibility distribution as shown in Example 1. In this study, if the membership function is preferred, then the characteristic function will be denoted as $\mu(x)$. On the other hand, if the possibility distribution is prefered, the characteristic function will be specified as $\pi(x)$.

Along with the expression of Equation (2.2), Zadeh [280] also proposed the following notations. When X is a finite set $\{x_1, x_2, ..., x_n\}$, a fuzzy set A is then expressed as:

$$A = \mu_A(x_1)/x_1 + ... + \mu_A(x_n)/x_n = \Sigma_i \, \mu_A(x_i)/x_i. \tag{2.3}$$

For Example 2, A can be expressed as:

$$A = \mu_A(\text{Hans})/\text{Hans} + \mu_A(\text{John})/\text{John} + \mu_A(\text{George})/\text{George} + \mu_A(\text{Young})/\text{Young}$$
$$= 0.2/\text{Hans} + 1/\text{John} + 0.8/\text{George} + 0/\text{Young}.$$

When X is not a finite set, A then can be written as:

$$A = \int_X \mu_A(x)/x \tag{2.4}$$

Sometimes, we might only need objects of a fuzzy set (but not its characteristic function), that is to transfer a fuzzy set into a crisp set. To do so, we need two concepts - support and α-level cut.

2.2.1.2 Support

The support of a fuzzy set A is the crisp subset of X and is presented as:

$$\text{supp } A = \{x \mid \mu_A(x) > 0 \text{ and } x \in X\}. \tag{2.5}$$

For Example 2, supp A = {Hans, John, George}, where Young does not belong to supp A because of $\mu(\text{Young}) \ngtr 0$.

2.2.1.3 α-level Set (α-cut)

The α-level set (α-cut) of a fuzzy set A is a crisp subset of X and is denoted by (Figure 2.4 shows a continuous case):

$$A_\alpha = \{x \mid \mu_A(x) \geq \alpha \text{ and } x \in X\}. \tag{2.6}$$

For Example 2, $A_{0.2}$ = {Hans, John, George} and $A_{0.7}$ = {John, George}.

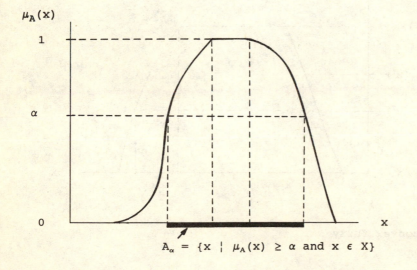

$$A_\alpha = \{x \mid \mu_A(x) \geq \alpha \text{ and } x \in X\}$$

Figure 2.4   An α-level set

Although characteristic function can be assigned by any number, a normalized value between 0 and 1 is always preferred. Thus let us introduce the normality as follows:

2.2.1.4 Normality

A fuzzy set A is normal if and only if $\text{Sup}_x\ \mu_A(x) = 1$, that is, the supreme of $\mu_A(x)$ over X is unity (see Figure 2.4 also). A fuzzy set is subnormal if it is not normal. A non-empty subnormal fuzzy set can be normalized by dividing each $\mu_A(x)$ by the factor $\text{Sup}_x\ \mu_A(x)$. (A fuzzy set is empty if and only if $\mu_A(x) = 0$ for $\forall x \in X$).

It is noted that a characteristic function is always a normalized function through this study.

2.2.1.5 Convexity and Concavity

A fuzzy set A in X is convex if and only if for every pair of point x^1 and x^2 in X, the membership function of A satisfies the inequality:

$$\mu_A(\delta x^1 + (1 - \delta)x^2) \geq \min (\mu_A(x^1), \mu_A(x^2)) \tag{2.7}$$

where $\delta \in [0, 1]$ (see Figure 2.5). Alternatively, a fuzzy set is convex if all α-level sets are convex.

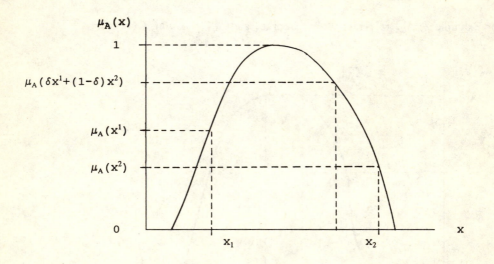

Figure 2.5   A convex fuzzy set

Dually, A is concave if its complement A^c is convex. It is easy to show that if A and B are convex, so is A ∩ B. Dually, if A and B are concave, so is A U B.

2.2.1.6 Extension Principle

Let X be a Cartesian product of universe, $X = X_1 \times X_2 \times ... \times X_n$, and $A_1, A_2, ..., A_n$ be n fuzzy subsets of $X_1, X_2, ..., X_n$, respectively. The Cartesian product of $A_1, A_2, ..., A_n$ is defined as:

$$A_1 \times ... \times A_n = \int_{X_1 x...xX_n} \min [\mu_{A1}(x_1), ..., \mu_{An}(x_n)]/(x_1, ..., x_n). \tag{2.8}$$

Let f be a mapping from $X_1 \times X_2 \times ... \times X_n$ to a universe Y such that $y = f(x_1, x_2, ..., x_n)$, where $x_1 \in X_1, ..., x_n \in X_n$. The extension principle (Zadeh [280]) allows us to induce a fuzzy subset B on Y through f from n fuzzy sets A_i such that:

$$\mu_B(y) = \sup_{y=f(x_1,...,x_n)} \min (\mu_{A1}(x_1), \mu_{A2}(x_2), ..., \mu_{An}(x_n)) \tag{2.9}$$

$$= 0 \quad \text{if } f^{-1}(y) = \emptyset$$

where $f^{-1}(y)$ is the universe image of y. $\mu_B(y)$ is the greatest among the membership values $\mu_{A1x...xAn}(x_1, ..., x_n)$ of the realization of y using n-ary $(x_1, ..., x_n)$.

For example, let $A = \Sigma \mu_A(x)/x = 0.6/-1 + 0.7/0 + 1/1 + 0.3/2$, and $f(x) = x^2$. Then, we have:

$$\mu_B(y) = \sup_{x \in f^{-1}(y)} \mu_A(x) \quad \text{if } f^{-1}(y) \neq \emptyset \tag{2.10}$$

$$= 0 \quad \text{otherwise.}$$

By applying the extension pronciple, $B = 0.7/0 + 1/1 + 0.3/4$ as shown in Figure 2.6:

2.2.1.7 Compatibility of Extension Principle with α-cuts

Denoting the image of $A_1, ... A_n$ by $B = f(A_1, ..., A_n)$, the following proposition holds:

$$\{f(A_1, ..., A_n)\}_\alpha = f(A_{1\alpha}, ..., A_{n\alpha}) \tag{2.11}$$

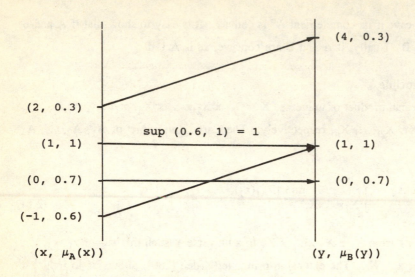

Figure 2.6 The extension principle

iff for $\forall y \in Y$, there exists $(x_1{}^*, ..., x_n{}^*)$ and $\mu_B(y) = \mu_{A1x...xAn}(x_1{}^*, ..., x_n{}^*)$. (The upper bound in Equation (2.9) is obtained.)

2.2.1.8 Relation

A fuzzy relation, R, in the product space $X \times Y = \{(x, y) \mid x \in X$ and $y \in Y\}$ is a fuzzy set in $X \times Y$ characterized by a membership function μ_R which associates with each ordered pair (x, y) a grade of membership $\mu_R(x, y)$ in R. More generally, an n-ary fuzzy relation in a product space $X = X_1 \times X_2 \times ... \times X_n$ is a fuzzy set in X characterized by an n-ary variate membership function $\mu_R(x_1, x_2, ..., x_n)$, $x_i \in X_i$, $i = 1, 2, ..., n$.

For example, let $X = Y = R^1$, where R^1 is the real line $(-\infty, \infty)$. Then $x >> y$ is a fuzzy relation in R^2. A subjective expression for μ_R, in this case, might be:

$$\mu_R(x, y) = 0 \qquad \text{for } x \leq y$$
$$= (1 + (x-y)^{-2})^{-1} \quad \text{for } x > y.$$

2.2.1.9 Decomposability

Let $X = \{x\}$, $Y = \{y\}$ and let C be a fuzzy subset in the product space $Z = X \times Y$ defined by a membership function $\mu_C(x, y)$. Then, C is decomposable along X and Y if and only if C admits of representation $C = A \cap B$ (A is a fuzzy subset in X and B is a fuzzy subset in

Y). That is:

$$\mu_C(x, y) = \mu_A(x) \wedge \mu_B(y) \tag{2.12}$$

where $\mu_A(x)$ and $\mu_B(x)$ are the membership functions of A and B. The same holds for a fuzzy subset in the product of any finite number spaces.

2.2.1.10 Decomposition Theorem

Any fuzzy subset A may be decomposed in the following form:

$$A = \max (\alpha_1 A_{\alpha 1}, \alpha_2 A_{\alpha 2}, \ldots, \alpha_n A_{\alpha n}) = \max_{\alpha i} \alpha_i A_{\alpha i} \tag{2.13}$$

where $\alpha_i \in (0, 1)$, $i = 1, 2, \ldots, n$. For example, $A = ((x_1, 0.2), (x_2, 0.7), (x_3, 1), (x_4, 0))$. Obviously, we should take $\alpha_1 = 0$, $\alpha_2 = 0.2$, $\alpha_3 = 0.7$ and $\alpha_4 = 1$. Then, we have the following results:

x	x_1	x_2	x_3	x_4
$\mu_A(x)$	0.2	0.7	1.0	0.0
$A_{0.0}$	1.0	1.0	1.0	1.0
$A_{0.2}$	1.0	1.0	1.0	0.0
$A_{0.7}$	0.0	1.0	1.0	0.0
$A_{1.0}$	0.0	0.0	1.0	0.0
$0.0 (A_{0.0})$	0.0	0.0	0.0	0.0
$0.2 (A_{0.2})$	0.2	0.2	0.2	0.0
$0.7 (A_{0.7})$	0.0	0.7	0.7	0.0
$1.0 (A_{1.0})$	0.0	0.0	1.0	0.0
$\max_{\alpha i} \alpha_i A_{\alpha i}$	0.2	0.7	1.0	0.0

2.2.1.11 Probability of Fuzzy Events

Let P be a probability measure on Ω. A fuzzy event A in Ω is defined to be a fuzzy subset A of Ω whose membership function μ_A is measurable. Then the probability of A is defined by the Lebesgue-Stieltjes integral:

$$P(A) = \int_\Omega \mu_A(x) \, dP. \tag{2.14}$$

Equivalently, $P(A) = E(\mu_A)$ where $E(\cdot)$ denotes the expectation operator. In the case of a normal crisp set, the above equation will reduce to the conventional definition of the probability of a crisp event.

Example 4[BM38]. Let per capita income be depicted by the fuzzy subset A = "around $a" and its membership function is defined as:

$$\mu_A(x) = \exp\{-(x-a)^2\}, \ x \in R.$$

Assume that the probability distribution of per capita income be normal distribution as depicted by the following probability density function:

$$p(x) = \frac{\exp\{-1/2[(x-m)/\sigma]^2\}}{\sigma\sqrt{2\pi}}$$

where m and σ are mean value and standard deviation, respectively, and $\pi = 0.341416...$ Then the probability of the fuzzy event A is:

$$P(A) = \frac{\int_{-\infty}^{+\infty} \exp\{-(x-a)^2\} \exp\{-1/2[(x-m)/\sigma]^2\} \ dx}{\sigma\sqrt{2\pi}}$$

where P(A) can be defined as the expectation of its membership function.

For a discrete case, let X be a set of events $(x_1, x_2, ..., x_n)$ with probabilities $P(x_i)$ where $P(x_i) \in [0, 1]$ and $\Sigma_i \ P(x_i) = 1$. A fuzzy event A is a fuzzy subset on X whose membership function is $\mu_A(x_i)$. The probability of the fuzzy event A is then defined as:

$$P(A) = \Sigma_i \ \mu_A(x_i)P(x_i). \tag{2.15}$$

Finally, it is worth noting that P(A) satisfies the following properties:

   (a) $0 \le P(A) \le 1$
   (b) $P(\emptyset) = 0$ and $P(\Omega) = 1$
   (c) $P(A) \le P(B)$ if $A \subset B$
   (d) $P(A^c) = 1 - P(A)$
   (e) $P(A \cup B) = P(A) + P(B) - P(A \cap B)$
   (f) $P(A \cup B) = P(A) + P(B)$ if $A \cap B = \emptyset$.

2.2.1.12 Conditional Fuzzy Sets

A fuzzy set B(x) in Y = {y} is conditioned on x if its membership function depends on x as a parameter. This dependence is expressed by $\mu_B(y|x)$.

Suppose that the parameter x ranges over a space X, so that to each x in X corresponds a fuzzy set B(x) in Y. Thus, we have a mapping, characterized by $\mu_B(y|x)$ - from X to the space of fuzzy sets in Y. Through this mapping, any given fuzzy set A in X induces a fuzzy set B in Y which is defined by:

$$\mu_B(y) = \sup_x \min (\mu_A(x), \mu_B(y|x))$$

$$= V_x (\mu_A(x) \wedge \mu_B(y|x)) \tag{2.16}$$

where $\mu_A(x)$ and $\mu_B(x)$ are the membership functions of A and B. Note that this equation is analogous - but not equivalent - to the expression for the marginal probability distribution of the joint distribution of two random variables, with $\mu_B(y|x)$ playing a role analogous to that of a conditional distribution.

2.2.2 Basic Operations

In this sub-section, we will summarize some basic set-theoretic operations which is useful in fuzzy mathematical programming and fuzzy multiple objective decision-making. These operations are based on the definitions from Bellman and Zadeh [12].

2.2.2.1 Inclusion

Let A and B be two fuzzy subsets of X. Then, A is included in B if and only if:

$$\mu_A(x) \leq \mu_B(x) \text{ for } \forall x \in X. \tag{2.17}$$

This will be denoted by A⊂B. For example, suppose X = {x_1, x_2, x_3, x_4}, A = $0.1/x_1 + 0.5/x_2 + 0/x_3 + 1/x_4$ and B = $0.1/x_1 + 0.6/x_2 + 0.3/x_3 + 1/x_4$. According to Equation (2.17), A is included in B (or A⊂B), since $0.1 \leq 0.1$, $0.5 \leq 0.6$, $0 \leq 0.3$ and $1 \leq 1$.

2.2.2.2 Equality

A and B are called equal if and only if:

$$\mu_A(x) = \mu_B(x) \text{ for } \forall x \in X. \tag{2.18}$$

This will be denoted by A = B. For example, suppose $X = \{x_1, x_2, x_3, x_4\}$, $A = 0.1/x_1 +$ $0.5/x_2 + 0/x_3 + 1/x_4$ and $B = 0.1/x_1 + 0.5/x_2 + 0/x_3 + 1/x_4$. According to Equation (2.18), A equals to B   (or A = B), since $0.1 = 0.1$, $0.5 = 0.5$, $0 = 0$ and $1 = 1$.

2.2.2.3 Complementation

A and B are complementary if and only if:

$$\mu_A(x) = 1 - \mu_B(x) \text{ for } \forall x \in X. \tag{2.19}$$

This will be denoted by $B = A^c$ or $A = B^c$, where A^c and B^c are the complements of A and B, respectively. For example, if $B = 0.1/x_1 + 0.6/x_2 + 0.3/x_3 + 1/x_4$, then $A = 0.9/x_1 + 0.4/x_2 + 0.7/x_3 + 0/x_4 = B^c$.

2.2.2.4 Intersection

The intersection of A and B may be denoted by A ∩ B which is the largest fuzzy subset contained in both fuzzy subsets A and B. When the min operator (aggregator) is used to express the logic "and," its corresponding membership is then characterized by:

$$\mu_{A \cap B}(x) = \min (\mu_A(x), \mu_B(x)) \text{ for } \forall x \in X$$
$$= \mu_A(x) \wedge \mu_B(x) \tag{2.20}$$

where $\wedge$ is conjunction here.

Example 5. Consider X = {Hans, John, George, Young}. Suppose A is the fuzzy subset of "good looking students" and B is the fuzzy subset of "intelligent students." The following data are also assumed:

$$A = 0.3/\text{Hans} + 0.4/\text{John} + 0.7/\text{George} + 0.8/\text{Young}$$
$$B = 0.7/\text{Hans} + 0.4/\text{John} + 0.1/\text{George} + 0.3/\text{Young}.$$

Then, the intersection of A and B (if the min-operator is adopted) A ∩ B = 0.3/Hans + 0.4/John + 0.1/George + 0.3/Young which is the fuzzy subset of "good looking and intelligent student" (see Table 2.1 also). Besides, $\wedge$ is considered as a hard "and."

2.2.2.5 Union

The union (A U B) of A and B is dual to the notion of intersection. Thus, the union of A and B is defined as the smallest fuzzy set containing both A and B. The membership function of A U B is given by:

$$\mu_{AUB}(x) = \max\ (\mu_A(x),\ \mu_B(x))\ \text{for}\ \forall x\ \epsilon\ X$$
$$= \mu_A\ V\ \mu_B \tag{2.21}$$

where V is disjunction. For Example 5, A U B = 0.3/Hans + 0.4/John + 0.1/George + 0.3/Young is the fuzzy subset of "good looking or intelligent students" (see Table 2.1 also). Besides, V is considered as a hard "or."

2.2.2.6 Algebraic Product

The algebraic product AB of A and B is characterized by the following membership function:

$$\mu_{AB}(x) = \mu_A(x)\mu_B(x)\ \text{for}\ \forall x\ \epsilon\ X. \tag{2.22}$$

This algebraic product is considered as a soft "and." For Example 5, AB = 0.21/Hans + 0.16/John + 0.07/George + 0.24/Young is the fuzzy subset of "good looking and intelligent students" (see Table 2.1 also).

2.2.2.7 Algebraic Sum

The algebraic sum A ⊕ B of A and B is characterized by the following membership function:

$$\mu_{A \oplus B}(x) = \mu_A(x) + \mu_B(x) - \mu_A(x)\mu_B(x). \tag{2.23}$$

This algebraic sum is considered as a soft "or." For Example 5, A ⊕ B = 0.79/Hans + 0.64/John + 0.73/George + 0.86/Young is the fuzzy subset of "good looking or intelligent students" (see Table 2.1 also).

2.2.2.8 Difference

The difference A - B of A and B is characterized by:

$$\mu_{A \cap B}{}^c(x) = \min\ (\mu_A(x),\ \mu_B{}^c(x)) \tag{2.24}$$

Table 2.1 Solutions of Example 5

X	Hans	John	George	Young
$\mu_A(x)$	0.3	0.4	0.7	0.8
$\mu_B(x)$	0.7	0.4	0.1	0.3
$\mu_{A\cap B}(x)$	0.3	0.4	0.1	0.3
$\mu_{A\cup B}(x)$	0.7	0.4	0.7	0.8
$\mu_{AB}(x)$	0.21	0.16	0.07	0.24
$\mu_{A\oplus B}(x)$	0.79	0.64	0.73	0.86

where B^c is the complement of B with $\mu_{B^c}(x) = 1 - \mu_B(x)$. For Example 5, we have A - B as shown in the following table:

X	Hans	John	George	Young
$\mu_A(x)$	0.3	0.4	0.7	0.8
$\mu_B(x)$	0.7	0.4	0.1	0.3
$\mu_{B^c}(x)$	0.3	0.6	0.9	0.7
$\mu_{A\cap B^c}(x)$	0.3	0.4	0.7	0.7

2.3 Membership Functions

While classical (crisp) mathematics are dichotomous in character, fuzzy set theory considers the situations involving the human factor with all its vagueness of perception, subjectivity, attitudes, goals and conceptions. By introducing vagueness and linguistics into the crisp set theory, fuzzy set theory becomes more robust and flexible than the original dichotomous classical set theory.

There are two essential features of the fuzzy set theory (Zimmermann [301]):

(I) Membership function of fuzzy sets and operators play a crucial role in fuzzy set theory.

(II) Fuzzy set theory is essentially a very general, flexible, formal theory. If it is to be applied to a real problem, it can and has to be adapted carefully. Neither the concept of membership nor the operator has a unique semantic interpretation. The context-dependent semantic interpretation will lead to different mathematical definitions and appropriate operators.

Thus membership functions and operators are actually the cornerstones of the fuzzy set theory. In this section, let us discuss some approaches to generate membership functions, while leaving the discussion of operators (aggregator) for the Section 2.4.

2.3.1 A Survey of Functional Forms

In this section, we will present a state-of-the-art survey of functional forms of membership (both preference-based membership functions and possibility distributions). This survey will provide researchers a direction of future studies in fuzzy and possibilistic mathematical programming problems.

Based on Dombi [61] (with modifications), we classify all existing membership functions into the following four classes:

I. Membership functions based on heuristic determination.

(a) Zadeh's unimodal functions:

$$\mu_{young}(x) = \begin{cases} 1/\{1+[(x-25)/5]^2\} & \text{if } x > 25 \\ 1 & \text{if } x \le 25 \end{cases} \tag{2.25}$$

$$\mu_{old}(x) = \begin{cases} 1/\{1+[(x-50)/5]^{-2}\} & \text{if } x \ge 50 \\ 0 & \text{if } x < 50. \end{cases} \tag{2.26}$$

(b) Dimitru and Luban's power functions:

$$\mu(x) = x^2/a^2 + 1, \ x \in [0, a] \tag{2.27}$$

$$\mu(x) = -x^2/a^2 - 2x/a + 1, \ x \in [0, a]. \tag{2.28}$$

(c) Svarowski's sin function:

$$\mu(x) = 1/2 + (1/2)[\sin\{[\pi/(b-a)][x-(a+b)/2]\}, \ x \in [a, b]. \tag{2.29}$$

II. Membership functions based on reliability concerns with respect to the particular problem.

(a) Zimmermann's linear function:

$$\mu(x) = 1 - x/a, \ x \ \epsilon \ [0, a]. \tag{2.30}$$

(b) Tanaka, Uejima and Asai's symmetric triangular function:

$$\mu(x) = \begin{cases} 1 - \mid b\text{-}x \mid /a, & \text{if } b\text{-}a \leq x \leq b\text{+}a \\ 0 & \text{otherwise.} \end{cases} \tag{2.31}$$

(c) Hannan's piecewise linear function:

$$\mu(x) = \Sigma_j \ \alpha_j \mid x\text{-}a_j \mid \ + \beta x + r, \ j = 1, 2, \ldots, N \tag{2.32}$$

$$\alpha_j = (t_{j+1} - t_j)/2$$
$$\beta = (t_{N+1} + t_1)/2$$
$$r = (s_{N+1} + s_1)/2$$

where $\mu(x) = t_i x + s_i$ for each segment i, $a_{i\text{-}1} \leq x \leq a_i$, $\forall i$. Here, t_i is the slope and s_i is the y-intercept for the section of the curve initiated at $a_{i\text{-}1}$ and terminated at a_i.

(d) Leberling's hyperbolic function:

$$\mu(x) = 1/2 + (1/2)\tanh(a(x\text{-}b)), \ -\infty \leq x \leq \infty \tag{2.33}$$

where a is a parameter.

(e) Sakawa and Yumine's exponential and hyperbolic inverse functions, respectively:

$$\mu(x) = c(1\text{-}e^{(b\text{-}x)/(b\text{-}a)}), \ x \ \epsilon \ [a, b] \tag{2.34}$$

$$\mu(x) = 1/2 + c\tanh^{-1}(d(x\text{-}b)) \tag{2.35}$$

where c and d are parameters.

(f) Dimitru and Luban's function:

$$\mu(x) = 1/(1+x/a) \tag{2.36}$$

where a is a parameter.

(g) Dubois and Prade's L-R fuzzy number:

$$\mu(x) = \begin{cases} L((a-x)/\alpha) & \text{if } x < a \\ R((x-b)/\beta) & \text{if } x > b \\ 1 & \text{if } a \le x \le b \end{cases} \tag{2.37}$$

where $L(\cdot)$ and $R(\cdot)$ are reference functions.

III. Membership functions based on more theoretical demand.

(a) Civanlar and Trussel's function:

$$\mu(x) = \begin{cases} ap(x) & \text{if } ap(x) \le 1 \\ 0 & \text{otherwise} \end{cases} \tag{2.38}$$

where $a \in [0, 1]$ is a parameter and $p(x)$ is the probability density function.

(b) Svarovski's function:

$$\mu(x) = \begin{cases} 0 & \text{if } x < a \\ K(x-a)^2 & \text{if } a \le x \le b \\ K_2 x^2 + K_1 x + K_0 & \text{if } b < x \le c \\ 1 & \text{if } x > c \end{cases} \tag{2.39}$$

where K, K_0, K_1 and K_2 are parameters.

VI. Membership functions as a model for human concepts.

(a) Hersh and Caramazza's function (which was experimentally formed in order to determine the nature of context effects upon the interpretation of a set of nature language terms, such as short, very short, sort of short, etc.):

$$\mu(x) = 1/2 + d(r/10) \tag{2.40}$$

where $d(x) = 1$ for "yes" reponses and $d(x) = -1$ for "no" responses and r is a confidence value.

(b) Zimmermann and Zysno's (see Section 2.3.2.1) function:

$$\mu(x) = 1/2 + (1/d)[1/(1+e^{-a(x-b)}) - c].$$ (2.41)

(c) Dombi's function:

$$\mu(x) = (1-s)x^2/\{(1-s)x^2+s(1-x)^2\}$$ (2.42)

where s (the characteristic value of the shape) is the intersection value of $y = \mu(x)$ and $y = x$.

As discussed in Chapter 1, membership functions can be distinguished into two classes of preference-based membership functions and possibility distributions. The preference-based membership function is constructed by eliciting the preference information from the decision makers. On the other hand, the possibility distribution, which is an analogous of probability distribution, is constructed by considering the possible occurrence of the events. However, this section is to show the existing functional forms, yet not to discern the differences between preference-based membership functions and possibility distributions.

The functions, which are most often used in modelling fuzzy and possibilistic mathematical programming problems, have been shown in Figures 1.1 and 1.2. Detailed discussions will be given in Chapters 3 and 4.

When modelling fuzziness and imprecision of the input data (A, b and c) in fuzzy and possibilistic mathematical programming problems, preference-based membership functions and possibility distributions are always assumed to be rationally given. Thus, how to generate preference-based membership functions and possibility distributions becomes a natural question. In the next section, we provide some examples of generating membership functions as references for interested readers.

2.3.2 Examples to Generate Membership Functions

There are two approaches to generate membership functions: axiomatic and semantic approaches (Giles [106]). The axiomatic approach, which is similar to the approaches used in utility theory (Zimmermann [301]), is centered on the mathematical consideration. On the other hand, the semantic approach is concentrated on the practical interpretation of the terms but not on the mathematical structure which is emphasized in the axiomatic concept. The semantic approach actually follows the perceptions of pragmatism in its insistence that all conclusions should be firmly based on the practical meaning of the concepts involved. There is no axiom or law - in principle not even the syntax - laid down in advance. The purpose is to start with an analysis of the immediate practical meanings of the concepts. These properties are then formulated in precise terms as axioms. Other concepts may then be defined in terms of the initial ones, and so a formal theory is gradually built up. Unlike the axiomatic approach, the semantical approach provides a derivation of a particular mathematical theory, and from the way the concepts are introduced, it yields a specific interpretation of this theory at the time.

In this study, we concentrate on the applications and not pure mathematics. The practical meaning and interpretation of membership are considered more important than mathematical terms. The semantical approaches, thus, are emphasized and discussed in the following subsections. However, this section is not required to understand fuzzy and possibilistic mathematical programming.

2.3.2.1 Distance Approach

When eliciting membership functions, the determination of the lowest necessary scale level of membership for a specific application is very important. Generally, the required scale level should be as low as possible in order to facilitate data acquisition which usually affords the participation of human beings. Besides, a suitable numerical handling is desirable in order to insure mathematically appropriate operations. Among the five classical scale levels (nominal, ordinal, interval, ratio, and absolute scale), Zimmermann and Zysno [305] proposed that the interval scale level seems to be most adequate, since the intended mathematical operations require at least interval scale quality.

The concept of using a distance $d(x)$ from a reference as an evaluation criterion has been popular in many fields, such as multiple attribute decision-making, multiple objective decision-making, goal decision-making, et al. Zimmermann and Zysno used this concept to derive membership functions. The rationality of this concept is that if the object has all the ideal features, the distance should be zero. By contrast, if no similarity between the object and the ideal exists, the distance then should be ∞. If this evaluation concept is formally represented by a fuzzy set A in X, then a certain degree of membership $\mu_A(x)$ (or simply μ) will be assigned to each element x. Membership can then be defined as a function of the distance between a given object x and a standard (ideal). Thus Zimmermann and Zysno proposed the following relation for μ and $d(x)$:

$$\mu = 1/(1 + d(x)) \qquad (2.43)$$

where $d(x) = 0 \rightarrow \mu = 1$, and $d(x) = \infty \rightarrow \mu = 0$. Obviously, Equation (2.43) is just a transformation rule (or a normalization process) which maps real number R into the interval [0, 1].

However, experience shows that ideals are very rarely ever fully realized and distance function is quite context dependent. The context-dependent parameters a and b are then created to represent the evaluation unit and the reference/standard, respectively. For instance, a may be the unit of length such as feet, meter, yards, etc., and b may be a fast and rough pre-evaluation such as "rather positive," "rather negative," etc. Since the relationship between physical units and perceptions is generally exponential, Zimmermann and Zysno proposed the following distance function as:

$$d(x) = 1/e^{a(x-b)} \tag{2.44}$$

and then:

$$\mu = 1/[1 + e^{-a(x-b)}] \tag{2.45}$$

where a, b can be considered as semantic parameters from a linguistic point of view.

Since concepts or categories, which are formally represented by sets, are normally linguistically described, the membership function is the formed representation of meaning. The vagueness of the concept is operated by the slope a and the identification threshold by b. For managerial terms such as "appropriate dividend" or "good utilization of capacities," the parameter a models the slope of the membership function in the tolerance interval and b represents the point at which the tendency of the subject's attitude changes from rather positive into rather negative.

Equation (2.45), however, is still too general to fit subjective models of different persons. Frequently, only a certain part of the logistic function is needed to represent a perceived situation. In order to allow for such a calibration, it is assumed that only a certain interval of the physical scale is mapped into the open interval (0, 1). And, the membership grades of the lower and upper bounds are assigned to be 0 and 1, respectively.

Since an interval scale is requested, the interval of the degrees of membership may be transformed linearly. On this scale level the ratios of two distances are in variant. Let μ^U and μ^L be the upper and lower bounds of the normalized membership scale, respectively, and μ_i is a degree of membership between these bounds, $\mu^L < \mu_i < \mu^U$, and let $\mu^{L'}$ and $\mu^{U'}$ be the corresponding values on the transformed scale. Then we have:

$$(\mu_i - \mu^L)/(\mu^U - \mu^L) = (\mu_i' - \mu^{L'})/(\mu^{U'} - \mu^{L'}).$$ (2.46)

For the normalized membership function, one may have $\mu^L = 0$ and $\mu^U = 1$. Hence,

$$\mu_i' = \mu_i(\mu^{U'} - \mu^{L'}) + \mu^{L'}.$$ (2.47)

Generally, it is preferable to define the range of validity by specifying the interval d with the center c (see Figure 2.7) as follows:

$$\mu^{U'} - \mu^{L'} = d \text{ and } (\mu^{U'} + \mu^{L'})/2 = c.$$ (2.48)

Then Equation (2.47) will become:

$$
\begin{aligned}
\mu_i' &= \mu_i(d) + \mu^{L'} \\
&= d\mu_i - d/2 + d/2 + \mu^{L'} \\
&= d\mu_i - d/2 + [(\mu^{U'} - \mu^{L'})/2 + \mu^{L'}] \\
&= d(\mu_i - 1/2) + c
\end{aligned}
$$ (2.49)

which can be specified by two parameters, a and b, if μ_i is replaced by μ_i'.

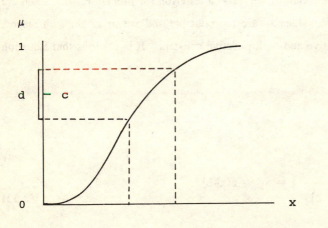

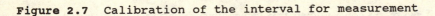

Figure 2.7 Calibration of the interval for measurement

Solving Equation (2.49) for μ_i will lead to the following complete membership model:

$$\mu_i = \left|\, (1/d)\left(\frac{1}{1 + e^{-a(x-b)}} - c \right) + 1/2 \,\right| \tag{2.50}$$

where $\mu_i \in [0, 1]$, $\mu_i(x) = 1$ for $x > x^U$, and $\mu_i(x) = 0$ for $x < x^L$.

The determination of the parameters from an empirical data base does not pose any difficulties in the general model of Equation (2.45). That is:

$$\mu = 1/[1 + e^{-a(x-b)}] \quad => \quad \ln\,[\mu/(1-\mu)] = a(x-b). \tag{2.51}$$

Now, suppose $y = \ln\,[\mu/(1-\mu)]$. A linear relationship between x and y is obvious. The straight line of the model is then defined by the least squares of deviations.

The estimation of the parameter c and d in the extended model still poses some problems. There is no direct way for a numerically optimal estimation. Thus Zimmermann and Zyson proposed the following interactive procedure. First, they assumed that a set of stimuli which is equally spread over the physical continuum was chosen such that the distance between any two of the neighboring stimuli is constant:

$$x_{i+1} - x_i = s \tag{2.52}$$

where s is a constant distance. This condition serves as a criterion for precision. If c and d are correctly estimated, then those scale values x_i' are reproducible and are invariant with respect to x_i with the exception of the additive and multiplicative constant. It is obvious that Equation (2.50) can be rewritten as:

$$d(\mu_i - 1/2) = 1/[1 + e^{-a(x-b)}]$$

or

$$\ln\left\{ \frac{d(\mu_i-1/2) + c}{1 - [d(\mu_i-1/2) + c]} \right\} = x_i' = a(x_i - b) \tag{2.53}$$

Let s' be the distance between the pair x_{i+1}' and x_i', and M' be their mean value. If the estimated values d^e and c^e are equal to their true values, then the estimated distance $s^{e'}$ and the mean $M^{e'}$ are equal to their respective true values and vice versa:

$$(d^e = d) \text{ and } (c^e = c) \quad <=> \quad (s^{e'} = s) \text{ and } (M^{e'} = M). \tag{2.54}$$

Our aim, therefore, is to reach the equivalence of $s^{e'}$ and s as well as $M^{e'} = M$.

Algorithm.

Step 0. Determine the initial appropriate values c_1 and d_1 by the following formula:

$$c_1 = (\Sigma_i \, \mu_i)/n \tag{2.55}$$
$$d_1 = \min (1-1/k, \, 2(1-c), \, 2c) \tag{2.56}$$

where n is the member of stimuli and k is the number of different degrees of membership. If only the values 0 and 1 occur, $d_1 = 1/2$. Only the linear interval in the middle of the logistic function is used. With increasing k, d converges to 1; the entire range of the function is then used. In any case d must not exceed the minor part of 2c or 2(1-c).

Step 1. Determine the x_i' which corresponds to the empirically determined μ_i:

$$M^{e'} = (\Sigma_i \, x_i')/n \tag{2.57}$$

$$s^{e'} = [\Sigma_i \, (x_{i+1}-x'_i)]/(n-1)$$
$$= (x_n'-x_1')/(n-1). \tag{2.58}$$

Step 2. Calculate the absolute difference between two estimates. That is:

$$| \, M^{e''} - M^{e'} \, | \leq \delta_M \tag{2.59}$$
$$| \, s^{e''} - s^{e'} \, | \leq \delta_s \tag{2.60}$$

where δ_M and δ_s are predetermined tolerances. If Equations (2.57) and (2.58) are satisfied, the last estimate is accepted as sufficiently exact and then STOP. Otherwise, go to the next step.

Step 3. Determine the interval of the base variable x_i' which corresponds to the (0, 1) interval of the membership value. That is to compute:

$$x^{U'} = M^{e'} + (n/2)s^{e'} \tag{2.61}$$

$$x^{L'} = M^{e'} - (n/2)s^{e'} \tag{2.62}$$

which is the upper and lower bounds, respectively.

Now, by Equation (2.53), the corresponding $\mu^{U'}$ and $\mu^{L'}$ can be computed, and then by Equation (2.48), new parameters c^e and d^e are estimated. Then go to Step 1.

Example. From the empirical evidence that 64 subjects (16 for each set) from 21 to 25 years of age individually rated 52 different statements of age concerning one of the four fuzzy sets: "very young man," "young man," "old man" and "very old man," Zimmermann and Zyson obtained monotonic membership functions as shown in Figure 2.8, and unimodal membership functions as shown in Figure 2.9.

The monotonic membership functions of "old man" and "very old man" (Figure 2.8) are rather similar. They differ only with respect to their inflection points, indicating a difference of about five years between "old man" and "very old man." The same holds for the monotonic membership functions of "very young man" and "young man"; their inflection points differ by nearly 15 years. It is interesting to note that the modifier "very" has a greater effect on "young" than "old," but in both cases it can be formally represented by a constant.

Finally, the meaning of "young" is less vague than that of "old" if the slope is an indicator for vagueness as shown in Figure 2.8. On the other hand, the variability of membership functions may be regarded as an indicator of ambiguity. Thus, though being less vague, "young" seems to be more ambiguous.

2.3.2.2 True-Valued Approach

Smets and Magrez [229] provided a definition of what is meant by a proposition with a truth value of 0.35 or any other intermediate value between True (or 1) and False (or 0). A canonical scale for the truth values is defined, and sets of fuzzy propositions are constructed for which the truth value has a unique numerical value on the canonical scale. Such a set of propositions with well-defined intermediate truth values is necessary to give meaning to the assertion "the truth P is 0.35."

Fuzzy logic has two characteristics: the truth domain is the whole [0, 1] interval, and the truth value can be a fuzzy subset of [0, 1]. Smets and Magrez considered the first characteristic, that is to reduce fuzzy logic to its multi-valued logic components.

According to Zadeh [274], fuzzy sets are those for which one can define, for each element, a grade of membership in [0, 1]. Smets and Magrez then postulated the relation between degree of truth and grade of membership so that the degree of membership $\mu_A(x)$ of an

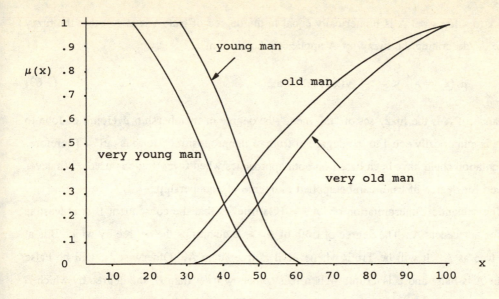

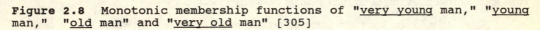

Figure 2.8 Monotonic membership functions of **"<u>very young</u> man," "<u>young</u> man," "<u>old</u> man" and "<u>very old</u> man"** [305]

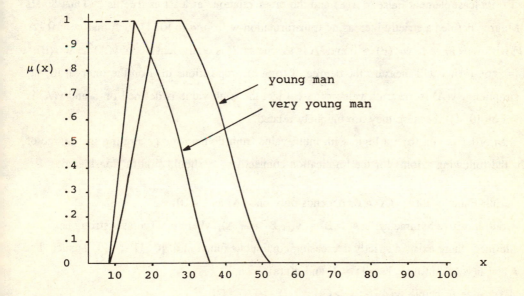

Figure 2.9 Unimodal membership functions of **"<u>very young</u> man" and "<u>young</u> man"** [305]

element x to a fuzzy set A is numerically equal to the degree of truth v(x is A') that the fuzzy predicate A' describing the fuzzy set A applies to the element x:

$$\mu_A(x) = a \quad <=> \quad v(x \text{ is } A) = a. \tag{2.63}$$

For instance, if A is the fuzzy set of "tall men," the degree of membership μ_A(John) of John to the set A is numerically equal to the degree of truth of the proposition "John is <u>tall</u>". Therefore, the presentation could have been based on both approaches when fuzzy logic is used. Whatever is deduced for degree of truth can be applied to degree of membership.

The semantical interpretation of "A -> B is true" is that the consequent B is at least as true as the antecedent A. The degree of truth of A -> B quantifies the degree by which B is at least as true as A. It will be Truth whenever B is truer than A. Otherwise, it will be False whenever A is False and B is Truth. When A is somehow truer than B, the degree by which B is at least as true as A might be intermediate between True and False.

Now, let v(A) be the truth value of proposition A with A ϵ Ω, where Ω is a Boolean algebra of propositions with T its tautology and F its contradiction. Then v: Ω -> σ is a mapping from Ω to a truth domain σ, where σ is a bounded set with its greatest element Truth = v(T), its least element False = v(F) and the order relation "at least as true as." Thus Smets and Magrez defined a strictly increasing transformation w from σ to [0, 1] such that w·v: Ω -> [0, 1] with w·v(T) = 1, w·v(F) = 0 and 'A is as true as B' is equivalent to "w·v(A) $\geq$ w·v(B)." If v = w·v, then v will measure the degree of truth of propositions in Ω on the interval [0, 1]. For simplicity, v(A) is the truth value of A (in fact the truth value is defined on σ and v(A) is defined on [0, 1]), because they are uniquely related.

In order to construct a logic with multi-value truth domain for generating the degree of truth, the following axioms for the implication connective -> should first be considered:

A1. Truth-Functionality: v(A->B) depends only on v(A) and v(B).

A2. Contrapositive Symmetry: v(A->B) = v($\neg$ B-> $\neg$ A), where the Trillas's strong negation $\neg$ is defined: there exists a strictly decreasing continuous function n: [0, 1] -> [0, 1] such that v($\neg$ A) = n(v(A)), n(0) = 1, n(1) = 0 and n(n(a)) = a.

A3. Exchange Principle: v(A->(B->C)) = v(B->(A->C)).

A4. Monotony: v(A->B) $\geq$ v(C->D) if v(A) $\leq$ v(C) and v(B) $\geq$ v(D).

A5. Boundary Condition: v(A->B) = 1 iff v(A) $\leq$ v(B).

A6. Neutrality Principle: v(T->A) = v(A).

A7. Continuity: v(A->B) is a continuous function of v(A).

With these axioms, Smets and Magrez have proved that the implication and the negation operators are necessary such that:

$$v(A->B) = f^{-1}(min\{f(1)-f(a)+f(b),f(1)\}) \tag{2.64}$$

$$v(\neg A) = f^{-1}(f(1)-f(a)) \tag{2.65}$$

where a = v(A), b = v(B), and the generator f is any bounded, continuous, monotonically increasing function from [0, 1] to [0, ∞) with f(0) = 0 and f(1) < ∞.

The f generator is defined up to any strictly monotone transformation. As the truth scale v(A) is also defined up to any strictly monotone transformation, we can then define the canonical scale for the truth value v(A) of a proposition by selecting the f generator such that f(v(T)) = f(1) = 1 and f(v(A)) = f(a) = a. In that case, we obtain the Lukasiewicz operator for the material implication connective:

$$v(A->B) = min \{1-v(A)+v(B), 1\} \tag{2.66}$$

$$v(\neg A) = 1 - v(A). \tag{2.67}$$

This conclusion is quite natural as f and v can always be adapted in order to obtain such a scale, and it is obviously the simplest we can construct.

However, this canonical scale does not explain what is meant by a truth value of 0.35. It must mean something more than just that it is between 0.34 and 0.36. Thus we need a set of propositions for which the truth value is uniquely defined. These propositions could be used later as a reference scale to measure the truth of other propositions.

According to Gaines, Smets and Magrez constructed a reference scale which defines the meaning of any truth value in [0, 1] as follows:

1. Let A = "John is _tall_" and ¬ A = "John is _not tall_." The truth of A and of ¬ A depends on the height h of John. If h = 110cm, A is false and ¬ A is true. If h = 190 cm, A is true and ¬ A is false (see Table 2.2). Thus, when h increases from 110 to 190, a = v(A) increases continuously from 0 = v(F) to 1 = v(T) (where T and F are the tautology and contradiction) and na = v(¬ A) decreases continuously from 1 to 0.

Table 2.2 Heights and true values of A = "John is <u>tall</u>"
and B = "Paul is <u>tall</u>" [372]

height	110cm	k'	h'	190cm
v(A)	0		0.5	1
v(¬A)	1		0.5	0
v(B)	0	0.25	0.5	1
v(¬B)	1	0.75	0.5	0
v(¬B -> B)	0	0.50		

There exists a value h' such that a = na. In such a case, ¬ A -> A is true, therefore
$v(\neg A\text{->}A)$ = 1 becomes $f^{-1}(f(1)\text{-}f(na)+f(a))$ = 1. As $f(na)$ = $f\bullet f^{-1}(f(1) - f(a))$ = $f(1) - f(a)$,
we have $f^{-1}(2f(a))$ = 1, thus $2f(a)$ = $f(1)$.

As $f(1) < \infty$, we can use an f scale such that $f(1) = 1$, in which case $f(a) = f(na) = 0.5$.
It is noted that the result would have been directly obtained by considering the strong negation
and the canonical scale as a = na is identical to a = 1-a.

2. Let B = "Paul is <u>tall</u>" and ¬ B = "Paul is <u>not tall</u>." The truth of B and of ¬ B
depend on the height k of Paul. Let us further postulate that k ≤ h', thus b = v(B) ≤ 0.5.

Consider the proposition ¬ B -> B, i.e., 'B is at least as true as ¬ B'. Its truth value c
= $v(\neg B\text{->}B)$ depends on k. If k = h', c = 1 as v(B) = v(¬ B). If k = 110cm, B is false and
¬ B is true, therefore c = 0. By increasing k from 110cm to h', c is increasing from 0 to 1.
Thus, there exists a k' such that $v(\neg B\text{->}B)$ = v(A) = 0.5 implies $f(1) - f(nb) + f(b) = 2f(b)$
= f(a) = 0.5 => f(b) = 0.25 and f(nb) = 0.75.

3. Let P_0 = A, P_1 = B, h_0 = h' and h_1 = k'. The procedure is iterated with the
propositions $P_i \equiv$ "X_i is <u>tall</u>" such that $v(\neg P_i\text{->}P_i) = v(P_{i\text{-}1})$, i = 1, 2, ..., and $v(P_i) < v(P_{i\text{-}1})$.
The values of the heights h_i of X_i, all i, are derived as above. For such h_i and with $p_i = v(P_i)$,
we obtain $f(p_i) = 2^{-i\text{-}1}$ and $f(np_i) = 1 - 2^{-i\text{-}1}$.

4. Other values of f can be obtained from expression based on the truth of ¬ P_i -> P_j,
as it corresponds to $f(p_i) + f(p_j) = 2^{-i\text{-}1} + 2^{-j\text{-}1}$. Further values are obtained from expressions like
$((\neg P_i\text{->}P_j)\text{->}P_k)\text{->}P_s$ whose true value is $f(p_i) + f(p_j) + f(p_k) + f(p_s)$, etc.

Appropriate sequences of implications based on propositions P_i can be constructed in order
to obtain any f value. For instance, let us construct the sequence to obtain f = 0.65625. As
0.65625 = 0.50 + 0.125 + 0.03125, 0.65625 = $f(p_0) + f(p_2) + f(p_4)$, it corresponds to the true
value of the implication $(\neg P_0\text{->}P_2)\text{->}P_4$ with P_i = "X_i is <u>tall</u>" and the height of X_i is h_i; thus,

thus, 0.65625 is the degree of truth of the proposition "P_4 is at least as true as not 'P_2 is at least as true as $\neg P_0$'" or "P_4 is at least as true as 'P_2 is less true than P_0'."

Finally, it is noted that f function is defined up to any strictly monotonic transformation such that $f(0) = 0$ and $f(1) = 1$. Thus we can state without loss of generality that $f(x) = x$, for computational efficiency. By use of the previous canonical scale, $v(P_i) = 2^{-i-1}$, and all other truth values can be equated to some sequence of implications based on propositions P_i. We have derived a reference scale for the truth value of any proposition Q by the use of the relation "Q is as true as" where P is a sequence of implications based on propositions P_j. Thus P and Q share the same numerical truth value.

2.3.2.3 Payoff Function

Giles [106] pointed out that the concepts of a set (fuzzy or not) are closely related to that of a property: any set determines a property which belongs to the set, and any property determines a set which contains all objects with the property. For example, to the property "tall," there corresponds the set of all tall people (objects). Then any statement about sets can be translated into a corresponding statement about properties, and vice versa. The concepts of the set and property thus appear to be equivalent. Nevertheless, the concept of a property is actually less abstract and closer to common usage than that of a set. Thus the analogue of degree of membership in a fuzzy set can be equivalent to that of the "degree of possession" of the corresponding property. For instance, the degree of membership of John in the set of all tall people may be identified with the degree of which John has the property, "tall."

There is a third notion related to these two concepts. This is "degree of truth" - for the previous example, the degree to which John is tall with the "degree of truth" of the statement, "John is tall." Every assignment of a degree of membership in a set or of a degree of possession of a property can be equally well-regarded as an assignment of a degree of truth to a corresponding statement. Indeed, we can hardly contemplate intermediate (other than 0 and 1) grades of membership in a set unless we are also prepared to consider intermediate (other than true and false) degrees of truth for a statement. Conversely, should we succeed in the task of attaching a well-defined meaning to intermediate degrees of truth, we would at the same time have given a meaning to intermediate grades of membership and intermediate "degrees of possession" of a property.

Once we have attached a clear meaning to statements with other than classical truth values, we can deduce how one should reason with such statements: the appropriate logic is determined as soon as a clear understanding of the meaning of the statements with which it has to operate has been reached. Thus the problem of the interpretation of grades of membership cannot be considered independently of the problem of fuzzy reasoning itself: it is part and parcel of the latter; a solution of one is a solution of both.

Giles further pointed out that the "assertions" concept, instead of the statements themselves, is fundamental, and the meaning of an assertion can be identified by its payoff function in terms of the decision theory. For example, consider the assertion "John is <u>tall</u>." The meaning of the fuzzy statement "John is <u>tall</u>" is that information specific to this fuzzy statement which in conjunction with one's state of belief regarding the relative possibility of different world states (here the relevant is the height of John in these various possible states) will allow him to decide whether (and how willingly) to assert "John is <u>tall</u>."

Now, let us use this example, "John is <u>tall</u>," to discuss how to construct the degree of membership from the "assertions" via the payoff concept.

Usually, one who asserts "John is <u>tall</u>" will receive approval from his peers to an extent dependent primarily on the height of John. In other words, the payoff of the assertion in any world state is approximately a function f(h) of the height h of John in that state. Insofar as this is the case, we can represent the payoff function of the assertion by the graph of the function f. Clearly, the utility will increase with height from some negative initial value to positive values. For example, the graph might be as shown in Figure 2.10. In this figure the payoff value f(h)

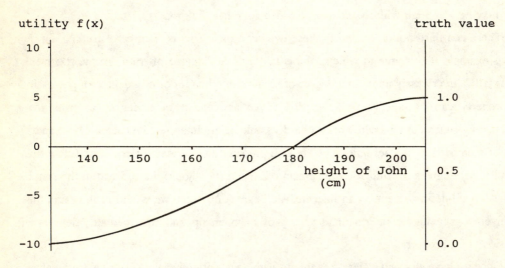

Figure 2.10 Payoff function of the assertion "John is tall" [106]

may be described as the degree of willingness of an agent to assert "John is <u>tall</u>." In particular, a positive (negative) value for the payoff $f(h)$ indicates that the agent will (won't) choose to assert "John is <u>tall</u>" when he knows John's height h. So one can tell where the graph crosses the axis just by asking for the least height at which John would be described as tall.

We can then determine other points on the graph of $f(h)$ by offering bribes. For instance, the graph suggests that the agent would be willing to assert "John is <u>tall</u>" even if $h = 170$, provided he was offered as a reward a prize worth 5 units of payoff. Actually, this argument is not quite correct, since when two events occur together their payoffs do not necessarily increase. Fortunately, one can flip a fair coin. If it shows a head, he will get the reward. Otherwise, he will be obligated to assert "John is <u>tall</u>." Therefore, if he agrees to this proposal, provided that he was sure John's height was at least 170cm, then $f(170) > -5$ is obtained. By using a lottery instead of the coin, one can similarly determine all other negative ordinates on the graph. On the other hand, the positive ordinates are also found by using a penalty instead of a reward.

It is noted that the previous discussion of the payoff function for the assertion "John is <u>tall</u>" has been simplified in the following two respects:

I. It is only approximately true that the payoff is a function of height, according to the following two reasons:

a) In a normal society the term "tall" does not solely represent geometrical height. For instance, a child is more likely to be considered tall than an adult of the same height.

b) Even the assertion that the payoff depends only on the nature of John must be qualified.

II. The meaning of an assertion, as represented by its payoff function, is determined by the society in which the assertion is used, and does not depend on the particular person who makes the assertion.

As a matter of fact, an assertion can be brought into standard form by a scaling and a shift. Here let us call the payoff function of the resulting standardized assertion "the truth function of the asserted fuzzy statement." The function is then the truth function of the sentence. Truth functions of fuzzy statements thus coincide with payoff functions of standardized assertions except that the extreme values 0 and 1 are necessarily attained in the latter case. Therefore, the graph of the truth function of the asserted fuzzy statement may be obtained from the graph of the payoff function of the assertion simply by changing the calibration on the y-axis. From Figure 2.10, the degree of the truth function for the fuzzy statement "John is <u>tall</u>" can be obtained by reading the right hand scale. The degree of membership of John in the fuzzy set of tall men is then identified with the _____ ent "John is <u>tall</u>."

For the payoff function, the following two significant points should be noted:

(1) It is clear that for any practical assertion, one value must be positive and the other negative. Otherwise, the assertion would either always or never be made, so that it would be useless for conveying the beliefs of a speaker.

(2) A scaling of the payoff function (multiplying it by a positive scale factor) corresponds to a change in emphasis of the assertion. For example, the payoff function for "John is <u>tall</u>" may be taken to differ only by a scale factor (>1) from that for the less emphatic "John is <u>tall</u>." Turning this observation around, one should regard any assertions a and a', whose payoff functions differ only by a positive scale factor, as assertions of the same fuzzy sentence except more (or less) emphasized.

2.3.2.4 Other Examples

Besides approaches presented in Sections 2.3.2.1, 2.3.2.2 and 2.3.2.3, there are several other considerations and approaches. Rapoport, Wallsten and Cox [213] dealt with the synthesis of subjective data for deriving membership functions. They provided direct (magnitude estimation) and indirect (graded pair-comparison) concepts to establish membership functions for probability phrases such as probable, rather likely, very likely, very unlikely, and so forth. Eliciting membership function is considered as eliciting utility function in the decision theory.

Bandemer [7] developed a fuzzy counterpart to regression analysis methods accounting for fuzzy observation. The fuzzy (functional) relations were then derived either directly, or via families of parametrized functions.

Pedrycz [203] solved an identification problem in terms of fuzzy relational equations from input/output fuzzy sets. Based on the clustering technique, Pedrycz proposed a general methodological scheme which includes the following three steps: (i) structure determination, (ii) parameter determination, and (iii) model valuation.

Bezdek and Hathaway [15] proposed the relational hard c-means (HCM) algorithm for the classification of objects when information about pairwise relationship between these objects is available. Their approach can be used to derive membership functions.

2.4 Fuzzy Decision and Operators

In a decision-making process, the principal ingredients include a set of alternatives/decisions, a set of constraints on the choice between different alternatives, and a performance/objective function which associates with each alternative the gain (or loss) resulting from the choice of that alternative. When this decision-making process is used in a fuzzy environment, Bellman and Zadeh [12] proposed that the symmetry between goals and constraints is the most important feature. A symmetry erases the differences between them and makes it possible to relate in a relatively simple way the concept of a decision to those of the goals and constraints of a decision process.

With this property of symmetry, Bellman and Zadeh proposed that a fuzzy decision is defined as the fuzzy set of alternatives resulting from the intersection of the goals/objectives and constraints.

2.4.1 Fuzzy Decision

Assume that objective(s) and constraints in an imprecise situation can be represented by fuzzy sets. For an illustration, suppose that we have a fuzzy goal G and a fuzzy constraint C in a decision space X expressed as follows:

G: x should be substantially larger than 10 , with

$$\mu_G(x) = [1 + (x\text{-}10)^{-2}]^{-1} \quad \text{if } x \geq 10$$
$$= 0 \quad \quad \text{if } x < 10; \text{ and}$$

C: x should be in the vicinity of 15,

$$\mu_C(x) = [1 + (x\text{-}15)^4]^{-1}.$$

Then, with the assumption of the symmetry we may make decisions which satisfy both the constraint "and" the goal. That is: G and C are connected to another by the operator "and" which corresponds to the intersection of fuzzy sets. This implies that in the example the combined effect of the fuzzy goal G and the fuzzy constraint C on the choice of alternatives may be represented by the intersection G∩C, with the membership function (see Figure 2.11):

$$\mu_{G \cap C}(x) = \mu_G(x) \wedge \mu_C(x) = \min \{\mu_G(x), \mu_C(x)\}. \tag{2.68}$$

Then Bellman and Zadeh [12] proposed that a fuzzy decision may be defined as the fuzzy set of alternatives resulting from the intersection of the goals and the constraints. That is: the decision D = G∩C is a fuzzy set resulting from the intersection of G and C, and has its membership function $\mu_D = \mu_G \cap \mu_C$ (see Figure 2.11).

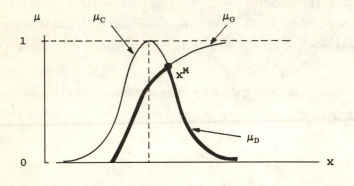

Figure 2.11 The relationship of G, C and D

A maximizing decision then can be defined as follows:

$$\mu_D(x^M) = \max \mu_D(x) \quad \text{for } x \in X$$
$$= 0 \qquad \text{elsewhere.} \tag{2.69}$$

If $\mu_D(x)$ has a unique maximum at x^M, then the maximizing decision is a uniquely defined crisp decision which can be interpreted as the action which belongs to all fuzzy sets representing either constraints or objective(s) with the highest possible degree of membership.

Suppose that the goals are defined as fuzzy sets G_1, G_2, ..., G_k in $Y = \{y\}$ while the constraints C_1, C_2, ..., C_m are defined as fuzzy sets in $X = \{x\}$. Now, given a fuzzy set G_i in Y, we can then find a fuzzy set G'_i in X which induces G_i in Y. Specifically, with $y = f(x)$ the membership function of G'_i is given by the equality:

$$\mu_{G'i}(x) = \mu_{Gi}(f(x)), \forall i. \tag{2.70}$$

The decision D, then, can be expressed as the intersection of G'_1, G'_2, ..., G'_k, and C_1, C_2, ..., C_m. By using Equation (2.70), we can then obtain the following:

$$\mu_D(x) = \mu_{G1}(f(x)) \wedge ... \wedge \mu_{Gk}(f(x)) \wedge \mu_{C1}(x) \wedge ... \wedge \mu_{Cm}(x) \tag{2.71}$$

where $f: X \to Y$. In this way, the case where the goals and the constraints are defined as fuzzy sets in different spaces can be reduced to the case where they are defined in the same space. The maximizing decision x^M as shown in Equation (2.69) can then be obtained.

From the above analysis, it is obvious that the decision D^M is obtained by using the max-min operator. However, except for the max-min operator, the decision D^M can also be obtained by using other operators. In the next section, we will discuss some properties of the max-min operator. Other operators will be discussed in Section 2.4.3.

2.4.2 Max-Min Operator

Zadeh [274] proposed the max and min operators to define intersection and union of fuzzy sets. Mathematically, Goguen [108] first considered the order structure of fuzzy sets and the properties of the max and min operator. The most complete arguments were provided by Bellman and Giertz [13] from a logic point of view, interpreting the intersection as "logical and," the union as "logical or" and the fuzzy set A as the statement. For an axiomatic justification, Bellman and Giertz considered: two statements, S and T, for which the truth values are $\mu(S)$ and $\mu(T)$, respectively. $\mu(S)$ and $\mu(T) \in [0, 1]$. The truth values of the "and" and "or" combination of these two statements are $\mu(S$ and $T)$ and $\mu(S$ or $T)$, which belong to [0, 1] and are interpreted as the values of the membership functions of the intersection and the union, respectively, of S and T statements.

Bellman and Giertz assumed that whenever truth values have been assigned to (arbitrary) statements S and T, then two real-valued functions f and g provide us with truth values for "S and T" and "S or T" such that:

$$\mu(S \text{ and } T) = f[\mu(S), \mu(T)] \tag{2.72}$$
$$\mu(S \text{ or } T) = g[\mu(S), \mu(T)]. \tag{2.73}$$

They feel that the following restrictions are reasonably imposed on f and g. The original statements are shown as follows:

(i) f and g are nondecreasing and continuous in both variables. Our willingness to accept "S and T" or to accept "S or T," should not decrease if our willingness to accept S(or T) increases. With suitably small changes in the level of acceptance of S or T, we should be able to change the truth value of "S and T," or of "S or T," by an arbitrarily small amount.

(ii) f and g are symmetric, i.e., $f(x, y) = f(y, x)$ and $g(x, y) = g(y, x)$. There is no reason to assign different truth values to "S and T" and "T and S."

(iii) $f(x, x)$ and $g(x, x)$ are strictly increasing in x. If $\mu(S_1) = \mu(S_2) > \mu(S_3) = \mu(S_4)$, we should be more willing to accept "S_1 and S_2" than to accept "S_3 and S_4."

(iv) $f(x, y) \leq \min \{x, y\}$ and $g(x, y) \geq \max \{x, y\}$. Accepting "S and T" requires more, and accepting "S or T" requires less than accepting S or T alone. Thus $\mu(S$ and $T) \leq \mu(S)$ and so on.

(v) $f(1, 1) = 1$ and $g(0, 0) = 0$. If both S and T are completely accepted, we must accept "S and T" completely as well, and if S and T are both completely rejected, then we must also reject "S or T."

(vi) Logically equivalent statements have equal truth values. To illustrate with an example, the statement:

$$S_1 \text{ and } (S_2 \text{ or } S_3)$$

is logically equivalent to:

$$(S_1 \text{ and } S_2) \text{ or } (S_1 \text{ and } S_3).$$

We have no reason to be more willing to accept one of these statements over the other. Condition (ii) is another example of this principle.

Bellman and Giertz then proved mathematically that

$$\mu(S \text{ and } T) = \min \{\mu(S), \mu(T)\} \tag{2.74}$$
$$\mu(S \text{ or } T) = \max \{\mu(S), \mu(T)\}. \tag{2.75}$$

The min and max operators also have been used in other fields, such as utility theory, game theory and so on, for a long time. Sometimes, the rationality of max and min operators may be reasonable for the human decision-making process. The decision maker may like to make a decision which maximizes the minimal possible return, or which minimizes the maximal possible penalty (or regret). Another good feature of the max and min operators is their computational tractability and simplicity. This is very important in practice. Freeling [93] pointed out that there are very strong reasons for the max-min operators to apply:

1) The computation of the fuzzy expected utilities is very simple.

2) The use of them allows the full machinery of fuzzy sets to be applied to the problem.

3) There is a very appealing interpretation of the fuzzy decision analysis as a multiple-level sensitivity.

2.4.3 Compensatory Operators

Even though the max-min operator has so many good properties, it is, however, not omnipotent. The following shows the other side of the argument.

Yager [267] showed that if the information with respect to membership grades is expected in the form of an ordinal preference function over the membership grades of the fuzzy subsets, then Zadeh's max and min operators are the only meaningful definitions of union and intersection. Furthermore, if the preference information is expressed in the form intensity information over each fuzzy subset individually, then Bellman and Zadeh's soft union and intersection ($\mu_{GUC}(x) = \mu_G(x) + \mu_C(x) - \mu_G(x)\mu_C(x)$ and $\mu_{G \cap C}(x) = \mu_G(x)\mu_C(x)$, respectively) becomes a meaningful way to combine fuzzy subsets without losing the information. However, if all we have is ratio information over each of the fuzzy subsets individually, then min and max are not meaningful at all.

Thole, Zimmermann and Zysno [245] conducted an experiment testing the suitability of minimum and product operators ($\mu_{G \cap C}(x) = \mu_G(x)\mu_C(x)$) for conjunction (here, "and" operator). The result, in contrast to the findings of Oden's studies [184] and the min operator, was a slight superiority, but was not perfect for the intersection operation in human categorizing processes. However, Thole et al. stated [245]:

"The fact that the empirical values have a tendency to exceed those predicated by means of the min-operator may be considered as a matter of no importance. However, from a slightly different point of view, the occurrence of rather systematic deviations is an experimental phenomenon, and possibly a lawful one. For it is conceivable that human beings use some kind of "compensation" when combining fuzzy sets in the sense of "and." This means that in rating objects with respect to a composite attribute they do not process the relevant information as if they were choosing the smaller of two grades of membership, but proceed internally as if they were using the smaller one only as an orientation and then modifying it in the direction of the higher value."

Therefore, searching compensatory operators for "and" and "or" operations has become one of the most important research fields in the empire of fuzzy set theory. Based on Chen and Hwang [BM3], we summarized the existing various compensatory min and max operators in Table 2.3.

Zimmermann and Zysno [299] made a further experiment to test the minimum, maximum, arithmetic mean and geometric mean operators. The results showed:

1) People use averaging operations when making judgments or evaluations resulting in membership values between minimum and maximum.

2) The geometric mean is an adequate model for human aggregation of fuzzy sets when a compensatory effect exists.

Table 2.3 A taxonomy of compensatory min and max operators

COMPENSATORY MIN OPERATORS:

1. **Algebraic Product:** $\mu_D(x) = \mu_G(x)\mu_C(x)$

2. **Bounded Product:** $\mu_D(x) = \max(0, \mu_G(x)+\mu_C(x)-1)$

3. **Hamacher's Min Operator (Γ-operator):** for $\Gamma \in [0, 1]$

$$\mu_D(x) = \frac{\mu_G(x)\mu_C(x)}{[\Gamma+(1-\Gamma)(\mu_G(x)+\mu_C(x)-\mu_G(x)\mu_C(x))]}$$

4. **Yager's Min Operator:** for $q \geq 0$

$$\mu_D(x) = 1 - \min\{1, [(1-\mu_G(x))^q + (1-\mu_C(x))^q]^{1/q}\}$$

5. **Dubois and Prade's Min Operator:** for $r \in [0, 1]$

$$\mu_D(x) = \frac{\mu_G(x)\mu_C(x)}{[\max(\mu_G(x), \mu_C(x), r)]}$$

6. **Werners's "Fuzzy And" Operator:** for $r \in [0, 1]$

$$\mu_D(x) = r\min((\mu_G(x), \mu_C(x)) + (1-r)(\mu_G(x)+\mu_C(x))/2$$

COMPENSATORY MAX OPERATORS:

1. **Algebraic Sum:** $\mu_D(x) = \mu_G(x)+\mu_C(x)-\mu_G(x)\mu_C(x)$

2. **Bounded Sum:** $\mu_D(x) = \min[1, \mu_G(x)+\mu_C(x)]$

3. **Hamacher's Max Operator:** for $\Gamma \in [0, 1]$

$$\mu_D(x) = \frac{[(1-\Gamma)\mu_G(x)\mu_C(x)+\Gamma(\mu_G(x)+\mu_C(x))]}{[\Gamma+\mu_G(x)\mu_C(x)]}$$

4. **Yager's Max Operator:** for $q \geq 1$

$$\mu_D(x) = \min\{1, [(\mu_G(x))^q + (\mu_C(x))^q]^{1/q}\}$$

5. **Dubois and Prade's Max Operator:** for $r \in [0, 1]$

$$\mu_D(x) = \frac{\{\mu_G(x)+\mu_C(x)-\mu_G(x)\mu_C(x)-\min[1-r, \mu_G(x), \mu_C(x)]\}}{\{\max[r, 1-\mu_G(x), 1-\mu_C(x)]\}}$$

6. **Werners's "Fuzzy Or" Operator:** for $r \in [0, 1]$

$$\mu_D(x) = r\max((\mu_G(x), \mu_C(x)) + (1-r)(\mu_G(x)+\mu_C(x))/2$$

---

3) Men use operators other than "and" and "or."

Thus, Zimmermann and Zysno provided a more general operator:

$$\mu_D(x) = \mu_{G \cap C}(x)^{1-r} \mu_{GUC}(x)^r \tag{2.76}$$

where $r \in [0, 1]$ is the degree of compensation between min and max operators. $\mu_{G \cap C}(x)$ and $\mu_{GUC}(x)$ are the results from using any (compensatory) min and max operators (as shown in Table 2.3), respectively.

Based on the recent development, the union and intersection operators should come from a class of triangular norms associated with certain modern algebras. These triangular norms (t-norms) share the following properties: (1) $T(1, \mu_G(x)) = \mu_G(x)$ for any $\mu_G(x)$ in $[0, 1]$; (2) monotonicity; (3) commutativity; and (4) associativity. $T(\mu_G(x), \mu_C(x))$ is affiliated with the interaction operator. A co-norm, $S(\mu_G(x), \mu_C(x))$, is affiliated with the union operator, and its properties are identical to those of $T(\mu_G(x), \mu_C(x))$ with the exception of (1) which is replaced with $S(0, \mu_G(x)) = \mu_G(x)$ for any $\mu_G(x)$ in $[0, 1]$. $T(\mu_G(x), \mu_C(x)) \leq \min(\mu_G(x), \mu_C(x))$ and $S(\mu_G(x), \mu_C(x)) \geq \max(\mu_G(x), \mu_C(x))$. These are not only Zadeh's original operators, but also are the operators that we would obtain in probability if G and C are overlapped (maximally dependent). Furthermore, $T(\mu_G(x), \mu_C(x)) \geq \max(\mu_G(x), \mu_C(x))$ and $S(\mu_G(x), \mu_C(x)) \leq \min(1, \mu_G(x) + \mu_C(x))$, which are operators called "bounded sum" pair in fuzzy set theory and also corresponding probability to conjunction and disjunction when two events are maximally exclusive [BM47]. In Table 2.3, compensatory min and max operators, except Werners's, belong to t-norm and t-conorm, respectively.

Finally, among these various operators, how can we select a suitable operator to obtain decisions? Zimmermann [BM64] proposed the following eight rules to justify a suitable operator for a particular fuzzy decision problem:

1. Axiomatic strength. An operator with less axiomatic restrictions is better.

2. Empirical fit. An operator must be an appropriate model of real system behavior which can normally only be proven by empirical testing.

3. Adaptability. An operator should be dependent on the context and the semantic interpretation.

4. Numerical efficiency. An operator should be computationally efficient.

5. Compensation. An operator should have compensation which is defined as: $\mu_D(x) = \mu_G(x) \cdot \mu_C(x)$, $\cdot$ is compensatory if a change in $\mu_G(x)$ can be counteracted by a change in $\mu_C(x)$.

6. Range of resulting membership. The larger the range of resulting membership the better the

operator.

7. <u>Aggregating behavior</u>. Since the degree of membership in the aggregated set depends very frequently on the number of sets combined, it is important that the resulting degree of membership be nonincreasing [108].

8. <u>Required scale level of membership functions</u>. An operator that requires the lowest scale level (nominal, interval, ratio, absolute) is the most preferable.

2.4.3.1 Numerical Examples for Operators

To illustrate computational procedure of various operators, let:

$$A = \{(4, 0), (5, 0.2), (6, 0.4), (7, 0.6), (8, 0.8), (9, 1.0), (10, 0)\}$$

and

$$B = \{(3, 0), (4, 0.5), (5, 0.7), (6, 1.0), (7, 0.7), (8, 0.5), (9, 0)\}.$$

With the assumption of $\Gamma = q = r = 0.5$, the results of intersection of A and B by using various min operators are summarized below:

x	3	4	5	6	7	8	9	10
$\mu_A(x)$	0	0	0.2	0.4	0.6	0.8	1.0	0
$\mu_B(x)$	0	0.5	0.7	1.0	0.7	0.5	0	0
Min Operator	0	0	0.2	0.4	0.6	0.5	0	0
COMPENSATORY MIN OPERATORS:								
Algebraic Product	0	0	0.14	0.4	0.42	0.4	0	0
Bounded Product	0	0	0	0.4	0.3	0.3	0	0
Γ-operator	0	0	0.16	0.4	0.45	0.42	0	0
Yager's Min Operator	0	0	0	0.4	0	0	0	0
Dubois and Prade's Min Operator	0	0	0.2	0.4	0.6	0.5	0	0
Werners's "Fuzzy And"	0	0.13	0.33	0.55	0.63	0.58	0.25	0

For example, at x = 7 we have:

Min Operator: $\mu_D(x=7) = \min [\mu_A(x=7), \mu_B(x=7)] = \min [0.6, 0.7] = 0.6$

Compensatory min operator:

1. Algebraic Product: $\mu_D(7) = \mu_A(7)\mu_B(7) = (0.6)(0.7) = 0.42$

2. Bounded Product: $\mu_D(7) = \max[0, \mu_A(7) + \mu_B(7) - 1] = \max(0, 0.3) = 0.3$

3. Γ-operator:

$$\mu_D(7) = \mu_A(7)\mu_B(7)/\{0.5+0.5[\mu_A(7)+\mu_B(7)-\mu_A(7)\mu_B(7)]\} = 0.42/0.94 = 0.45$$

4. Yager's Min Operator:

$$\mu_D(7) = 1 - \min\{1, [(\mu_A(7))^{0.5} + (\mu_B(7))^{0.5}]^{1/0.5}\} = 1 - \min(1, 2.60) = 0$$

5. Dubois and Prade's Min Operator:

$$\mu_D(7) = \mu_A(7)\mu_B(7)/\{\max[\mu_A(7), \mu_B(7), 0.5]\} = 0.42/\{\max[0.6, 0.7, 0.5]\} = 0.6$$

6. Werners's "Fuzzy And": $\mu_D(7) = 0.5 \min(0.6, 0.7) + 0.5(0.6+0.7)/2 = 0.625$.

With the assumption of $\Gamma = q = r = 0.5$, the results of union of A and B by using various max operators are summarized below:

x	3	4	5	6	7	8	9	10
$\mu_A(x)$	0	0	0.2	0.4	0.6	0.8	1.0	0
$\mu_B(x)$	0	0.5	0.7	1.0	0.7	0.5	0	0
max operator	0	0.5	0.7	1.0	0.7	0.8	1.0	0
COMPENSATORY MAX OPERATORS:								
algebraic sum	0	0.5	0.76	1.0	0.88	0.90	1.0	0
bounded sum	0	0.5	0.9	1.0	1.0	1.0	1.0	0
Hamacher's max operator	0	0.5	0.96	1.0	0.93	0.94	1.0	0
Yager's max operator	0	0.5	1.0	1.0	1.0	1.0	1.0	0
Dubois and Prade's max operator	0	0.5	0.7	1.0	0.76	0.8	1.0	0
Werners's "fuzzy or"	0	0.38	0.58	0.85	0.68	0.73	0.75	
Zimmermann and Zysno's r-operator	0	0	0.33	0.63	0.61	0.6	0	0

For example, at $x = 7$ we have:

Max Operator: $\mu_D(x=7) = \max[\mu_A(x=7), \mu_B(x=7)] = \min[0.6, 0.7] = 0.7$

Compensatory Max Operator:

1. Algebraic Sum: $\mu_D(7) = \mu_A(x) + \mu_B(x) - \mu_A(7)\mu_B(7) = 0.6 + 0.7 - (0.6)(0.7) = 0.88$

2. Bounded Sum: $\mu_D(7) = \min[1, \mu_A(7) + \mu_B(7)] = \min(1, 1.3) = 1.0$

3. Hamacher's Max Operator:

$$\mu_D(7) = \{(0.5)\mu_A(7)\mu_B(7)+0.5[\mu_A(7)+\mu_B(7)]\}/[0.5+\mu_A(7)\mu_B(7)]$$
$$= 0.86/0.92 = 0.93$$

4. Yager's Max Operator:

$$\mu_D(7) = \min\{1, [(\mu_A(7))^{0.5} + (\mu_B(7))^{0.5}]^{1/0.5}\} = \min(1, 2.60) = 1.0$$

5. Dubois and Prade's Max Operator:

$$\mu_D(7) = \frac{\{\mu_A(7)+\mu_B(7)-\mu_A(7)\mu_B(7)-\min[1-0.5, 0.6, 0.7]\}}{\{\max[0.5, 1-\mu_A(7), 1-\mu_B(7)]\}} = 0.38/0.5 = 0.76$$

6. Werners's "Fuzzy Or": $\mu_D(7) = 0.5 \max(0.6, 0.7) + 0.5 (0.6+0.7)/2 = 0.675$.

As to Zimmermann and Zysno's operator of $\mu_D(x) = \mu_{G\cap C}(x)^{1-r}\mu_{GUC}(x)^r$, we assume that r = 0.5, $\mu_{G\cap C}(x) = \mu_A(x)\mu_B(x)$ and $\mu_{GUC}(x) = \{1 - [1-\mu_A(x)][1-\mu_B(x)]\}$. Then, at x = 7 we have the following result:

$$\mu_D(7) = [\mu_A(7)\mu_b(7)]^{0.5}\{1 - [1-\mu_A(7)][1-\mu_B(7)]\}^{0.5} = (0.65)(0.94) = 0.61.$$

2.5 Fuzzy Arithmetic

In the previous section, we have discussed some major operations for handling fuzzy sets and obtaining decisions. However, those operators are not enough to calculate all the situations. For example, they cannot solve an equation involving fuzzy numbers which is always composed of addition, subtraction, multiplication and division. Equations involving fuzzy numbers are the most important ingredients in mathematical programming problems and many other fields. Thus, in this section we will discuss the operations of addition, subtraction, multiplication and division of fuzzy numbers. More detailed discussion can be found in Kaufmann and Gupta [BM31][BM32], and Chen and Hwang [BM3].

2.5.1 Addition of Fuzzy Numbers

Assume that A and B are two fuzzy numbers. The addition of X and Y can be accomplished by using α-level cut and max-min convolution.

I. α-level cut. By use of the concept of confidence intervals, the α-level sets of X and Y are

$X_\alpha = [x_\alpha{}^L, x_\alpha{}^U]$ and $Y_\alpha = [y_\alpha{}^L, y_\alpha{}^U]$ where $x_\alpha{}^L$, $x_\alpha{}^U$, $y_\alpha{}^L$ and $y_\alpha{}^U \epsilon R$. Then the result Z of the addition of the fuzzy numbers X and Y can be defined by the α-level sets. That is:

$$Z_\alpha = X_\alpha (+) Y_\alpha = [x_\alpha{}^L + y_\alpha{}^L, x_\alpha{}^U + y_\alpha{}^U] \tag{2.77}$$

for every $\alpha \epsilon [0, 1]$. Equation (2.77) obviously shows that the lower bound of the resulting fuzzy set Z is the sum of the lower bounds of the fuzzy numbers X and Y, and the upper bound of the resulting fuzzy set Z is the sum of the upper bounds of the fuzzy numbers X and Y.

Example 6. Assume the membership functions of fuzzy numbers X and Y are (see Figure 2.12):

$$\mu_X(x) = \begin{cases} 0 & \text{if } x \leq 0 \text{ or } x > 4 \\ x/2 & \text{if } 0 < x \leq 2 \\ (4-x)/2 & \text{if } 2 < x \leq 4 \end{cases}$$

$$\mu_Y(y) = \begin{cases} 0 & \text{if } y \leq 3 \text{ or } y > 11 \\ (y-3)/5 & \text{if } 3 < y \leq 8 \\ (11-y)/3 & \text{if } 8 < y \leq 11. \end{cases}$$

Then, by using α-level cut, we have $\mu_X(x) \geq \alpha$ and $\mu_Y(y) \geq \alpha$ for $\alpha \epsilon (0, 1]$. That is:

$$x/2 \geq \alpha \text{ and } (4-x)/2 \geq \alpha \quad => \quad 2\alpha \leq x \leq 4-2\alpha.$$

Therefore, we have $X_\alpha = [x_\alpha{}^L, x_\alpha{}^U] = [2\alpha, 4-2\alpha]$. Similarly, we have $Y_\alpha = [y_\alpha{}^L, y_\alpha{}^U] = [3+5\alpha, 11-3\alpha]$. Then:

$$Z_\alpha = X_\alpha (+) Y_\alpha = [(2\alpha)+(3+5\alpha), (4-2\alpha)+(11-3\alpha)] = [7\alpha+3, 15-5\alpha].$$

Now, let $Z_\alpha = [z_\alpha{}^L, z_\alpha{}^U] = [7\alpha+3, 15-5\alpha]$ and we have $z_\alpha{}^L = 7\alpha + 3$ and $z_\alpha{}^U = 15 - 5\alpha$, or $(z_\alpha{}^L - 3)/7 = \alpha$ and $(15 - z_\alpha{}^U)/5 = \alpha$, where $\alpha \epsilon (0, 1]$. Finally, we obtain the following membership function for the fuzzy number Z:

$$\mu_Z(z) = \begin{cases} 0 & \text{if } z \leq 3 \text{ or } z > 15 \\ (z-3)/7 & \text{if } 3 < z \leq 10 \\ (15-z)/5 & \text{if } 10 < z \leq 15. \end{cases}$$

II. Max-min convolution. By using the extension principle, the result Z of the addition of the fuzzy numbers X and Y can be defined as:

$$Z(z) = \max_{z=x+y} \{\min [\mu_X(x), \mu_Y(y)]\} \tag{2.78}$$

where x, y and z ϵ R.

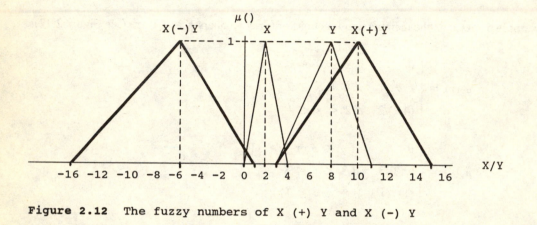

Figure 2.12 The fuzzy numbers of X (+) Y and X (-) Y

Example 7. Assume that the fuzzy numbers X and Y are:

X	x	1	2	3	4	5	6	
	$\mu_X(x)$	0.1	0.5	1.0	0.7	0.1	0.0	
Y	y	1	2	3	4	5	6	7
	$\mu_Y(y)$	0.0	0.1	0.7	1.0	0.8	0.5	0.0

By using Equation (2.78), we have:

$$\mu_Z(z=6) = \max [\min(0.1, 0.8), \min (0.5,1.0), \min (1.0,0.7), \min (0.7,0.1),$$
$$\min (0.1,0.0)]$$
$$= \max [0.1, 0.5, 0.7, 0.1, 0.0] = 0.7$$

where $6 = 1 + 5 = 2 + 4 = 3 + 3 = 4 + 2 = 5 + 1$. Other possible values of z are summarized in Table 2.4. From Table 2.4, we can easily and systematically obtain the resulting fuzzy number Z. For example, the value of $\mu_Z(z=6)$ is the maximum of the diagonally blocked numbers which are the minimum values of the corresponding row and column values. Thus we have the following solution:

Z	z	2	3	4	5	6	7	8	9	10	11	12	13
$\mu_Z(z)$		0.0	0.1	0.1	0.5	0.7	1.0	0.8	0.7	0.5	0.1	0.1	0.0

Table 2.4 Results of X (+) Y for Example 7

X x	$\mu(x)$	Y 	y $\mu(y)$	1 0.0	2 0.1	3 0.7	4 1.0	5 0.8	6 0.5	7 0.0
1	0.1		(z=2)	0.0	0.1	0.1	0.1	0.1	0.1	0.1
2	0.5		(3)	0.0	0.1	0.5	0.5	0.5	0.5	0.0
3	1.0		(4)	0.0	0.1	0.7	1.0	0.8	0.5	0.0
4	0.7		(5)	0.0	0.1	0.7	0.7	0.7	0.5	0.0
5	0.1		(6)	0.0	0.1	0.1	0.1	0.1	0.1	0.1
6	0.0		(7)	0.0	0.0	0.0	0.0	0.0	0.0	0.0
				(z=8)	(9)	(10)	(11)	(12)	(13)	

2.5.2 Subtraction of Fuzzy Numbers

Similar to the addition operation, we have the following two definitions of substraction of fuzzy numbers:

I. α-level cut. By using α-level cut, we can define the resulting fuzzy number Z of the subtraction of the fuzzy numbers X and Y as:

$$Z_\alpha = X_\alpha (-) Y_\alpha = [x_\alpha^L - y_\alpha^U, x_\alpha^U - y_\alpha^L] \tag{2.79}$$

for every $\alpha \in [0, 1]$. The calculation is quite similar to the addition operation.

Example. Consider Example 6. Obviously, we have $Z_\alpha = [5\alpha-11, 1-7\alpha]$. The membership function of the fuzzy number Z is then (see Figure 2.12):

$$\mu_Z(z) = \begin{cases} 0 & \text{if } z \leq -11 \text{ or } z > 1 \\ (z+11)/5 & \text{if } -11 < z \leq -6 \\ (1-z)/7 & \text{if } -6 < z \leq 1. \end{cases}$$

II. Max-min convolution. By using the extension principle, we can also define the subtraction operation as:

$$\mu_Z(z) = \max_{z=x-y} \{\min [\mu_X(x), \mu_Y(y)]\} \tag{2.80}$$

$$= \max_{z=x+y} \{\min [\mu_X(x), \mu_Y(-y)]\}$$

$$= \max_{z=x+y} \{\min [\mu_X(x), \mu_{-Y}(y)]\}.$$

As we can see, the subtraction is equivalent to the addition of image of Y to X.

Example. Consider Example 7. Now, let $Z = X (-) Y$. The calculation will be similar to the addition operation. Thus we can directly compute Table 2.5.

Table 2.5 Results of X (-) Y for Example 7

			(z=0)	(−1)	(−2)	(−3)	(−4)	(−5)	(−6)
X	Y	y	1	2	3	4	5	6	7
x	$\mu(x)$	$\mu(y)$	0.0	0.1	0.7	1.0	0.8	0.5	0.0
1	0.1		0.0	0.1	0.1	0.1	0.1	0.1	0.1
2	0.5	(z=1)	0.0	0.1	0.5	0.5	0.5	0.5	0.0
3	1.0	(2)	0.0	0.1	0.7	1.0	0.8	0.5	0.0
4	0.7	(3)	0.0	0.1	0.7	0.7	0.7	0.5	0.0
5	0.1	(4)	0.0	0.1	0.1	0.1	0.1	0.1	0.1
6	0.0	(5)	0.0	0.0	0.0	0.0	0.0	0.0	0.0

From the above table, we can easily and systematically obtain the resulting fuzzy number Z. For example, the value of $\mu_Z(z=0) = 0.7$ is the maximum of the diagonally blocked numbers which are the minimum values of the corresponding row and column values. Thus we have the following solution:

Z z	-6	-5	-4	-3	-2	-1	0	1	2	3	4	5
$\mu_Z(z)$	0.1	0.1	0.5	0.8	0.7	1.0	0.7	0.7	0.1	0.1	0.0	0.0

2.5.3 Multiplication of Fuzzy Numbers

For the potential sign effect, we first assume that $\mu_X(x) = 0$ for $x < 0$ and $\mu_Y(y) = 0$ for $y < 0$. Then, as previously discussed, we have the following two definitions of the multiplication of fuzzy numbers:

I. α-level cut. By using α-level cut, we can define the resulting fuzzy number Z of the multiplication of the fuzzy numbers X and Y as:

$$Z_\alpha = X_\alpha \, (\cdot) \, Y_\alpha = [[x_\alpha^L y_\alpha^L, \, x_\alpha^U y_\alpha^U] \tag{2.81}$$

for every $\alpha \, \epsilon \, [0, 1]$.

Example. Consider Example 6. $X_\alpha = [2\alpha, 4-2\alpha]$ and $Y_\alpha = [3+5\alpha, 11-3\alpha]$. By using Equation (2.81), we obtain:

$$Z_\alpha = [(2\alpha)(3+5\alpha), \, (4-2\alpha)(11-3\alpha)] = [10\alpha^2+6\alpha, \, 6\alpha^2-34\alpha+44]$$

which is not a simple triangular fuzzy number. The membership function of the fuzzy number Z is then (see Figure 2.13 also):

$$\mu_Z(z) = \begin{cases} 0 & \text{if } z \leq 0 \text{ or } z > 44 \\ -6 + (36+40z)^{1/2}/20 & \text{if } 0 < z \leq 16 \\ 34 - (100+24z)^{1/2}/12 & \text{if } 16 < z \leq 44. \end{cases}$$

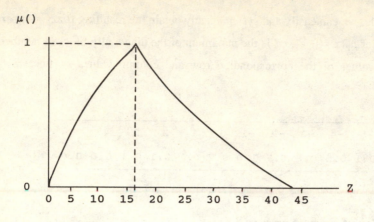

Figure 2.13 Fuzzy number Z = X (·) Y

II. Max-min convolution. When using the extension principle, the multiplication operation becomes more complex. Thus Kaufmann and Gupta [BM31] proposed the following procedure:

1. First, find z^1 (the peak of the fuzzy number Z) such that $\mu_Z(z^1) = 1$. Then we calculate the left and right legs.

2. The left leg of $\mu_Z(z)$ is defined as:

$$\mu_Z(z) = \max_{xy \leq z} \ \{\min [\mu_X(x), \mu_Y(y)]\}. \tag{2.82}$$

3. The right leg of $\mu_Z(z)$ is defined as:

$$\mu_Z(z) = \max_{xy \geq z} \ \{\min [\mu_X(x), \mu_Y(y)]\}. \tag{2.83}$$

Example. Consider Example 7. We can directly compute Table 2.6. From Table 2.6, we can obtain the peak at x = 3 and y = 4. That is $\mu_Z(12) = 1$. Next, we should find the left and right legs of the fuzzy number Z by using Equations (2.82) and (2.83). The results are:

Z	z	1	2	...	5	6	...	8	9	...	11	12	13	14
	$\mu_z(z)$	0.0	0.1		0.1	0.5		0.5	0.7		0.7	1.0	0.8	0.8
	z	15	16	...	20	21	...	24	25	...	35	36	...	42
	$\mu_z(z)$	0.8	0.7		0.7	0.5		0.5	0.1		0.1	0.0		0.0

Table 2.6 Results of X $(\cdot)$ Y for Example 7

X x	$\mu(x)$	Y y $\mu(y)$	1 0.0	2 0.1	3 0.7	4 1.0	5 0.8	6 0.5	7 0.0
1	0.1		**0.0**	**0.1**	**0.1**	**0.1**	**0.1**	**0.1**	0.1
2	0.5		**0.0**	**0.1**	**0.5**	0.5	0.5	0.5	0.0
3	1.0		**0.0**	**0.1**	0.7	<u>1.0</u>	0.8	0.5	0.0
4	0.7		**0.0**	0.1	0.7	0.7	0.7	0.5	0.0
5	0.1		**0.0**	0.1	0.1	0.1	0.1	0.1	0.1
6	0.0		**0.0**	0.0	0.0	0.0	0.0	0.0	0.0

Note: when z = 6, we choose the maximum among the bold numbers

2.5.4 Division of Fuzzy Numbers

Like the relation between addition and subtraction, the division operation is essentially an extension of the multiplication operation. Thus we can define the division of fuzzy numbers as follows:

I. α-level cut. First, by using α-level cut, we have the following result:

$$Z_\alpha = X_\alpha \, (:) \, Y_\alpha = [[x_\alpha^L/y_\alpha^U, \, x_\alpha^U/y_\alpha^L] \tag{2.84}$$

for every $\alpha \in [0, 1]$.

Example. Consider Example 6. $X_\alpha = [2\alpha, 4-2\alpha]$ and $Y_\alpha = [3+5\alpha, 11-3\alpha]$. $Z_\alpha = [2\alpha/(11-3\alpha), (4-2\alpha)/(3+5\alpha)]$ or

$$\mu_Z(z) = \begin{cases} 0 & \text{if } z \leq 0 \text{ or } z > 4/3 \\ 11z/(3z+2) & \text{if } 0 < z \leq 1/4 \\ (4-3z)/(2+5z) & \text{if } 1/4 < z \leq 4/3. \end{cases}$$

II. Max-min convolution. As discussed in the previous section, we must first find the peak of the fuzzy number $Z = X \ (:) \ Y$ when the max-min convolution is used. Then the left leg is defined as:

$$\mu_Z(z) = \max_{x/y \leq z} \ \{\min [\mu_X(x), \mu_Y(y)]\} \tag{2.85}$$

$$= \max_{xy \leq z} \ \{\min [\mu_X(x), \mu_Y(1/y)]\}$$

$$= \max_{xy \leq z} \ \{\min [\mu_X(x), \mu_{1/Y}(y)]\}.$$

The right leg of $\mu_Z(z)$ is defined as:

$$\mu_Z(z) = \max_{x/y \geq z} \ \{\min [\mu_X(x), \mu_Y(y)]\} \tag{2.86}$$

$$= \max_{xy \geq z} \ \{\min [\mu_X(x), \mu_Y(1/y)]\}$$

$$= \max_{xy \geq z} \ \{\min [\mu_X(x), \mu_{1/Y}(y)]\}.$$

From the above equations, we can see that the division operation can be completed by using the multiplication operation.

2.5.5 Triangular and Trapezoid Fuzzy Numbers

Among the various types of fuzzy numbers, triangular and trapezoid fuzzy numbers are of the most importance. They are especially useful in solving possibilistic mathematical programming problems. Thus, in this section we will provide some tables to show the results of applying fuzzy arithmetic on the triangular and trapezoid fuzzy numbers.

First, the triangular fuzzy numbers can be denoted as $X = (x, \alpha, \beta)$ and $Y = (y, r, \delta)$, where x and y are the central values ($\mu_X(x) = 1$ and $\mu_Y(y) = 1$), α and r are the left spreads, and β and δ are the right spreads (see Figure 2.14). Then the results of applying fuzzy arithmetic on the fuzzy numbers X and Y are shown in Table 2.7.

Similarly, we can denote the trapezoid fuzzy numbers as $X = (x_1, x_2, \alpha, \beta)$ and $Y = (y_1, y_2, r, \delta)$ as shown in Figure 2.14. The results of applying fuzzy arithmetic on the fuzzy numbers X and Y are shown in Table 2.8.

On the other hand, the triangular fuzzy numbers can also be denoted as $X = (x^m, x^p, x^o)$ and $Y = (y^m, y^p, y^o)$ where $x^m = x$ ($y^m = y$), $x^p = x - \alpha$ ($y^p = y - r$) and $x^o = x + \beta$ ($y^o = y + \delta$). Then the results of applying fuzzy arithmetic on the fuzzy numbers X and Y are shown in Table 2.9.

Similarly, the trapezoid fuzzy numbers can also be denoted as $X = (x_1^m, x_2^m\ x^p, x^o)$ and $Y = (y_1^m, y_2^m, y^p, y^o)$ where $x_1^m = x_1$ ($y_1^m = y_1$), $x_2^m = x_2$ ($y_2^m = y_2$), $x^p = x_1 - \alpha$ ($y^p = y_1 - r$) and $x^o = x_2 + \beta$ ($y^o = y_2 + \delta$). Then the results of applying fuzzy arithmetic on the fuzzy numbers X and Y are shown in Table 2.10.

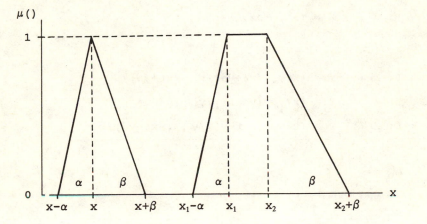

Figure 2.14 The triangular and trapezoid fuzzy numbers

Table 2.7 Fuzzy arithmetic on triangular fuzzy numbers
$X = (x, \alpha, \beta)$ & $Y = (y, r, \delta)$

Image of Y: $-Y = (-y, \delta, r)$

Inverse of Y: $Y^{-1} = (y^{-1}, \delta y^{-2}, r y^{-2})$

Addition: $X \; (+) \; Y = (x+y, \alpha+r, \beta+\delta)$

Subtraction: $X \; (-) \; Y = X \; (+) \; -Y = (x-y, \alpha+\delta, \beta+r)$

Multiplication:

$X > 0, \; Y > 0: \; X \; (\cdot) \; Y = (xy, \; xr+y\alpha, \; x\delta+y\beta)$

$X < 0, \; Y > 0: \; X \; (\cdot) \; Y = (xy, \; y\alpha-x\delta, \; y\beta-xr)$

$X < 0, \; Y < 0: \; X \; (\cdot) \; Y = (xy, \; -x\delta-y\beta, \; -xr-y\alpha)$

Scalar Multiplication:

$a > 0, \; a \in R: \; a \; (\cdot) \; X = (ax, \; a\alpha, \; a\beta)$

$a < 0, \; a \in R: \; a \; (\cdot) \; X = (ax, \; -a\beta, \; -a\alpha)$

Division:

$X > 0, \; Y > 0: \; X \; (:) \; Y = (x/y, \; (x\delta+y\alpha)/y^2, \; (xr+y\beta)/y^2)$

$X < 0, \; Y > 0: \; X \; (:) \; Y = (x/y, \; (y\alpha-xr)/y^2, \; (y\beta-x\delta)/y^2)$

$X < 0, \; Y < 0: \; X \; (:) \; Y = (x/y, \; (-xr-y\beta)/y^2, \; (-x\delta-y\alpha)/y^2)$

Table 2.8 Fuzzy arithmetic on trapezoid fuzzy numbers
$X = (x_1, x_2, \alpha, \beta)$ & $Y = (y_1, y_2, r, \delta)$

Image of Y: $-Y = (-y_1, -y_2, \delta, r)$

Inverse of Y: $Y^{-1} = (y_2^{-1}, y_1^{-1}, \delta/[y_2(y_2+\delta)], r/[y_1(y_1-r)])$

Addition: $X (+) Y = (x_1+y_1, x_2+y_2, \alpha+r, \beta+\delta)$

Subtraction: $X (-) Y = (x_1-y_2, x_2-y_1, \alpha+\delta, \beta+r)$

Multiplication:

$X > 0, Y > 0: X (\cdot) Y = (x_1y_1, x_2y_2, x_1r+y_1\alpha-\alpha r, x_2\delta+y_2\beta-\beta\delta)$

$X < 0, Y > 0: X (\cdot) Y = (x_1y_2, x_2y_1, y_2\alpha-x_1\delta+\alpha\delta, y_1\beta-x_2r-\beta r)$

$X < 0, Y < 0: X (\cdot) Y = (x_2y_2, x_1y_1, -x_2\delta-y_2\beta-\beta\delta, -x_1r-y_1\alpha+\alpha r)$

Scalar Multiplication:

$a > 0, a \in R: a (\cdot) X = (ax_1, ax_2, a\alpha, a\beta)$

$a < 0, a \in R: a (\cdot) X = (ax_1, ax_2, -a\beta, -a\alpha)$

Division:

$X > 0, Y > 0:$

$\quad X (:) Y = (x_1/y_2, x_2/y_1, (x_1\delta+y_2\alpha)/[y_2(y_2+\delta)], (x_2r+y_1\beta)/[y_1(y_1-r)])$

$X < 0, Y > 0:$

$\quad X (:) Y = (x_1/y_1, x_2/y_2, (y_1\alpha-x_1r)/[y_1(y_1-r)], (y_2\beta-x_2\delta)/[y_2(y_2+\delta)])$

$X < 0, Y < 0:$

$\quad X (:) Y = (x_2/y_1, x_1/y_2, (-x_2r-y_1\beta)/[y_1(y_1-r)], (-x_1\delta-y_2\alpha)/[y_2(y_2+\delta)])$

Table 2.9 Fuzzy arithmetic on triangular fuzzy numbers
$X = (x^m, x^p, x^o)$ & $Y = (y^m, y^p, y^o)$

Image of Y: $-Y = (-y^m, -y^o, -y^p)$

Inverse of Y: $Y^{-1} = (1/y^m, 1/y^o, 1/y^p)$

Addition: $X (+) Y = (x^m+y^m, x^p+y^p, x^o+y^o)$

Subtraction: $X (-) Y = X (+) -Y = (x^m-y^m, x^p-y^o, x^o-y^p)$

Multiplication:

  $X > 0, Y > 0$: $X (\cdot) Y = (x^m y^m, x^p y^p, x^o y^o)$

  $X < 0, Y > 0$: $X (\cdot) Y = (x^m y^m, x^p y^o, x^o y^p)$

  $X < 0, Y < 0$: $X (\cdot) Y = (x^m y^m, x^o y^o, x^p y^p)$

Scalar Multiplication:

  $a > 0, a \in R$: $a (\cdot) X = (ax^m, ax^p, ax^o)$

  $a < 0, a \in R$: $a (\cdot) X = (ax^m, ax^o, ax^p)$

Division:

  $X > 0, Y > 0$: $X (:) Y = (x^m/y^m, x^p/y^o, x^o/y^p)$

  $X < 0, Y > 0$: $X (:) Y = (x^m/y^m, x^o/y^o, x^p/y^p)$

  $X < 0, Y < 0$: $X (:) Y = (x^m/y^m, x^o/y^p, x^p/y^o)$

Table 2.10   Fuzzy arithmetic on trapezoid fuzzy numbers
$X = (x_1^m, x_2^m x^p, x^o)$ & $Y = (y_1^m, y_2^m, y^p, y^o)$

Image of Y: $-Y = (-y_2^m, -y_1^m -y^o, -y^p)$

Inverse of Y: $Y^{-1} = (1/y_2^m, 1/y_1^m, 1/y^o, 1/y^p)$

Addition: $X (+) Y = (x_1^m+y_1^m, x_2^m+y_2^m, x^p+y^p, x^o+y^o)$

Subtraction: $X (-) Y = (x_1^m-y_2^m, x_2^m-y_1^m, x^p-y^o, x^o-y^p)$

Multiplication:

$X > 0, Y > 0$: $X (\cdot) Y = (x_1^m y_1^m, x_2^m y_2^m, x^p y^p, x^o y^o)$

$X < 0, Y > 0$: $X (\cdot) Y = (x_2^m y_1^m, x_1^m y_2^m, x^p y^o, x^o y^p)$

$X < 0, Y < 0$: $X (\cdot) Y = (x_2^m y_2^m, x_1^m y_1^m, x^o y^o, x^p y^p)$

Scalar Multiplication:

$a > 0, a \in R$: $a (\cdot) X = (ax_1^m, ax_2^m, ax^p, ax^o)$

$a < 0, a \in R$: $a (\cdot) X = (ax_2^m, ax_1^m, ax^o, ax^p)$

Division:

$X > 0, Y > 0$: $X (:) Y = (x_1^m/y_2^m, x_2^m/y_1^m, x^p/y^o, x^o/y^p)$

$X < 0, Y > 0$: $X (:) Y = (x_2^m/y_2^m, x_1^m/y_1^m, x^o/y^o, x^p/y^p)$

$X < 0, Y < 0$: $X (:) Y = (x_2^m/y_1^m, x_1^m/y_2^m, x^o/y^p, x^p/y^o)$

2.6 Fuzzy Ranking

Fuzzy ranking is an important topic in fuzzy mathematical programming. Consider the constraints of the fuzzy mathematical programming: $Ax \leq b$. If A, b and/or x are fuzzy, the inequalities $Ax \leq b$ will become a ranking problem. Chapter 4 will discuss how to use fuzzy ranking approaches in solving fuzzy/imprecise constraints.

Fuzzy ranking approaches have been systematically and thoroughly classified and well presented in Chen and Hwang [BM3] as shown in Table 2.11. For detailed discussions, readers are invited to refer to Chen and Hwang's Fuzzy *Multiple Attribute Decision Making: Methods and Applications* [BM3]. Most of these approaches can be used to handle the constraints (of mathematical programming) involving fuzzy numbers. However, we should pay attention to the meaning and usefulness of these approaches for various possible mathematical programming problems.

Table 2.11 A taxonomy of fuzzy ranking methods [BM3]

	Comparison Medium	Technique Involved	Approach
Fuzzy ranking	Preference relation	Degree of optimality	Baas & Kwakernaak Watson, Weiss & Donnell Baldwin & Guild
		Hamming distance	Yager Kerre Nakamura Kolodzijezyk
		α-cut	Adamo Buckley & Chanas Mabuchi
		Comparison function	Dubois and Prade Tsukamoto et al. Delgado et al.
	Fuzzy mean and spread	Probability distribution	Lee & Li
	Fuzzy scoring	Proportion to optimal	McCahon
		Left/right scores	Jain Chen Chen and Hwang
		Centroid index	Yager Murakami et al.
		Area measurement	Yager
	Linguistic expression	Intuition	Efstathiou & Tong
		Linguistic approximation	Tong & Bonissone

3 FUZZY MATHEMATICAL PROGRAMMING

Fuzzy set theory was first developed for solving the imprecise/vague problems in the field of artificial intelligence, especially for imprecise reasoning and modelling linguistic terms. In solving decision making problems, the pioneer work came from Bellman and Zadeh [12], Tanaka, Okuda and Asai [236][237], Negoita et al. [176][178][179], Zimmermann [296][298], Orlovsky [192], Yager [266] and Freeling [92].

Since Tanaka et al. [236], there have been a number of fuzzy linear programming models. In this monograph, we classify linear programming models with imprecise information into two main classes: fuzzy linear programming and possibilistic linear programming. In this chapter we first discuss fuzzy linear programming problems and models. Possibilistic linear programming problems and models will be discussed in the next chapter. Dyson [79] argued that the membership function is equivalent to the utility function when the membership functions of fuzzy linear programming are based on a preference concept like the utility theory. But, this is not true for the possibility distributions in possibilistic linear programming.

Some important fuzzy linear programming problems and methods are discussed in Section 3.1. In this section we introduce two major models: symmetric and nonsymmetric [304]. The symmetric models are based on the definition of fuzzy decision proposed by Bellman and Zadeh [12]. They assumed that objective(s) and constraints in an imprecise situation can be represented by fuzzy sets. A decision, then, may be stated as the confluence of the fuzzy objective(s) and constraints, and may be defined by a max-min operator. That is: assume that we are given a fuzzy objective set G and a fuzzy constraint set C in a space of alternatives X. Then, G and C combine to form a decision D which is a fuzzy set resulting from the intersection of G and C, and corresponding $\mu_D = \mu_G \cap \mu_C$. This relation between G, C, D is depicted in Figure 3.1.

The nonsymmetric models are based on the following two approaches [304]:

(1) The determination of the fuzzy set decision.

(2) The determination of a crisp maximizing decision by aggregating the objective function, after appropriate transformations, with the constraints.

In Section 3.2 we will present our integrated linear programming decision support system. Finally, we will discuss some extension of the fuzzy linear programming in Section 3.3.

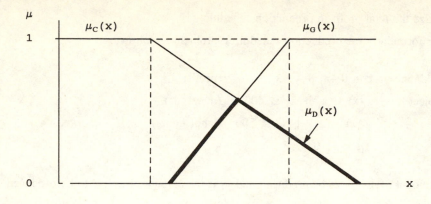

Figure 3.1 The relation of the fuzzy sets C, G and D

3.1 Fuzzy Linear Programming Models

In this section, five main fuzzy linear programming problems will be discussed. The methodology for each problem is presented under the pre-assumption of linear membership function forms, and max-min operators. Other membership function forms and operators are also assumed in some specified cases. For simplicity, we assume that these membership functions and operators are consistent to the decision maker's judgement and rationality in the decision processes.

It is worthy to note that the pre-assumed membership functions of fuzzy linear programming models are based on the preference concept, but not the possibility concept which will be discussed in Chapter 4.

3.1.1 Linear Programming Problem with Fuzzy Resources

In order to understand as clearly as possible how and where the fuzziness is applied in linear programming problems with fuzzy resources available, a simplified linear programming case is demonstrated.

The Hardee Toy Company makes two kinds of toy dolls. Doll A is a high quality toy with a \$0.40 per unit profit and doll B is of lower quality with a \$0.30 per unit profit. Each doll A requires twice as many labor hours as each doll B, and the total available labor hours are 500 hours per day. The supply of material is sufficient for only 400 dolls per day (both A and B combined). It is assumed that all the dolls the factory has made can be sold. The manager, then, wishes to maximize the total profit in production scheduling.

The manager formulates the production scheduling problem as follows:

$$
\begin{array}{lll}
\text{maximize} & z = 0.4x_1 + 0.3x_2 & \text{(profit)} \\
\text{subject to} & g_1(x) = x_1 + x_2 \leq 400 & \text{(materials)} \\
& g_2(x) = 2x_1 + x_2 \leq 500 & \text{(labor hours)} \\
& x_1, x_2 \geq 0 &
\end{array}
$$

where x_1 is the amount of doll A to be produced and x_2 is the amount of doll B to be produced. After solving the above LP problem by using the graphic method (see Figure 3.2), the manager gets the optimal solution: he produces 100 units of doll A and 300 units of doll B, and earns $130 profit. Subsequently, the manager feels that the available material may change in some range because he can get additional materials from suppliers, and that available time may also change because he can ask the workers to work overtime. Therefore, he begins studying the fuzzy set theory and introduces it into his production scheduling problem. Now the manager decides that the profit should be substantially larger than $130 because of the increased resources.

The membership functions of the material and time shown in Table 3.1 and Figure 3.3 are provided by the manager. He explains that the values of the membership functions mean the degree of his satisfaction (preference) on the constraints g_1 (material) and g_2 (labor hours). For example, when the total amount of material used is less than or equal to 400, the constraint "$g_1(x) \leq 400$" is absolutely satisfied. When the total amount of material used is greater than or equal to 500, the constraint "$g_1(x) \leq 400$" is absolutely violated, so the degree of satisfaction is equal to 0. Between 400 and 500, the degree of membership function is monotonic linear decreasing. The membership function of the time availability can be explained in the same way. Thus the manager considers that the only feasible solutions should be those satisfying the membership functions of the material, labor hours, and profit at the same time. The feasible alternative with the maximum degree of satisfaction is then chosen as the optimal solution of the production scheduling problem.

From Figure 3.4, when the available material and labor hours are 400 and 500, respectively, the optimal solution is $z = 130$ with the satisfaction level $= 1$, at $x_1 = 100$ and $x_2 = 300$ as indicated by point b. For $g_1(x) = 450$ and $g_2(x) = 550$ the optimal solution is $z' = 140$ with the satisfaction level $= 0.5$, at $x_1 = 100$ and $x_2 = 350$ as indicated by point b'. Similarly, for $g_1(x) = 500$ and $g_2(x) = 600$, the optimal solution is $z'' = 160$ with the satisfaction level $= 0$, at $x_1 = 100$ and $x_2 = 400$ as indicated by point b''. Obviously, the optimal solutions are fuzzy and their membership values are determined by the membership values (for example, 0, .5, 1) of constraints.

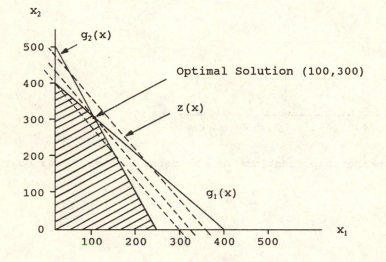

Figure 3.2 The graphic approach for a linear programming problem

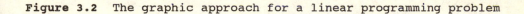

Table 3.1 The membership functions of the material and labor hours

material	$g_1(x)$	150	200	250	300	350	400	450	500	
available	$\mu_{g1}(x)$	1	1	1	1	1	1	0.5	0	
time	$g_2(x)$	200	250	300	350	400	450	500	550	600
available	$\mu_{g2}(x)$	1	1	1	1	1	1	1	0.5	0

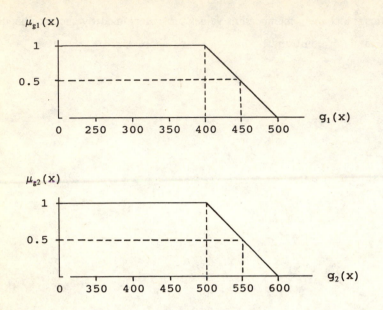

Figure 3.3 The membership functions of the materials and labor hours

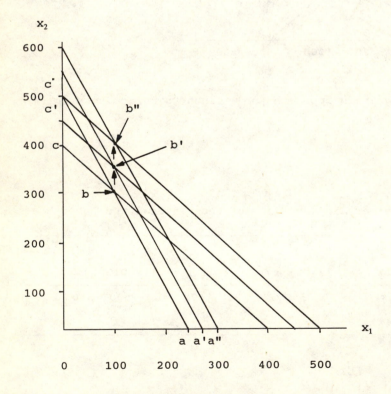

Figure 3.4 Graphic solutions of the fuzzy linear programming problem

In the above example, we can see that in fuzzy linear programming, the fuzziness of available resources is characterized by the membership function over the tolerance range (ambiguous range). The membership function of the optimal solutions can then be constructed by the convolution of the membership functions of constraints.

The general model of this linear programming with fuzzy resources available is formulated as:

$$\max \quad z = cx$$
$$\text{s.t.} \quad (Ax)_i \leq \tilde{b}_i, \, i = 1, 2, ..., m$$
$$x \geq 0 \qquad\qquad (3.1.a)$$

where $\tilde{b}_i$, $\forall i$, are in $[b_i, b_i+p_i]$ with given p_i. We may also consider the following fuzzy inequality constraints:

$$\max \quad z = cx$$
$$\text{s.t.} \quad (Ax)_i \lesseqgtr b_i, \quad i = 1, 2, ..., m$$
$$x \geq 0 \qquad\qquad (3.1.b)$$

where $\lesseqgtr$ is called fuzzy less than or equal to. Assume that tolerance p_i for each fuzzy constraint is known. Then $(Ax)_i \leq \tilde{b}_i$ will be equivalent to $(Ax)_i \leq (b_i + \theta p_i)$, $\forall i$, where θ is in $[0, 1]$. If the membership functions of both cases are the same, then Equations (3.1.a) and (3.1.b) will be the same problem [137]. Thus we will consider both problems as equivalent in this study.

Verdegay [250] first proved that the problem of Equation (3.1) is equivalent to the crisp parametric programming problem. Thus we can use the parametric programming methods to solve our first fuzzy linear programming problem. Werners [257], on the other hand, considered that the objective should be fuzzy because of the fuzzy constraints. Both approaches are discussed as follows.

3.1.1.1 Verdegay's Approach - A Nonsymmetric Model

For Equation (3.1), Verdegay [250] considered that if the membership functions of the fuzzy constraints:

$$\mu_i(x) = \begin{cases} 1 & \text{if } (Ax)_i < b_i \\ 1 - [(Ax)_i - b_i]/p_i & \text{if } b_i \leq (Ax)_i \leq b_i + p_i \\ 0 & \text{if } (Ax)_i > b_i + p_i \end{cases} \qquad (3.2)$$

are continuous and monotonic functions, and trade-off between those fuzzy constraints are allowed, then Equation (3.1) will be equivalent to:

$$\text{max} \quad cx$$
$$\text{s.t.} \quad x \in X_{\alpha} \quad (3.3)$$

where $X_{\alpha} = \{x \mid \mu_i(x) \geq \alpha, \forall i, x \geq 0\}$, for each $\alpha \in [0, 1]$. The α-level cut concept is based on the previous works of Tanaka et al. [236] and Orlovski [192].

The membership functions indicate that: (1) if $(Ax)_i \leq b_i$ then the ith constraint is absolutely satisfied, $\mu_i(x) = 1$; (2) if $(Ax)_i \geq b_i + p_i$ where p_i is the maximum tolerance from b_i and determined by the decision maker in any systematic or nonsystematic way, then the ith constraint is absolutely violated, $\mu_i(x) = 0$; (3) if $(Ax)_i \in (b_i, b_i + p_i)$; then the membership functions are monotonically decreasing. That is, the more resources consumed, the less satisfaction the decision maker feels.

We can then substitute membership functions of Equation (3.2) into Equation (3.3) and obtain the following problem:

$$\text{max} \quad cx$$
$$\text{s.t.} \quad (Ax)_i \leq b_i + (1 - \alpha)p_i, \forall i$$
$$x \geq 0 \text{ and } \alpha \in [0, 1] \quad (3.4)$$

which is equivalent to a parametric programming, while $\alpha = 1 - \theta$. Thus the fuzzy linear programming problem given by Equation (3.1) can be equivalent to a crisp parametric linear programming problem when some proper forms of membership functions of the fuzzy constraints are assumed. It is noted that for each α, we have an optimal solution, so the solution with α grade of membership is actually fuzzy.

For illustrating this approach, let consider the following two examples.

3.1.1.1a Example 1: The Knox Production-Mix Selection Problem

Let us consider a product-mix selection problem [BM41][137]. Suppose that the Knox Mix company has the option of using one or more of four different types of production processes. The first and second processes yield items of product A, and the third and fourth yield items of product B. The inputs for each process are labor measured in man-weeks, pounds of material Y, and boxes of material Z. Since each process varies in its input requirements, the profitabilities of the process differ, even for processes producing the same item. The manufacturer's decision on a week's production schedule is limited in the range of possibilities by the available amounts of manpower and both kinds of raw materials. The full technology and input restrictions are given in Table 3.2. Suppose that production levels in processes 1, 2, 3, and 4 are x_1, x_2, x_3 and x_4, respectively. The problem can then be formulated as the following linear programming problem:

$$
\begin{aligned}
\max \quad & 4x_1 + 5x_2 + 9x_3 + 11x_4 && \text{(profit)} \\
\text{s.t.} \quad & g_1(x) = x_1 + x_2 + x_3 + x_4 \le 15 && \text{(man-weeks)} \\
& g_2(x) = 7x_1 + 5x_2 + 3x_3 + 2x_4 \le 120 && \text{(material Y)} \\
& g_3(x) = 3x_1 + 5x_2 + 10x_3 + 15x_4 \le 100 && \text{(material Z)} \\
& x_1, x_2, x_3 \text{ and } x_4 \ge 0.
\end{aligned}
$$

We then solve this linear programming problem by use of the simplex method. The final tableau is shown in Table 3.3. The optimal solution is: $x^* = (50/7, 0, 55/7, 0) = (7.14, 0, 7.86, 0)$ and $z^* = \$695/7 = \$99.29$. The actual resources used are 15, 73.57 and 100 units for man-weeks, material Y and material Z, respectively.

Let us assume that the available total man-weeks and material Z are imprecise and their maximum tolerances are 3 and 20 units, respectively. Then, the membership functions (see Figure 3.5) of the fuzzy constraints are:

Table 3.2   The input data for the Knox product-mix selection problem

Item	Man-weeks	Material-Y (pounds)	Material-Z (boxes)	Unit profit
One item of product A				
process 1	1	7	3	4
process 2	1	5	5	5
One item of product B				
process 3	1	3	10	9
process 4	1	2	15	11
Total available	≤ 15	≤ 120	≤ 100	maximize

Table 3.3 The final tableau of the simplex method for the Knox product-mix selection problem

Basic variable		4 x1	5 x2	9 x3	11 x4	0 s1	0 s2	0 s3	RHS
4	x1	1	5/7	0	-5/7	10/7	0	-1/7	50/7
0	s2	0	-6/7	0	13/7	-61/7	1	4/7	325/7
9	x3	0	2/7	1	12/7	-3/7	0	1/7	55/7
zj - cj		0	3/7	0	11/7	13/7	0	5/7	695/7

$$\mu_1(x) = \begin{cases} 1 & \text{if } g_1(x) < 15 \\ 1 - [g_1(x)-15]/3 & \text{if } 15 \leq g_1(x) \leq 18 \\ 0 & \text{if } g_1(x) > 18 \end{cases}$$

$$\mu_3(x) = \begin{cases} 1 & \text{if } g_3(x) < 100 \\ 1 - [g_3(x)-100]/20 & \text{if } 100 \leq g_3(x) \leq 120 \\ 0 & \text{if } g_3(x) > 120. \end{cases}$$

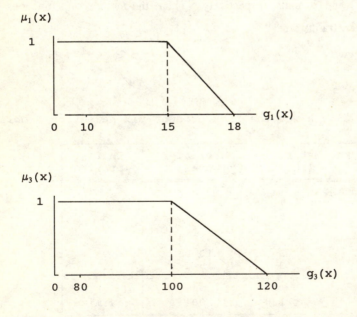

Figure 3.5 The membership functions of man-weeks and material Y constraints

Then we have the following problem:

$$\max \quad 4x_1 + 5x_2 + 9x_3 + 11x_4$$

$$\text{s.t.} \quad \mu_1(x) \geq \alpha$$

$$g_2(x) = 7x_1 + 5x_2 + 3x_3 + 2x_4 \leq 120$$

$$\mu_3(x) \geq \alpha$$

$$\alpha \in [0, 1], \text{ and } x_1, x_2, x_3 \text{ and } x_4 \geq 0$$

or

$$\max \quad 4x_1 + 5x_2 + 9x_3 + 11x_4$$

$$\text{s.t.} \quad g_1(x) = x_1 + x_2 + x_3 + x_4 \leq 15+3(1-\alpha)$$

$$g_2(x) = 7x_1 + 5x_2 + 3x_3 + 2x_4 \leq 120$$

$$g_3(x) = 3x_1 + 5x_2 + 10x_3 + 15x_4 \leq 100+20(1-\alpha)$$

$$\alpha \in [0, 1], \text{ and } x_1, x_2, x_3 \text{ and } x_4 \geq 0.$$

Set $\theta = 1 - \alpha$. The following parametric programming problem is obtained:

$$\max \quad 4x_1 + 5x_2 + 9x_3 + 11x_4$$

$$\text{s.t.} \quad g_1(x) = x_1 + x_2 + x_3 + x_4 \leq 15 + 3\theta$$

$$g_2(x) = 7x_1 + 5x_2 + 3x_3 + 2x_4 \leq 120$$

$$g_3(x) = 3x_1 + 5x_2 + 10x_3 + 15x_4 \leq 100 + 20\theta$$

$$x_1, x_2, x_3 \text{ and } x_4 \geq 0.$$

where $\theta \in [0, 1]$ is a parameter. By use of the parametric technique and the final table of the simplex method shown in Table 3.3, we can obtain the following results:

$$(10/7, 0, -1/7) \begin{pmatrix} 3\theta \\ 0 \\ 20\theta \end{pmatrix} = 10\theta/7$$

$$(-61/7, 1, 4/7) \begin{pmatrix} 3\theta \\ 0 \\ 20\theta \end{pmatrix} = -103\theta/7$$

$$(-3/7, 0, 1/7) \begin{pmatrix} 3\theta \\ 0 \\ 20\theta \end{pmatrix} = 11\theta/7.$$

The final simplex table is shown in Table 3.4. Since the RHS $50/7 + 10\theta/7$, $325/7 - 103\theta/7$ and $55/7 + 11\theta/7$, for $\theta \in [0, 1]$, are always greater than zero, the optimal solution is then: $x^* = (7.14 + 1.43\theta, 0, 7.86 + 1.57\theta, 0)$, and $z^* = \$(99.29 + 19.86\theta)$. To display these results clearly, Table 3.5 is provided for the decision maker.

Table 3.4 The final parametric tableau of the simplex method for the Knox product-mix selection problem

Basic variable	4 x1	5 x2	9 x3	11 x4	0 s1	0 s2	0 s3	RHS
4 x1	1	5/7	0	-5/7	10/7	0	-1/7	$50/7+10\theta/7$
0 s2	0	-6/7	0	13/7	-61/7	1	4/7	$325/7+103\theta/7$
9 x3	0	2/7	1	12/7	-3/7	0	1/7	$55/7+11\theta/7$
zj - cj	0	3/7	0	11/7	13/7	0	5/7	$695/7+139\theta/7$

Table 3.5 The solutions of the parametric programming problem

θ	z^*	resources actually used		
		man-weeks	material Y	material Z
0.0	99.29	15.00	73.57	100.00
0.1	101.28	15.30	75.04	102.00
0.2	103.27	15.60	76.51	104.00
0.3	105.26	15.90	77.98	106.00
0.4	107.25	16.20	79.45	108.00
0.5	109.24	16.50	80.92	110.00
0.6	111.23	16.80	82.39	112.00
0.7	113.22	17.10	83.86	114.00
0.8	115.21	17.40	85.33	116.00
0.9	117.20	17.70	86.80	118.00
1.0	119.19	18.00	88.27	120.00

3.1.1.1b Example 2: A Transportation Problem

Consider a transportation problem with fuzzy demand [252]:

$$\min \quad cx$$

$$\text{s.t.} \quad \sum_{i \in I} x_{ij} \geq b_j, \quad j \in J = \{1, 2, ..., n\text{-}1\}$$

$$\sum_{j \in J} x_{ij} \leq a_i, \quad i \in I = \{1, 2, ..., m\}$$

$$x_{ij} \geq 0, \quad \forall i, j.$$

The membership functions of the fuzzy constraints, $\mu_j()$, $j \in J$, and $\mu_i()$, $i \in I$, are assumed to be continuous and strictly monotonic, respectively. Then it can be proved that:

$$\sum_{i \in I} \mu_i^{-1}(\alpha) \geq \sum_{j \in J} \mu_j^{-1}(\alpha), \quad \alpha \in [0, 1].$$

Therefore, the fuzzy transportation problem can be solved by means of the crisp parametric transportation problem. That is:

$$\min \quad cx$$

$$\text{s.t.} \quad \sum_{i \in I} x_{ij} \geq \mu_j^{-1}(\alpha), \quad j \in J$$

$$\sum_{j \in J} x_{ij} \leq \mu_i^{-1}(\alpha), \quad i \in I$$

$$x_{ij} \geq 0, \quad \forall i \text{ and } j, \quad \alpha \in [0, 1].$$

In order to get $\sum_i \mu_i^{-1}(\alpha) = \sum_j \mu_j^{-1}(\alpha)$, an n-th dummy destination is introduced. Thus, the following parametric linear programming problem is obtained:

$$\min \quad cx$$

$$\text{s.t.} \quad \sum_{i \in I} x_{ij} = \mu_j^{-1}(\alpha), \quad j \in J = \{1, 2, ..., n\}$$

$$\sum_{j \in J} x_{ij} = \mu_i^{-1}(\alpha), \quad i \in I$$

$$x_{ij} \geq 0, \; \forall \; i \text{ and } j, \; \alpha \in [0, 1].$$

Now, consider the following numerical illustration of the fuzzy transportation problem:

4	5	2	1	$(-\infty, 8, 13)$
6	2	4	3	$(-\infty, 6, 9)$
3	1	1	1	$(-\infty, 5, 6)$

$(3, 4, \infty) \quad (3, 5, \infty) \quad (3, 6, \infty) \quad (1, 4, \infty)$

where the triangular shaped notation of (s_1, s_2, s_3) is introduced for simplicity (see the following figure).

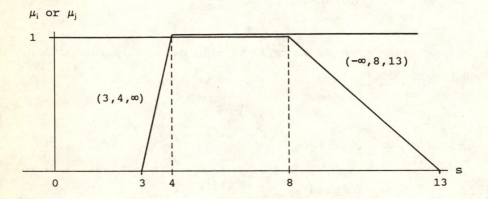

By introducing the 5-th dummy destination with its membership function $(-\infty, 0, 18)$, we obtain:

4	5	2	1	0	$13-5\alpha$
6	2	4	3	0	$9-3\alpha$
3	1	1	1	0	$6-\alpha$

$3+\alpha \qquad 3+2\alpha \qquad 3+3\alpha \qquad 1+3\alpha \qquad 18-18\alpha$

The optimal solution that provides a fuzzy solution to the fuzzy transportation is:

For $\alpha \in [0, 3/5]$:

$3+\alpha$		6α	$1+3\alpha$	$9-15\alpha$
				$9-3\alpha$
	$3+2\alpha$	$3-3\alpha$		

For $\alpha \in [3/5, 12/13]$:

$3+\alpha$		$9-9\alpha$	$1+3\alpha$	
	$-9+15\alpha$			$18-18\alpha$
	$12-13\alpha$	$-6+12\alpha$		

For $\alpha \in [12/13, 1]$:

$3+\alpha$		$-3+4\alpha$	$13-10\alpha$	
	$3+2\alpha$		$-12+13\alpha$	$18-18\alpha$
		$6-\alpha$		

For transportation problems, OhEigeartaigh [185] considered the case where the membership functions of the fuzzy demands are triangular forms. Sometimes the decision maker may prefer to have some particular amount of demand. More or less than this amount is less preferred. Thus, the triangular membership function is rational. Chanas and Kulej [35] also provided an approach to solve a fuzzy linear programming problem with triangular membership functions of fuzzy resources.

3.1.1.2 Werners's Approach - A Nonsymmetric Model

Werners [257, 258] proposed that the objective function of Equation (3.1) should be fuzzy because of fuzzy total resources or fuzzy inequality constraints. Similar to the previous section, let us assume that the tolerances p_i for the fuzzy resources are available and given. Then for solving Equation (3.1), Werners first defined z^0 and z^1 as follows:

$$z^0 = \max \quad cx$$
$$\text{s.t.} \quad (Ax)_i \leq b_i, \ \forall i, \text{ and } x \geq 0$$

$$z^1 = \max \quad cx$$
$$\text{s.t.} \quad (Ax)_i \leq b_i+p_i, \ \forall i, \text{ and } x \geq 0. \tag{3.5}$$

Thus, we can construct a continuously nondecreasing linear membership function for the objective by use of z^0 and z^1. Since the optimal solution will be in between z^0 and z^1, the satisfaction of the optimal solution will increase when its value increases. The membership function μ_0 (see Figure 3.6) of the objective is:

$$\mu_0(x) = \begin{cases} 1 & \text{if } cx > z^1 \\ 1 - (z^1 - cx)/(z^1 - z^0) & \text{if } z^0 \leq cx \leq z^1 \\ 0 & \text{if } c\,x < z^0. \end{cases} \tag{3.6}$$

With the above membership function, we can then use the max-min operator to obtain optimal decision. Then Equation (3.1) can be solved by solving " $\max_{x \geq 0} \alpha$, where $\alpha = \min [\mu_0(x), \mu_1(x), \ldots, \mu_m(x)]$." That is:

$$
\begin{aligned}
\max \quad & \alpha \\
\text{s.t.} \quad & \mu_0(x) \geq \alpha \\
& \mu_i(x) \geq \alpha, \forall i \\
& \alpha \in [0, 1] \text{ and } x \geq 0.
\end{aligned} \tag{3.7}
$$

Equation (3.7) is essentially a symmetrical model which is similar to the model proposed by Zimmermann (see Section 3.1.2.1). The difference is the membership function of the objective function.

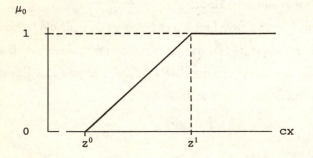

Figure 3.6   The membership function of the objective function

3.1.1.2a Example 1: The Knox Production-Mix Selection Problem

Let us consider the previous example. From the solution of the crisp linear programming problem, we can see that 120 units for the constraint of material Y will have too many idle resources. We now assume that the constraint of material Y will be 80 units, and that the tolerances for man-weeks, and material Y and Z are 5, 40, 30, respectively. Our problem then becomes:

$$\max \quad z = 4x_1 + 5x_2 + 9x_3 + 11x_4$$
$$\text{s.t.} \quad g_1(x) = x_1 + x_2 + x_3 + x_4 \leq \widetilde{15}$$
$$g_2(x) = 7x_1 + 5x_2 + 3x_3 + 2x_4 \leq \widetilde{80}$$
$$g_3(x) = 3x_1 + 5x_2 + 10x_3 + 15x_4 \leq \widetilde{100}$$
$$x_1, x_2, x_3 \text{ and } x_4 \geq 0$$

where the membership function $\mu_i(x)$ for the ith fuzzy constraints, $\forall i$, are:

$$\mu_1(x) = \begin{cases} 1 & \text{if } g_1(x) < 15 \\ 1 - [g_1(x)-15]/5 & \text{if } 15 \leq g_1(x) \leq 20 \\ 0 & \text{if } g_1(x) > 20 \end{cases}$$

$$\mu_2(x) = \begin{cases} 1 & \text{if } g_2(x) < 80 \\ 1 - [g_2(x)-80]/40 & \text{if } 80 \leq g_2(x) \leq 120 \\ 0 & \text{if } g_2(x) > 120 \end{cases}$$

$$\mu_3(x) = \begin{cases} 1 & \text{if } g_3(x) < 100 \\ 1 - [g_3(x)-100]/30 & \text{if } 100 \leq g_3(x) \leq 130 \\ 0 & \text{if } g_3(x) > 130. \end{cases}$$

Then, solving Equation (3.5), we obtain:

$$z^0 = z^*(\theta = 0) = 99.29$$
$$z^1 = z^*(\theta = 1) = 130$$

where θ is the parameter in the new parametric programming problem. The symmetric linear programming problem can then be formulated as follows:

max α

s.t. $\mu_0 \geq \alpha$

$\mu_i \geq \alpha$, i = 1, 2, 3

$\alpha \in [0, 1]$

where the membership function μ_0 of the fuzzy objective is defined as:

$$\mu_0(x) = \begin{cases} 1 & \text{if } z > 130 \\ 1 - (130\text{-}z)/(130\text{-}99.29) & \text{if } 99.29 \leq z \leq 130 \\ 0 & \text{if } z < 99.29. \end{cases}$$

This problem is actually equivalent to:

min θ

s.t. $z = 4x_1 + 5x_2 + 9x_3 + 11x_4 \geq 130 - 30.71\theta$

$g_1(x) = x_1 + x_2 + x_3 + x_4 \leq 15 + 5\theta$

$g_2(x) = 7x_1 + 5x_2 + 3x_3 + 2x_4 \leq 80 + 40\theta$

$g_3(x) = 3x_1 + 5x_2 + 10x_3 + 15x_4 \leq 100 + 30\theta$

x_1, x_2, x_3 and $x_4 \geq 0$ and $\theta \in [0, 1]$

where $\theta = 1 - \alpha$. The solution is: $x^* = (8.57, 0, 8.93, 0)$ and $z^* = \$114.65$ at $\theta = 0.5$, while the actually used resources are 17.5, 86.78 and 115.01 for man-weeks, material Y and material Z, respectively.

3.1.1.2b Example 2: An Air Pollution Regulation Problem

Consider an air pollution regulation problem [231]. It is intended to improve the quality of life indicated by the purity of air in a given area, where each point is expressed by the coordinates (x_1, x_2) of a cartesian coordinate system. The pollutant's concentration was modified so that (a) given air quality standards are strictly complied to, (b) people's desire to reach an even better concentration rate than the adapted standard is taken into consideration, and (c) the reduction of the emission does not entail too large a decrease of the plant's production (d) by minimizing the reduction cost.

Because of these requirements, the fuzzy optimization technique is needed to solve the air pollution regulation problem.

First, let us make the following assumptions:

(1) Consider a single kind of pollutant which is released from several resources in the control area.

(2) The given air quality standards are to be complied with at scattered so-called receptor-points in the control area.

(3) There is no change in weather and climate conditions. This implies that we do not need parameters expressing quantities like wind-speed and direction, pressure change, moisture, temperature, etc. This can be considered in a more elaborate model. A known "transfer function" s_j exists for each source j, j = 1, 2, ..., J, which indicates the concentration of the pollutant at the ground level with the coordinates (x_1, x_2). This function may be of generalized Gauss type:

$$s_j(x_1, x_2) = \alpha_j \exp(\beta_j)$$

where the sources' special parameters α_j and β_j are known functions of x_1, x_2. The concentration rate at a receptor-point i (i = 1, 2, ..., I) caused by the source j is:

$$s_{ji} = s_j(x_1^i, x_2^i)$$

where (x_1^i, x_2^i) are the coordinates of the receptor-point i. s_{ji} is one of the given data and expresses the actual concentration rates.

The decision variables $E_j \epsilon$ [0, 1] serve as an indicator for the reduction of the pollutant at the source j, so that:

E_j is the reduction rate, and

$(1-E_j)s_{ji}$ is the concentration rate after reduction.

Finally, assuming that the total concentration at one receptor-point is the sum of the pollutant's concentrations that are released from all sources, we determine:

$$\sum_{j=1}^{J} (1-E_j)s_{ji}$$

as the receptor-point's total concentration (after reduction, if there exist a j so that $E_j > 0$).

Now, the first type of fuzzy restriction is "Let, for each receptor-point i, Σ_j $(1-E_j)s_{ji}$ not exceed the standard d but try to go below e_i which is the desirable quantity at the receptor-point i." This can be written as:

$$\sum_{j=1}^{J} (1-E_j)s_{ji} \leq e_i; \; d, \; \forall i$$

with the membership functions:

$$\mu_i = 1 - \{[\, \Sigma_j \, (1-E_j)s_{ji}]-e_i\}/(d-e_i), \; \forall i.$$

The second type of fuzzy restriction is "Ensure, for each source j, that the reduction rate does not exceed E_j^u and even try to go below w_j!" This is:

$$E_j \leq w_j; \; E_j^u, \; \forall j$$

with the membership functions

$$\mu_j = 1 - (E_j-w_j)/(E_j^u-w_j), \; \forall j$$

where w_j and $E_j^u \in [0, 1]$, $\forall j$. The objective function is to minimize $\Sigma_j \, c_j E_j$ where c_j is the total reduction cost at the source j, assuming (or having approximately determined) that the cost function is linear on $[0, E_j^u]$. Thus, we obtain:

$$\begin{aligned}
\min \quad & \Sigma_j \, c_j E_j \\
\text{s.t.} \quad & \Sigma_j \, (1-E_j)s_{ji} \leq e_i + \theta(d - e_i), \; \forall i \\
& E_j \leq w_j + \theta(E_j^u - w_j), \; \forall j \\
& E_j \geq 0, \; \forall j
\end{aligned}$$

where s_{ji}, e_i, d, w_j, E_j^u, c_j are given data.

Now, let us consider the following numerical example. Suppose that the transfer functions are:

$$s_1(x_1, x_2) = \exp\{-[1+(x_2-1)^2]/x_1\}/x_1, \; x_1 > 0$$
$$s_2(x_1, x_2) = \exp\{-[2+x_2^2/4]/x_1\}/x_1, \; x_1 > 0$$
$$s_3(x_1, x_2) = \exp\{-[1+(x_2+1)^2]/(x_1-2)\}/(x_1-2), \; x_1 > 2$$

for the three sources. Beyond this, assume that there are only six scattered receptor points as depicted in Figure 3.7.

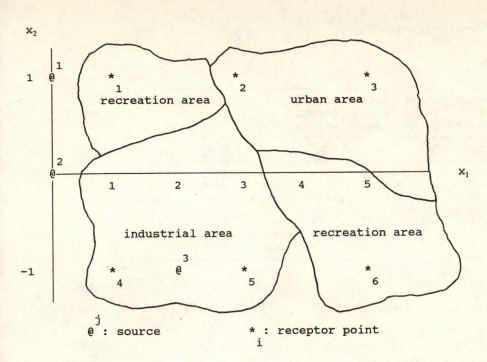

Figure 3.7 The control area of the air pollution regulation problem

The control area is divided into three different parts, so that the desirable concentration rate e_i depends on the location but the maximum permitted concentration d is a constant standard over the entire control area. e_i and d are:

area	e_i	d
recreation	0.42	0.5
urban	0.44	0.5
industrial	0.49	0.5

The receptor points 2, 3, and 4 are not taken into account because they already have a smaller sum of the plant's concentration rates than the desirable rates. The data of the sources' assumed maximum permitted reduction rate E_j, the desirable rate w_j and the reduction cost c_j are:

j	c_j	E_j	w_j
1	2	0.2	0.04
2	4	0.35	0.15
3	1	0.55	0.3

Other data are shown as follows:

receptor points and its coordinates			actual concen. rates (before reduction)				desirable concen. rate
i	x_1^i	x_2^i	s_{1i}	s_{2i}	s_{3i}	$\Sigma_i\ s_{ji}$	e_i
1	1	1	.3679	.1054	--	.4733	.42
2	3	1	.2388	.1757	.0067	.4030	.44
3	5	1	.1638	.1275	.0630	.3543	.44
4	1	-1	.0067	.1054	--	.1121	.49
5	3	-1	.0630	.1575	.3679	.5884	.49
6	5	-1	.0736	.1275	.2388	.4399	.42

The problem becomes:

$$\min \quad z = \Sigma_j\ c_j E_j$$

$$\text{s.t.} \quad .3679E_1 + .1054E_2 \geq .0533 - .08\theta$$

$$.0630E_1 + .1575E_2 + .3679E_3 \geq .0984 - .01\theta$$

$$.0736E_1 + .1275E_2 + .2388E_3 \geq .0199 - .08\theta$$

$$E_1 \leq .04 + .16\theta$$

$$E_2 \leq .15 + .2\theta$$

$$E_3 \leq .3 + .25\theta$$

$$E_j \geq 0, \forall j.$$

We first obtain: $z^0 = .2403$ and $z^1 = .87643$. We can then establish Werners's membership function for the objective function and solve the fuzzy linear programming problem. The optimal decision is: $E_1^* = .04$, $E_2^* = .00$ and $E_3^* = .26$.

3.1.2 Linear Programming Problem With Fuzzy Resources and Objective

For the same reason as mentioned in the previous section, the following two problems are equivalent here:

$$\widetilde{\max} \quad z = cx$$
$$\text{s.t.} \quad (Ax)_i \leq \widetilde{b_i}, \quad i = 1, 2, ..., m$$
$$x \geq 0 \tag{3.8.a}$$

and

$$\widetilde{\max} \quad z = cx$$
$$\text{s.t.} \quad (Ax)_i \lesssim b_i, \quad i = 1, 2, ..., m$$
$$x \geq 0 \tag{3.8.b}$$

To solve Equation (3.8), the methods proposed by Zimmermann [296] and Chanas [34] are presented in this section.

3.1.2.1 Zimmermann's Approach - A Symmetric Model

In this approach, the goal b_0 and its corresponding tolerance p_0 of the fuzzy objective are given initially, and also are for the fuzzy resources: b_i and its corresponding tolerances p_i, $\forall i$. The fuzzy objective and the fuzzy constraints are then considered without difference, and their corresponding regions can be described in the intervals $[b_i, b_i+p_i]$, $\forall i$. Thus, Equation (3.8) can be considered as [296]:

$$\text{find} \quad x$$
$$\text{such that} \quad cx \geq b_0$$
$$(Ax)_i \leq b_i, \quad \forall i,$$
$$x \geq 0 \tag{3.9}$$

In the fuzzy set theory, the fuzzy objective function and the fuzzy constraints are defined by their corresponding membership functions. For simplicity, let us assume that the membership function μ_0 of the fuzzy objective is a non-decreasing continuous linear function, and the membership functions μ_i, $\forall i$, of the fuzzy constraints are non-increasing continuous linear membership functions as follows (see Figure 3.8):

$$\mu_0(x) = \begin{cases} 1 & \text{if } cx > b_0 \\ 1 - (b_0 - cx)/p_0 & \text{if } b_0 - p_0 \le cx \le b_0 \\ 0 & \text{if } cx < b_0 - p_0 \end{cases}$$

$$(3.10)$$

$$\mu_i(x) = \begin{cases} 1 & \text{if } (Ax)_i < b_i \\ 1 - [(Ax)_i - b_i]/p_i & \text{if } b_i \le (Ax)_i \le b_i + p_i \\ 0 & \text{if } (Ax)_i > b_i + p_i, \forall i. \end{cases}$$

$$(3.11)$$

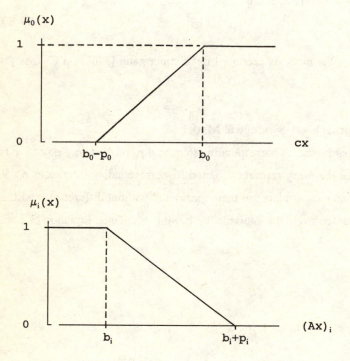

Figure 3.8 The membership functions for the fuzzy objective constraint $c\,x \ge \tilde{b}_0$ and the ith fuzzy constraint $(Ax)_i \le \tilde{b}_i$

Zimmermann then used Bellman and Zadeh's max-min operator to solve the Equation (3.9). Thus the optimal solution can be obtained by:

$$\max \mu_D = \max \{ \min [\mu_0(x), \mu_1(x), \dots, \mu_m(x)] \}$$

$$(3.12)$$

where μ_D is the membership function of the decision space D, and $\mu_D = \min (\mu_0, \mu_1, \dots, \mu_m)$.

If $\alpha = \mu_D$, the Equation (3.9), via the Equation (3.12), will be equivalent to:

$$\max \quad \alpha$$

$$\text{s.t.} \quad \mu_0(x) = 1 - (b_0 - cx)/p_0 \geq \alpha$$

$$\mu_i(x) = 1 - [(Ax)_i - b_i]/p_i \geq \alpha, \, i = 1, \, ..., \, m$$

$$\mu_i(x), \, \forall i, \, \text{and} \, \alpha \in [0, 1] \tag{3.13}$$

or

$$\max \quad \alpha$$

$$\text{s.t.} \quad cx \geq b_0 - (1 - \alpha)p_0$$

$$(Ax)_i \leq b_i + (1 + \alpha)p_i, \, \forall i$$

$$x \geq 0 \, \text{and} \, \alpha \in [0, 1] \tag{3.14}$$

where c, A, b_0, p_0, b_i and p_i, $\forall i$, are given initially.

Obviously, Equation (3.14) is a crisp linear programming technique. A unique optimal solution can be obtained. It should be noted that this approach is considered as the first practical method to solve a linear programming problem with fuzzy resources and objective.

3.1.1.2a Example 1: The Knox Production-Mix Selection Problem

Consider the above product-mix selection problem and assume that $b_0 = 111.57$ and $p_0 = 10$. Our problem becomes:

$$\text{find} \quad x$$

$$\text{such that} \quad g_0(x) = z = 4x_1 + 5x_2 + 9x_3 + 11x_4 \gtrsim 111.57$$

$$g_1(x) = x_1 + x_2 + x_3 + x_4 \lesssim 15$$

$$g_2(x) = 7x_1 + 5x_2 + 3x_3 + 2x_4 \lesssim 80$$

$$g_3(x) = 3x_1 + 5x_2 + 10x_3 + 15x_4 \lesssim 100$$

$$x_1, \, x_2, \, x_3 \, \text{and} \, x_4 \geq 0.$$

where the membership functions of the fuzzy constraints are described in previous subsections and the membership function μ_0 of the fuzzy objective is defined as:

$$\mu_0(x) = \begin{cases} 1 & \text{if } g_0(x) > 111.57 \\ 1 - [111.57 - g_0(x)]/10 & \text{if } 101.57 \leq g_0(x) \leq 111.57 \\ 0 & \text{if } g_0(x) > 101.57 \end{cases}$$

which is then equivalent to:

$$\max \quad \alpha$$

$$\text{s.t.} \quad g_0(x) = 4x_1+5x_2+9x_3+11x_4 \geq 111.57 - 10(1-\alpha)$$

$$g_1(x) = x_1+x_2+x_3+x_4 \leq 15 + 5(1-\alpha)$$

$$g_2(x) = 7x_1+5x_2+3x_3+2x_4 \leq 80 + 40(1-\alpha)$$

$$g_3(x) = 3x_1+5x_2+10x_3+15x_4 \leq 100 + 30(1-\alpha)$$

$$x_1, x_2, x_3, x_4 \geq 0 \text{ and } \alpha \, \epsilon \, [0, 1].$$

The solution is: $x^* = (8.01, 0, 8.50, 0)$ and $z^* = \$108.54$ at $\alpha = 0.70$, while the resources used are 16.51, 81.57 and 109.03 for man-weeks, material Y and material Z, respectively.

3.1.2.1b Example 2: A Regional Resource Allocation Problem

Consider a simplified regional resource allocation problem which involves the communal farming communities (the system of kibbutzim, in Israel) [148].

It is common for groups of kibbutzim to join together as a confederation to share common technical services and to coordinate their production. Within the confederation, the over-all planning is carried out by a coordinating technical office. The function of the office is to plan agricultural production of the confederation for the coming year.

The allocation problem concerns a confederation of three kibbutzim. The agricultural output of each kibbutz is limited by the amount of available irrigable land and the quantity of water (see Table 3.6) (which can only be imprecisely specified) allocated for irrigation by the Water Commissioner, a national government official.

Table 3.6   Irrigable land and water capacity

| Kibbutz | Fuzzy Interval | |
	Irrigable Land (acres)	Water Capacity (acre feet)
1	[400, 440]*	[600, 660]
2	[600, 630]	[800, 840]
3	[300, 320]	[375, 450]

* The fuzzy specification for the fuzzy interval is corresponding to "should be less than 400 or not much greater than 400" and the membership function form is defined as equation (3.11).

The crops under consideration are sugar beets, cotton and sorghum. These crops differ in their expected net return and their consumption of water. Imprecise capacity of land that can be devoted to each of the crops is also imposed by the Ministry of Agriculture as follows:

Table 3.7 Crop data for the confederation of kibbutzim

Crop	Net Return ($/acre)	(acre feet) / (acre)	Maximum Capacity (acres) (fuzzy interval)
Sugar beets	400	3	[600, 650].
Cotton	300	2	[500, 540]
Sorghum	100	1	[325, 350]

* The fuzzy specification for the fuzzy interval is corresponding to "should be less than 600 or not much greater than 600" and the membership function form is defined as equation (3.11).

The three kibbutzim of the Confederation have agreed that every kibbutz will plant a pre-specified proportion of its available irrigable land. Nevertheless, any combination of the crops may be grown at any of the kibbutzim.

For the coming year, the coordinating technical office has to determine the amount of land to be allocated to each crop in the respective kibbutzim so that the above prescribed restrictions are satisfied. In place of taking the maximization of total net return of the confederation as an objective, it is decided that a target value, \$ 350,000, should be employed and the total net return is required to exceed the target value. In case the target value is too optimistic, the total net return is allowed to fall below it. The bottom line is \$ 250,000. Thus:

"Total net return should be greater than \$350,000 or not much smaller than \$ 350,000," and is defined by the membership function form of equation (3.10) with the associated fuzzy interval [250,000, 350,000].

Based on the above information, the resource allocation problem of the Confederation can be formulated as a fuzzy linear programming problem as follows:

Objective (fuzzy):

$$z = 400(x_{11}+x_{12}+x_{13})+300(x_{21}+x_{22}+x_{23})+100(x_{31}+x_{32}+x_{33}) \geq 350,000; 250,000$$

Constraints:

Water consumption (fuzzy):

$$g_{11}(x) = 3x_{11} + 2x_{21} + x_{31} \leq 600; 660$$
$$g_{12}(x) = 3x_{12} + 2x_{22} + x_{32} \leq 800; 840$$

$$g_{13}(x) = 3x_{13} + 2x_{23} + x_{33} \leq 375; 450$$

Crop-land (fuzzy):

$$g_{21}(x) = x_{11} + x_{12} + x_{13} \leq 600; 650 \quad \text{(Sugar beets)}$$

$$g_{22}(x) = x_{21} + x_{22} + x_{23} \leq 500; 540 \quad \text{(Cotton)}$$

$$g_{23}(x) = x_{31} + x_{32} + x_{33} \leq 325; 350 \quad \text{(Sorghum)}$$

Irrigable land (fuzzy):

$$g_{31}(x) = x_{11} + x_{21} + x_{31} \leq 400; 440$$

$$g_{32}(x) = x_{12} + x_{22} + x_{32} \leq 600; 630$$

$$g_{33}(x) = x_{13} + x_{23} + x_{33} \leq 300; 320$$

Land appropriation (precise):

$$g_{41}(x) = (x_{11}+x_{21}+x_{31})/400 = (x_{12}+x_{22}+x_{32})/600$$

$$g_{42}(x) = (x_{12}+x_{22}+x_{32})/600 = (x_{13}+x_{23}+x_{33})/300$$

$$g_{43}(x) = (x_{13}+x_{23}+x_{33})/300 = (x_{11}+x_{21}+x_{31})/400$$

Non-negativity:

$$x_{ij} \geq 0, \text{ i and j} = 1, 2, 3$$

In the above table, x_{ij} = the number of acres of land in kibbutz j to be allocated to plant crop i, i = 1 (sugar beets), 2 (cotton) and 3 (sorghum).

Since the membership functions of the fuzzy objective and the fuzzy constraints are defined as equations (3.10) and (3.11), respectively, we can then obtain the following model:

max $\quad \alpha$

s.t. $\quad 1 - (350000-z)/100000 \geq \alpha$

$\qquad 1 - [g_{11}(x)-600]/60 \geq \alpha$

$\qquad 1 - [g_{12}(x)-800]/40 \geq \alpha$

$\qquad 1 - [g_{13}(x)-375]/75 \geq \alpha$

$\qquad 1 - [g_{21}(x)-600]/50 \geq \alpha$

$\qquad 1 - [g_{22}(x)-500]/40 \geq \alpha$

$\qquad 1 - [g_{23}(x)-325]/25 \geq \alpha$

$\qquad 1 - [g_{31}(x)-400]/40 \geq \alpha$

$\qquad 1 - [g_{32}(x)-600]/30 \geq \alpha$

$\qquad 1 - [g_{33}(x)-300]/20 \geq \alpha$

$\qquad 3(x_{11}+x_{21}+x_{31})-2(x_{12}+x_{22}+x_{32}) = 0$

$$(x_{12}+x_{22}+x_{32})-2(x_{13}+x_{23}+x_{33}) = 0$$
$$4(x_{13}+x_{23}+x_{33})-3(x_{11}+x_{21}+x_{31}) = 0$$
$$\alpha \in [0, 1], \text{ and } x_{ij} \geq 0, \forall i,j.$$

The optimal solution is: $x^*_{11} = 152.78$, $x^*_{12} = 48.61$, $x^*_{13} = 128.47$, $x_{21} = 104.94$, $x^*_{22} = 337.96$, $x^*_{23} = x^*_{31} = x_{32} = 0$, and $x^*_{33} = 64.82$ with $\alpha^* = 0.213$.

3.1.2.1c Example 3: A Fuzzy Resource Allocation Problem

Consider a fuzzy resource allocation problem [172]. The problem is that there are J resources, indexed by $j \in \{1, 2, ..., J\}$, which are allocated to I tasks, $i \in \{1, 2, ..., I\}$. Each task contributes to the completion of one or more activities, indexed by $k \in \{1, 2, ..., K\}$. The quantity of resources j allocated to task i is denoted by x_{ji} for all i and j, where x_{ji} is a non-negative real number, and $x = (x_{ji})$ is the JxI allocation matrix. The total consumption y_j of the resource j is then given by

$$y_j = \Sigma_i \, x_{ji}$$

and y_j should not be larger than the available quantity h_j of resource j, where $h_j > 0$ for all j. Accordingly, for each $j \in \{1, ..., J\}$ introduce a fuzzy set A_j describing the intuitive requirement that: "The consumption of resource j should be as small as possible and not greater than or equal to h_j." Mathematically, A_j is defined by $A_j = \{y_j, \mu_j(y_j)\}$ where $y_j \in [0, \infty)$ for all j. It is assumed that μ_j is continuous and that:

$$\mu_j(y_j^1) > \mu_j(y_j^2) \quad \text{if } y_j^1 < y_j^2 \text{ and } y_j^1, y_j^2 \in [0, h_j)$$
$$\mu_j(0) > 0$$
$$\mu_j(y_j) = 0 \qquad \text{if } y_j \geq h_j.$$

For a given activity k, it is assumed that the total effectiveness on that activity of all the allocation is given by:

$$x_k = \Sigma_j \Sigma_i \, \Gamma_{jik} x_{ji}$$

where, for each (j, i, k), $\Gamma_{jik} \geq 0$ is a real number describing the effectiveness of an allocation for each $x_{ji} = 1$.

A fuzzy set is introduced for each activity k in order to describe the requirement that:

$[0, \infty)$ for all k, where r_k is the membership function of B_k, assuming that r_k is continuous and

$$r_k(0) \geq 0$$
$$r_k(x_k{}^1) < r_k(xj_k{}^2) \quad \text{if } x_k{}^1 < x_k{}^2 \text{ and } x_k{}^1, x_k{}^2 \geq 0$$
$$\min_k r_k(0) < \min_j \mu_j(0).$$

By using the max-min operator, an allocation matrix $x = (x_{ji})$ is selected on the basis of the fuzzy set D (see Figure 3.1 also) defined by $D = \{x, z(x)\}$ and $x = (x_{ji}) \geq 0$, where $z(x)$ $= \min \{\min_j \mu_j(y_j), \min_k r_k(x_k)\}$. A decision x^* with the highest value z^* of the membership function $z(x)$ is then given by the solution of maximizing $z(x)$ where $x = (x_{ji}) \geq 0$. That is to solve the following problem:

$$\max \quad \alpha$$
$$\text{s.t.} \quad \mu_j(\Sigma_i \, x_{ji}) \geq \alpha, \; \forall j$$
$$r_k(\Sigma_j \, \Sigma_i \; \Gamma_{jik} x_{ji}) \geq \alpha, \; \forall k$$
$$x = (x_{ji}) \geq 0.$$

To solve the above problem, let us extend the definition of r_k as $r_k(\delta_k) = r_k(0)(\delta_k + 1)$ for $\delta_k \in [-1, 0]$, showing that r_k is a strictly increasing function of δ_k.

It is observed that the continuity and strict monotonicity assumptions on the functions μ_j and r_k demand the existence of inverse functions defined by the requirements that:

$$\mu_j(y_j) = \alpha \quad <=> \quad \mu_j^{-1}(\alpha) = y_j, \; y_j \epsilon[0, h_j), \; \alpha\epsilon(0, \mu_j(0)]$$

$$r_k(\delta_k) = \alpha \quad <=> \quad r_k^{-1}(\alpha) = \delta_k, \; \delta_k\epsilon[-1, 0], \; \alpha\epsilon[0, r_k(\infty)].$$

It is noted that $\mu_j^{-1}(0)$ is not uniquely defined, while $\mu_j(y_j) = 0$ and $y_j \geq h_j$.

It follows that μ_j^{-1} and r_k^{-1} are respectively strictly decreasing and strictly increasing functions. For $\alpha \epsilon [0, \min_j \mu_j(0)]$ $(z^* \leq \min_j \mu_j(0))$, our problem becomes:

$$\max \quad \delta$$
$$\text{s.t.} \quad a_j\delta - \Sigma_i \, x_{ji} \geq a_j\alpha - \mu_j^{-1}(\alpha), \; \forall j$$
$$b_k\delta - \Sigma_j \, \Sigma_i \; \Gamma_{jik} x_{ji} \leq b_k\alpha - r_k^{-1}(\alpha), \; \forall k$$
$$x = (x_{ji}) \geq 0$$

where $a_j < 0$ for $\forall$ j and $b_k > 0$ for $\forall$k are given (selected) real numbers. Thus the optimal solution of our fuzzy resource allocation problem can be determined by the following procedure:

Step 1. Start with $n = 1$ and a selected value α_1 of α to solve the above problem. If $\delta^*(\alpha_1) = \alpha_1$, terminate the procedure; otherwise go to Step 2.

Step 2. Choose $\alpha_{n+1} > \alpha_n$ if $\delta^*(\alpha_n) > \alpha_n$ and choose $\alpha_{n+1} < \alpha_n$ if $\delta^*(\alpha_n) < \alpha_n$, and then solve the problem. Terminate the procedure if $\delta^*(\alpha_{n+1}) = \alpha_{n+1}$ or if $z^* \epsilon (\alpha_n, \alpha_{n+1})$ where $\alpha_n < \alpha_{n+1}$, or $z^* \epsilon (\alpha_{n+1}, \alpha_n)$ where $\alpha_{n+1} < \alpha_n$, and the difference $| \alpha_{n+1} - \alpha_n |$ is sufficiently small. Otherwise replace n by $(n+1)$ and return to Step 2.

Now, suppose that the following data are given ($J = I = K = 2$):

$$(\Gamma_{ji1}) = \begin{pmatrix} 2 & 1 \\ 0 & 0 \end{pmatrix} \qquad (\Gamma_{ji2}) = \begin{pmatrix} 0 & 0 \\ 0 & 1 \end{pmatrix}$$

$$\mu_1(y_1) = \begin{cases} 3 - y_1 & \text{if } y_1 \epsilon [0, 1] \\ 7/2 - (3/2)y_1 & \text{if } y_1 \epsilon (1, 2) \\ 0 & \text{if } y_1 \epsilon [2, \infty) \end{cases}$$

$$\mu_2(y_2) = \begin{cases} 2 - y_2 & \text{if } y_2 \epsilon [0, 2) \\ 0 & \text{if } y_2 \epsilon [2, \infty) \end{cases}$$

$$r_1(x_1) = \begin{cases} 3x_1 & \text{if } x_1 \epsilon [0, 1) \\ 2 + x_1 & \text{if } x_1 \epsilon [1, \infty) \end{cases}$$

$$r_2(x_2) = \begin{cases} x_2 & \text{if } x_2 \epsilon [0, 1) \\ -1 + 2x_2 & \text{if } x_2 \epsilon [1, \infty) \end{cases}$$

and let $a_1 = a_2 = -1$ and $b_1 = b_2 = 1$. This problem can then be formulated as the following linear programming problem:

$$
\begin{aligned}
\max \quad & \delta \\
\text{s.t.} \quad & -\delta - x_{11} - x_{12} - s_1 = -\alpha - \mu_1^{-1}(\alpha) \\
& -\delta - x_{21} - x_{22} - s_2 = -\alpha - \mu_2^{-1}(\alpha) \\
& \delta - 2x_{11} - x_{12} - s_3 = \alpha - r_1^{-1}(\alpha) \\
& \delta - x_{22} - s_4 = \alpha - r_2^{-1}(\alpha) \\
& x_{ji} \geq 0, \ \forall j, i.
\end{aligned}
$$

The choice $\alpha_1 = 0$ in Step 1 of the algorithm and the solution of this problem gives the basic variable δ, s_1, x_{12}, x_{22} with the values 1, 1, 1, 1 and $\delta^*(0)$ (= 1) $> \alpha_1$ (= 0). So, let $\alpha_2 = 2$. The solution is then $\delta^* = 1/2$, $x_{11} = 1/2$, $x_{12} = 2/3$ and $x_{22} = 3/2$. Now, $\delta^*(2) = 1/2 > 2$ and demand $\alpha_3 < 2$. The choice of $\alpha_3 = (\alpha_1 + \alpha_2)/2$ gives the solution $(\delta^*, x_{11}, x_{12}, x_{22}) = (1, 1, 1, 1)$. $\delta^*(1) = 1$ shows that an optimal solution is given by $z^* = 1$ and $(x_{11}, x_{12}, x_{21}, x_{22})^* = (1, 2, 0, 1)$.

3.1.2.2 Chanas's Approach - A Nonsymmetric Model

For Equation (3.8), Chanas considered that the goal b_0 and its tolerance p_0 cannot be specified initially because of lack of knowledge about the fuzzy feasible region. Therefore, in order to help the decision maker to determine b_0 and p_0, Chanas [34] first solves the following problem, and then presents the results to the decision maker to determine b_0 and p_0:

$$\max \quad cx$$

$$\text{s.t.} \quad (Ax)_i \leq \tilde{b}_i, \forall i,$$

$$x \geq 0. \tag{3.15}$$

where the tolerances p_i for each b_i is given and the membership functions of the fuzzy constraints are assumed as Equation (3.11). Then we obtain $\mu_i \geq 1 - [(Ax)_i) - b_i]/p_i \geq \alpha$, $\forall i$, which are equivalent to $(Ax)_i \leq b_i + (1 - \alpha)p_i$, $\forall i$. If $\theta = 1 - \alpha$, then we will have $(Ax)_i \leq b_i + \theta p_i$. Obviously, Equation (3.15) is equivalent to the following parametric programming:

$$\max \quad cx$$

$$\text{s.t.} \quad (Ax)_i \leq b_i + \theta p_i, \forall i$$

$$x \geq 0 \text{ and } \theta \in [0, 1] \tag{3.16}$$

where c, A, b_i and p_i, $\forall i$, are given and θ is a parameter. It is easy to see that for every admissible solution $x^*(\theta)$ with a fixed parameter θ, the condition:

$$\mu_i(Ax^*(\theta)) \geq 1 - \theta, \forall i$$

is valid. On the other hand, for every non-zero basic solution (if $p_i > 0$), there exists at least one active constraint, i, such that $\mu_i(Ax^*(\theta)) \geq 1 - \theta$ (or $(Ax)_i = b_i + \theta p_i$) and consequently the common degree of satisfaction of the constraints is:

$$\mu_c(Ax^*(\theta)) = \Lambda_i \; \mu_i((Ax^*(\theta)) = 1 - \theta. \tag{3.17}$$

Hence, for every θ we obtain a solution (if one exists) which satisfies jointly the constraints with degree $1 - \theta$.

The optimal solutions $z^*(\theta)$ and $x^*(\theta)$ of Equation (3.16) are then presented to the decision maker. The decision maker can now choose b_0 and its corresponding p_0. According to these values, we can then construct the membership function μ_0 of the objective function as (3.10). Since the final optimal solution will exist at $x^*(\theta)$, μ_0 becomes:

$$\mu_0(x^*(\theta)) = \begin{cases} 1 & \text{if } cx^*(\theta) > b_0 \\ 1 - [b_0 - cx^*(\theta)]/p_0 & \text{if } b_0 - p_0 \le cx^*(\theta) \le b_0 \\ 0 & \text{if } cx^*(\theta) < b_0 - p_0 \end{cases} \tag{3.18}$$

which is always piecewise linear function with respect to θ.

In Equation (3.16), the overall membership function μ_c of the fuzzy constraint set is $1 - \theta$, by use of the min-operator. Therefore, the final optimal solution $x^*(\theta^*)$ and $z^*(\theta^*)$ will exist at:

$$\max \mu_D(\theta) = \max \{ \min [\mu_0(\theta), \mu_c(\theta)] \} = \max [\mu_0(\theta) \, A \, \mu_c(\theta)] \tag{3.19}$$

as depicted in Figure 3.9. To illustrate this approach, we proposed the algorithm as shown in Figure 3.10.

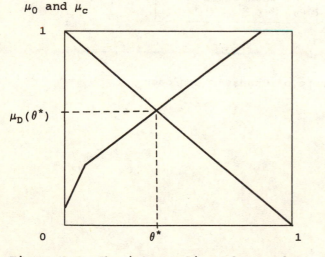

Figure 3.9 The intersection of μ_0 and μ_c

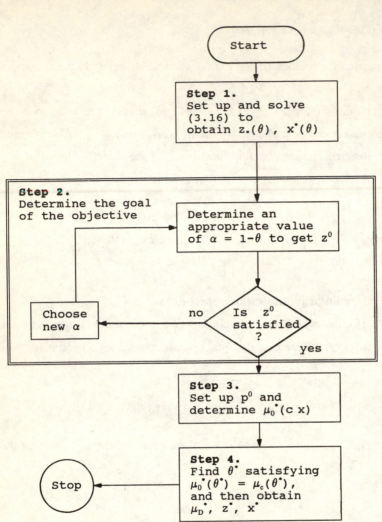

Figure 3.10 Solution procedure of Chanas's approach

3.1.2.2a Example 1: An Optimal System Design Problem

Consider designing an optimal system problem proposed by Zeleny [293]. However, we modify it as an example of a fuzzy linear programming problem.

Zeleny has proposed that modern production systems are increasingly characterized by novel high-productivity attributes: minimal buffers, close-to-zero inventories, full utilization of resources, flexibility of layouts, resource portfolios and so on. Therefore, the crisp linear programming, with its mathematically "built-in" under-utilization of resources, single-mindedness of purpose, and fixation on the "fixed," does not appear to be a sufficiently rich and flexible tool to match these novel requirements.

Here, a simplified example of an optimal design system in a fuzzy environment is considered and solved. The constraints and the objective are fuzzified.

A small company has been producing a highly profitable decorative material for the last several Christmas seasons. This decorative material came in two versions, x and y. Five different component ingredients were needed: golden thread, silk, velvet, silver thread and nylon. The prices of these inputs and their technological contributions to both x and y are given in Table 3.8. The profit margins are $400 per unit of x and $300 per unit of y.

Table 3.8 Inputs and technological coefficients

Resource	Technological Coefficients		Price/unit ($)
	x	y	
golden thread	4	0	30
silk	2	6	40
velvet	12	4	9.5
silver thread	0	3	20
nylon	4	4	10

In order to maintain these margins, the company does not allow "substantially more than $2600" to be spent on the purchase of the components. If the decision maker spends $2600 to buy 20, 24, 60, 10.5, 26 units of golden thread, silk, velvet, silver thread, nylon, respectively, we can solve the following linear programming and get the optimal production scheduling in the given situation:

max   400x + 300y   (profits)

s.t.   4x   ≤ 20   (golden thread)

       2x + 6y   ≤ 24   (silk)

       12x + 4y   ≤ 60   (velvet)

              3y   ≤ 10.5   (silver thread)

       4x + 3y   ≤ 26   (nylon).

By graphic technique, the solution obtained is: $x^* = (3.125, 2.625)$ and the optimal profit = 2437.5.

Now let us formulate the above problem into the following fuzzy linear high-productivity-system programming model (Lai [139]):

max   400x + 300y   (profits)

s.t.   4x $= b_1$ (golden thread)

       $2x + 6y = b_2$ (silk)

       $12x + 4y = b_3$ (velvet)

              $3y = b_4$ (silver thread)

       $4x + 3y = b_5$ (nylon)

       $30b_1 + 40b_2 + 9.5b_3 + 20b_4 + 10b_5 \leq \widetilde{2600}$ (budget)

       $x, y, b_1, b_2, b_3, b_4, b_5 \geq 0$

where b_i, i = 1, ..., 5, are the optimal resource levels to achieve high-productivity systems with close-to-zero inventory. The solution procedure is shown as follows:

Step 1. Suppose the maximal tolerance for the budget constraint is $200, then the parametric programming version becomes:

max 400x + 300y

s.t. $30(4x) + 40(2x+6y) + 9.5(12x+4y) + 20(3y) + 10(4x+3y) = 354x + 368y \leq 2600 + 200\theta$

x, y ≥ 0, and $\theta \epsilon [0, 1]$.

The optimal solution $x^* = ((2600 + 200\theta)/354 , 0)$ and $z^*(\theta) = (400/354) (2600 + 200\theta)/350$.

Step 2. For simplicity, assume that $z_0 = 3100$ is the goal of the profit objective.

Step 3. Assume that $p_0 = 200$ and the membership function of the profit objective is linear as Equation (3.10). Thus, one can get the following membership of the objective:

$$\mu_0^*(\theta) = 1 - \cfrac{3100 - 400(2600+200\theta)/354}{200} = 0.189 + 1.13\theta$$

Step 4. Set $\mu_C = \mu_0$ as shown in Figure 3.11. We obtain $0.189 + 1.13\theta = 1 - \theta$ and then θ^* = 0.38. Thus, $\mu_D^* = \mu_0 = \mu_c = 1 - 0.38 = 0.62$. The optimal solution is then: $x^* = [2600 + 200(0.38)]/354 = 7.56$, $y^* = 0$, and $z^* = \$3023.73$. The necessary resources are: $b_1^* = 30.24$, $b_2^* = 15.12$, $b_3^* = 90.72$, $b_4^* = 0$ and $b_5^* = 30.24$ and the budget is $\$ 2676.24$.

Table 3.9 represents the differences between the two solutions of the crisp linear programming and fuzzy linear high-productivity programming.

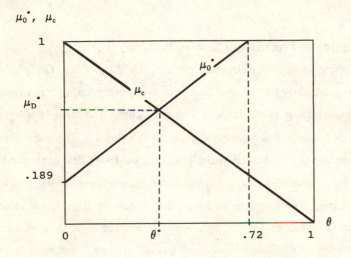

Figure 3.11 The optimal solution of the fuzzy optimal system design problem

Table 3.9 The comparison of the crisp solution and fuzzy high-productivity solution

Crisp standard system	Fuzzy high-productivity system
$x^* = 4.125$ $y^* = 2.625$	$x^* = 7.56$ $y^* = 0$
Profit $z^* = \$2437.5$	$z^* = 3023.73$
Budget = \$2600	\$2676.24
Resources $b_1 = 20$ $b_2 = 24$ $b_3 = 60$ $b_4 = 20.5$ $b_5 = 26$	$b_1^* = 30.24$ $b_2^* = 15.12$ $b_3^* = 90.72$ $b_4^* = 0$ $b_5^* = 30.24$

3.1.2.2b Example 2: An Aggregate Production Planning Problem

Consider a single objective aggregate production planning (APP) problem [145]. APP deals with matching supply and demand of forecasted and fluctuated customers' orders over the medium time range, up to approximately twelve months into the future. The term "aggregate" implies that the planning is done for a few aggregate product categories. The purposes of aggregate production planning are (1) to set up overall production levels for each product category to meet the fluctuating or uncertain demands in the near future; (2) to set up decisions and policies on the issues of hiring, layoff, overtime, backorder, subcontracting, and inventory level, which means that the APP will determine the appropriate resources to be used as well.

When using any of the aggregate production planning models, it is often assumed that the goals and the model inputs (resources and demands) are deterministic/crisp. In practice, demands, resources and costs are usually imprecise/fuzzy. The current APP model represents the information in a fuzzy environment where the objective function and the parameters are not completely defined and cannot be accurately measured. The best compromise APP will balance the cost of building and holding inventory against the cost of adjusting activity levels to meet the fluctuations in demands. The forecasted demand in a particular period could either be satisfied or backordered. However, the backorder must be fulfilled within the next period. The single objective linear programming model for the APP can then be formulated as:

$$\underset{\sim}{\min} \quad \overset{T}{\underset{t=1}{\Sigma}} \; [C_{pt}(P_{rt}+P_{ot})+C_{rt}W_t+C_{ot}(kP_{ot})+C_{it}I_t+C_{bt}B_t+C_{ht}H_t+C_{lt}L_t] \qquad (ex.1)$$

$$\text{s.t.} \quad W_t \le \widetilde{W}_{tmax} \qquad (ex.2)$$

$$W_t = W_{t-1} + H_t - L_t \qquad (ex.3)$$

$$kP_{rt} \le \delta W_t \qquad (ex.4)$$

$$kP_{ot} \le \beta_t\delta W_t \qquad (ex.5)$$

$$P_{rt} + P_{ot} + I_{t-1} - B_{t-1} \ge \widetilde{F}_{tmin} \qquad (ex.6)$$

$$I_t - B_t = I_{t-1} - B_{t-1} + P_{rt} + P_{ot} - F_t \qquad (ex.7)$$

$$P_{rt}, P_{ot}, W_t, I_t, B_t, H_t, L_t \ge 0, \; \forall t \qquad (ex.8)$$

where the parameters and decision variables are explained in the table:

Parameters and constants:

C_{pt} : production cost excluding labor cost in period t (\$/unit)

C_{ot} : overtime labor cost in period t (\$/man-hour)

C_{rt} : labor cost in period t (\$/man-day)

C_{it} : inventory carrying cost in period t (\$/unit-period)

C_{ht} : cost to hire one worker in period t (\$/man-day)

C_{lt} : cost to layoff one worker in period t (\$/man-day)

C_{bt} : stockout cost in period t (\$/unit-period)

k : conversion factor in hours of labor per unit of production

δ : regular working hours per worker per day

β_t : fraction of working hours available for overtime production

T : planning horizon or number of periods

I_0 : initial inventory level (units)

W_0 : initial work force level (man-day)

B_0 : initial backorder level (units)

W_{tmax} : maximum work force available in period t (man-day)

F_t : forecasted demand in period t (units)

F_{tmin} : minimum demand in period t (units)

Decision variables:

P_{rt} : regular time production in period t (units)

P_{ot} : overtime production in period t (units)

W_t : work force level in period t (man-day)

I_t : inventory level in period t (units)

B_t : backorder level in period t (units)

H_t : worker hired in period t (man-day)

L_t : worker layoff in period t (man-day)

The fuzzy linear programming model for the single objective APP problem has a fuzzy goal, fuzzy work force level allowed in each period, and fuzzy demands. Equation (ex.1) is the objective of minimizing total cost. Equation (ex.2) is a policy that sets the limit of maximum available labor force in any period. Equation (ex.3) represents the constraint that the work force in period t is equal to the work force in period t-1 plus new hires minus layoffs. Equations (ex.4) and (ex.5) represent the constraints that limit the regular and overtime production to the available hours. Equation (ex.6) stands for the constraint that a production backorder in any period is not carried for more than one period. Equation (ex.7) requires a solution that the inventory level or backorder level in period t be equal to what it was in period t-1 plus the regular time production, overtime production and minus the forecasted demand. Equation (ex.8) is a non-negative constraint.

Now, let us consider the following numeric data:

a) There is a six-period planning horizon with fuzzy demands of intervals [250, 275], [285, 330], [430, 500], [260, 320], [300, 350], [270, 340] units. Here, it is assumed that the trend of future demand is optimistic in periods 1 and 2. The decision maker has the highest satisfaction level at demand levels 275 and 330, and the lowest satisfaction level at demand 250 and 285 for periods 1 and 2, respectively. The decision maker feels that the demands in periods 5 and 6 are pessimistic, and in-between in periods 3 and 4. The minimum demands are 250, 285, 430, 260, 300 and 270 units, respectively.

b) The production cost other than labor cost is $20 per unit. Three hours of labor are needed for each unit produced. The regular labor force works an eight-hour day.

c) The initial work force is 100 workers. The maximum numbers of workers allowed are [90, 100] and [100, 115], for periods 1 and 2, [120, 150] for periods 3, 4, 5, and 6. The costs associated with the regular payroll, hiring and firing are $64, $30 and $40 per worker per day, respectively.

d) Overtime production is limited to no more than 30% of regular time production. The overtime charge is based on \$15 per worker per hour.

e) There is no beginning inventory. The inventory carrying cost is \$2 per unit per period. Backorder must not be carried over more than one period. The backorder cost is \$3 per unit per period.

The formulation of this APP model is:

$$\min \ \sum_{t=1}^{6} [20(P_{rt} + P_{ot}) + 64W_t + 15(3P_{ot}) + 2I_t + 3B_t + 30H_t + 40L_t]$$

s.t.

$$W_1 \leq \widetilde{90}, \quad W_2 \leq \widetilde{100}, \quad W_3 \leq \widetilde{120}$$

$$W_4 \leq \widetilde{120}, \quad W_5 \leq \widetilde{120}, \quad W_6 \leq \widetilde{120}$$

$$W_1 = W_0 + H_1 - L_1, \quad W_2 = W_1 + H_2 - L_2$$

$$W_3 = W_2 + H_3 - L_3, \quad W_4 = W_3 + H_4 - L_4$$

$$W_5 = W_4 + H_5 - L_5, \quad W_6 = W_5 + H_6 - L_6$$

$$3P_{r1} \leq 8W_1, \quad 3P_{r2} \leq 8W_2, \quad 3P_{r3} \leq 8W_3$$

$$3P_{r4} \leq 8W_4, \quad 3P_{r5} \leq 8W_5, \quad 3P_{r6} \leq 8W_6$$

$$3P_{o1} \leq (0.3)8W_1, \quad 3P_{o2} \leq (0.3)8W_2$$

$$3P_{o3} \leq (0.3)8W_3, \quad 3P_{o4} \leq (0.3)8W_4$$

$$3P_{o5} \leq (0.3)8W_5, \quad 3P_{o6} \leq (0.3)8W_6$$

$$P_{r1} + P_{o1} + I_0 - B_0 \geq 250$$

$$P_{r2} + P_{o2} + I_1 - B_1 \geq 285$$

$$P_{r3} + P_{o3} + I_2 - B_2 \geq 430$$

$$P_{r4} + P_{o4} + I_3 - B_3 \geq 260$$

$$P_{r5} + P_{o5} + I_4 - B_4 \geq 300$$

$$P_{r6} + P_{o6} + I_5 - B_5 \geq 270$$

$$I_1 - B_1 = I_0 - B_0 + P_{r1} + P_{o1} - \widetilde{275}$$

$$I_2 - B_2 = I_1 - B_1 + P_{r2} + P_{o2} - \widetilde{330}$$

$$I_3 - B_3 = I_2 - B_2 + P_{r3} + P_{o3} - \widetilde{450}$$

$$I_4 - B_4 = I_3 - B_3 + P_{r4} + P_{o4} - \widetilde{300}$$

$$I_5 - B_5 = I_4 - B_4 + P_{r5} + P_{o5} - \widetilde{300}$$

$$I_6 - B_6 = I_5 - B_5 + P_{r6} + P_{o6} - \widetilde{270}$$

$$P_{rt}, \ P_{ot}, \ W_t, \ I_t, \ B_t, \ H_t, \ L_t \geq 0, \ t = 1, \ldots 6$$

where the membership functions of the fuzzy demands and work force levels are given in Figure 3.12. The membership function of work force level in period 1, for example, can be represented as follows:

$$\mu_{W1max} = \begin{cases} 1 & \text{if } W_{1max} < 90 \\ 1 - (W_{1max} - 90)/10 & \text{if } 90 \leq W_{1max} \leq 100 \\ 0 & \text{if } 100 < W_{1max}. \end{cases}$$

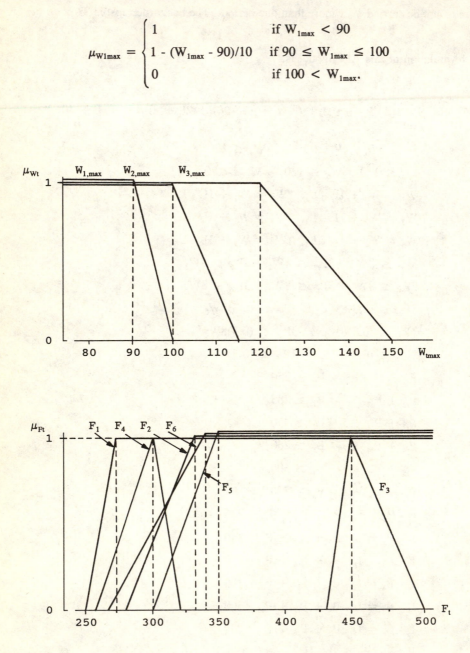

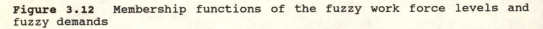

Figure 3.12 Membership functions of the fuzzy work force levels and fuzzy demands

According to Chanas's concept [34, 35], we can obtain the following equivalent constraints for the first fuzzy resource constraint $W_1 \leq 90 + 10\theta$. The other fuzzy resources of the work force level are $W_2 \leq 100 + 15\theta$, and W_3, W_4, W_5 and $W_6 \leq 120 + 30\theta$. For the equality constraints of fuzzy demands, we have:

$$I_0 + B_1 - I_1 - B_0 + P_{r1} + P_{o1} + x_1 = 275$$
$$x_1 \leq 275 - 275 + (250+0)\theta = 25\theta$$
$$I_1 + B_2 - I_2 - B_1 + P_{r2} + P_{o2} + x_2 = 330$$
$$x_2 \leq 45\theta$$
$$I_2 + B_3 - I_3 - B_2 + P_{r3} + P_{o3} + x_3 = 450 + 50\theta$$
$$x_3 \leq 70\theta$$
$$I_3 + B_4 - I_4 - B_3 + P_{r4} + P_{o4} + x_4 = 300 + 20\theta$$
$$x_4 \leq 60\theta$$
$$I_4 + B_5 - I_5 - B_4 + P_{r5} + P_{o5} + x_5 = 300 + 50\theta$$
$$x_5 \leq 50\theta$$
$$I_5 + B_6 - I_6 - B_5 + P_{r6} + P_{o6} + x_6 = 270 + 70\theta$$
$$x_6 \leq 70\theta$$

where x_t is an auxiliary variable, $t = 1, 2, \ldots, 6$. Our problem becomes:

$$\min \quad z = \sum_{t=1}^{6} [20(P_{rt} + P_{ot}) + 64W_t + 15(3P_{ot}) + 2I_t + 3B_t + 30H_t + 40L_t]$$

s.t.
$$W_1 \leq 90 + 10\theta, \quad W_2 \leq 100 + 15\theta$$
$$W_i \leq 120 + 30\theta, \ i = 3, 4, 5 \text{ and } 6$$
$$W_1 = W_0 + H_1 - L_1, \quad W_2 = W_1 + H_2 - L_2$$
$$W_3 = W_2 + H_3 - L_3, \quad W_4 = W_3 + H_4 - L_4$$
$$W_5 = W_4 + H_5 - L_5, \quad W_6 = W_5 + H_6 - L_6$$
$$3P_{r1} \leq 8W_1, \quad 3P_{r2} \leq 8W_2, \quad 3P_{r3} \leq 8W_3$$
$$3P_{r4} \leq 8W_4, \quad 3P_{r5} \leq 8W_5, \quad 3P_{r6} \leq 8W_6$$
$$3P_{o1} \leq (0.3)8W_1, \quad 3P_{o2} \leq (0.3)8W_2$$
$$3P_{o3} \leq (0.3)8W_3, \quad 3P_{o4} \leq (0.3)8W_4$$
$$3P_{o5} \leq (0.3)8W_5, \quad 3P_{o6} \leq (0.3)8W_6$$
$$P_{r1} + P_{o1} + I_0 - B_0 \geq 250$$
$$P_{r2} + P_{o2} + I_1 - B_1 \geq 285$$

$$P_{r3} + P_{o3} + I_2 - B_2 \geq 430$$
$$P_{r4} + P_{o4} + I_3 - B_3 \geq 260$$
$$P_{r5} + P_{o5} + I_4 - B_4 \geq 300$$
$$P_{r6} + P_{o6} + I_5 - B_5 \geq 270$$
$$I_0 + B_1 - I_1 - B_0 + P_{r1} + P_{o1} + x_1 = 275$$
$$I_1 + B_2 - I_2 - B_1 + P_{r2} + P_{o2} + x_2 = 330$$
$$I_2 + B_3 - I_3 - B_2 + P_{r3} + P_{o3} + x_3 = 450 + 50\theta$$
$$I_3 + B_4 - I_4 - B_3 + P_{r4} + P_{o4} + x_4 = 300 + 20\theta$$
$$I_4 + B_5 - I_5 - B_4 + P_{r5} + P_{o5} + x_5 = 300 + 50\theta$$
$$I_5 + B_6 - I_6 - B_5 + P_{r6} + P_{o6} + x_6 = 270 + 70\theta$$
$$x_1 \leq 25\theta, \ x_2 \leq 45\theta, \ x_3 \leq 70\theta$$
$$x_4 \leq 60\theta, \ x_5 \leq 50\theta, \ x_6 \leq 70\theta$$
$$P_{rt}, P_{ot}, W_t, I_t, B_t, H_t, L_t \text{ and } x_t \geq 0, \ \forall \, t.$$

Now, we can use the parametric programming approach proposed by Chanas [35] to solve this problem. However, we provided 11 possible solutions, as in Table 3.10, which are considered most helpful for decision makers.

The total costs associated with the production plans are between \$81,915 and \$91,213. For $\alpha = 1 - \theta = 1$, the decision maker has the highest satisfaction level in the fuzzy resources utilized. The results indicate that each fuzzy parameter has achieved the maximum satisfaction level in the decision maker's mind. The production plan is to produce more by using overtime production, hiring fewer people, and establishing a lower inventory level. On the other hand, the decision maker will hire more people to produce more through regular time production and build up inventory during the optimistic periods when a lower aspiration level (α) is presented.

Suppose that the total cost should be in [\$81,000, \$88,000] by referring to Table 3.10. The goal is \$81,000 with the tolerance \$7,000. Then we can obtain the following membership function of the objective function (z = total cost):

$$\mu_z(z^*(\theta)) = \begin{cases} 1 & \text{if } z^*(\theta) < 81000 \\ 1 - [z^*(\theta)-81000]/7000 & \text{if } 81000 \leq z^*(\theta) \leq 88000 \\ 0 & \text{if } z^*(\theta) > 88000. \end{cases}$$

Therefore, the following relation is obtained:

Table 3.10 Fuzzy solutions of the fuzzy APP problem

θ	$z^*(\theta)$	θ	$z^*(\theta)$	θ	$z^*(\theta)$	θ	$z^*(\theta)$
1.0	81915	0.7	84625	0.4	87419	0.1	90262
0.9	82809	0.6	85545	0.3	88360	0.0	91213
0.8	83717	0.5	86482	0.2	89311		

Aggregate production plan:

θ	1.0	0.9	0.8	0.7	0.6	0.5	0.4	0.3	0.2	0.1	0.0
P_{r1}	267	264	261	259	256	253	251	248	245	243	240
P_{r2}	307	303	299	295	291	287	283	279	275	271	267
P_{r3}	307	303	299	295	292	295	298	302	308	314	320
P_{r4}	280	283	286	289	292	295	298	302	308	314	320
P_{r5}	280	283	286	389	292	295	298	300	300	300	300
P_{r6}	270	270	270	270	270	270	270	270	270	270	270
P_{o1}	0	0	0	0	4	9	14	20	25	30	35
P_{o2}	0	12	31	50	63	67	72	75	76	77	77
P_{o3}	85	91	90	88	88	89	89	91	92	94	96
P_{o4}	0	0	0	0	0	0	0	0	0	0	0
P_{o5}	0	0	0	0	0	0	0	0	0	0	0
P_{o6}	0	0	0	0	0	0	0	0	0	0	0
I_1	17	12	6	1	0	0	0	0	0	0	0
I_2	38	37	42	47	50	47	43	37	30	22	14
I_3	0	0	0	0	0	0	0	0	0	0	0
I_4	20	17	14	11	8	5	2	0	0	0	0
I_5	0	0	0	0	0	0	0	0	0	0	0
I_6	0	0	0	0	0	0	0	0	0	0	0
B_1	0	0	0	0	0	0	0	0	0	0	0
B_2	0	0	0	0	0	0	0	0	0	0	0
B_3	0	2	4	6	8	10	12	14	16	18	20
B_4	0	0	0	0	0	0	0	0	0	0	0
B_5	0	0	0	0	0	0	0	0	0	0	0
B_6	0	0	0	0	0	0	0	0	0	0	0
W_1	100	99	98	97	96	95	94	93	92	91	90
W_2	115	114	112	111	109	108	106	105	103	102	100
W_3	115	114	112	111	110	111	112	113	116	118	120
W_4	105	106	107	108	110	111	112	113	116	118	120
W_5	105	106	107	108	110	111	112	113	113	113	112
W_6	101	101	101	101	101	101	101	101	101	101	101
H_1	0	0	0	0	0	0	0	0	0	0	0
H_2	15	15	14	14	13	13	12	12	11	11	10
H_3	0	0	0	0	1	3	6	8	13	16	20
H_4	0	0	0	0	0	0	0	0	0	0	0
H_5	0	0	0	0	0	0	0	0	0	0	0
H_6	0	0	0	0	0	0	0	0	0	0	0
L_1	0	1	2	3	4	5	6	7	8	9	10
L_2	0	0	0	0	0	0	0	0	0	0	0
L_3	0	0	0	0	0	0	0	0	0	0	0
L_4	10	8	5	3	0	0	0	0	0	0	0
L_5	0	0	0	0	0	0	0	0	3	5	8
L_6	4	5	6	7	9	10	11	12	12	12	11

$$z^*(\theta) \leq 81000 + 70000\theta.$$

By use of Chanas's approach, the solution is presented in Table 3.11 and Figure 3.13.

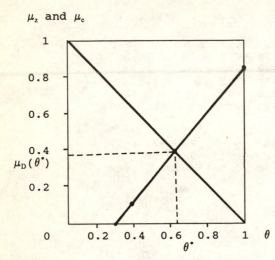

μ_z and μ_c

Figure 3.13 Intersection of μ_0 and μ_c

Table 3.11 Chanas's optimal solution of the APP problem

$\theta = 0.62$ $\quad$ $(\alpha = 0.38)$ $\quad\quad$ $z^*(0.62) = 85348$

Production Level		Inventory Level		Work Force Level	
P_{r1}	257	I_1	0	W_1	96
P_{r2}	292	I_2	51	W_2	109
P_{r3}	292	I_3	0	W_3	109
P_{r4}	291	I_4	9	W_4	109
P_{r5}	291	I_5	0	W_5	109
P_{r6}	270	I_6	0	W_6	101
P_{o1}	3	B_1	0	H_1	0
P_{o2}	62	B_2	0	H_2	13
P_{o3}	87	B_3	8	H_3	0
P_{o4}	0	B_4	0	H_4	0
P_{o5}	0	B_5	0	H_5	0
P_{o6}	0	B_6	0	H_6	0

3.1.3 Linear Programming Problem with Fuzzy Parameters in the Objective Function

It often happens that the unit profit or cost cannot be precisely determined. Therefore, considering the linear programming with imprecise coefficients in the objective function is valuable. This problem is formulated as:

$$\max \quad \tilde{c}x$$
$$\text{s.t.} \quad (Ax)_i \leq b_i, \forall i$$
$$x \geq 0. \tag{3.20}$$

For Equation (3.20), Verdegay [251, 252] proposed the following equivalent parametric linear programming:

$$\max \quad cx$$
$$\text{s.t.} \quad \mu(c) \geq 1 - \alpha$$
$$(Ax)_i \leq b_i, \forall i$$
$$\alpha \in [0, 1] \text{ and } x \geq 0 \tag{3.21}$$

where $c = (c_1, c_2, \ldots, c_n)$, $\mu(c) = \inf_j \mu_j(c_j)$, and $\mu(c)$ is the membership function of (cx) and $\mu_j(c_j)$, $\forall j$, are the membership functions of c_j. The problem given by Equation (3.21) is equivalent to:

$$\max \quad cx$$
$$\text{s.t.} \quad \mu_j(c_j) \geq 1 - \alpha, \forall j$$
$$(Ax)_i \leq b_i, \forall i$$
$$\alpha \in [0, 1] \text{ and } x \geq 0$$

which is equivalent to:

$$\max \quad \Sigma_j c_j x_j$$
$$\text{s.t.} \quad c_j \geq \mu_j^{-1}(1 - \alpha), \forall j$$
$$(Ax)_i \leq b_i, \forall i$$
$$\alpha \in [0, 1] \text{ and } x \geq 0$$

or

$$\max \quad \Sigma_j \, \mu_j^{-1}(1 - \alpha)x_j$$

$$\text{s.t.} \quad (Ax)_i \leq b_i, \, \forall i$$

$$\alpha \in [0, 1] \text{ and } x \geq 0. \tag{3.22}$$

Obviously, Equation (3.22) is a parametric programming, so we can obtain the solution of Equation (3.20) by solving Equation (3.22). The solution is fuzzy and for each $(1 - \alpha)$-level cut of the fuzzy objective set, the solution has an α grade of preference.

On the other hand, Verdegay proposed that for a given fuzzy linear programming Equation (3.20) or (3.1), there always exists another one that is dual and has the same fuzzy solution. Therefore, we can solve the dual problem of Equation (3.20) by using the approaches discussed in Section 3.1.1.

The dual problem of Equation (3.4) is:

$$\min \quad [b+p(1-\alpha)]u$$

$$\text{s.t.} \quad uA^T \geq c, \, u \geq 0 \text{ and } \alpha \in [0, 1] \tag{3.23}$$

which is equivalent to:

$$\min \quad au$$

$$\text{s.t.} \quad a = b + p(1 - \alpha)$$

$$uA^T \geq c, \, u \geq 0 \text{ and } \alpha \in [0, 1].$$

Let $\beta = 1 - \alpha$. The problem becomes:

$$\min \quad au$$

$$\text{s.t.} \quad a \leq b + p\beta \quad (<=> 1 - [b_i\text{-}a_i]/p_i \geq 1\text{-}\beta, \, \forall i)$$

$$uA^T \geq c, \, u \geq 0 \text{ and } \beta \in [0, 1]$$

or

$$\min \quad au$$

$$\text{s.t.} \quad \mu_i(a_i) \geq 1 - \beta, \, \forall i$$

$$uA^T \geq c, \, u \geq 0 \text{ and } \beta \in [0, 1].$$

Finally, we obtain:

$$\min \quad \text{au}$$
$$\text{s.t.} \quad \text{uA}^T \geq \text{c}, \; \text{u} \geq 0. \tag{3.24}$$

In conclusion, the linear programming with fuzzy parameters in the objective function can be solved by the approaches discussed in Section 3.1.1.

3.1.4 Linear Programming with All Fuzzy Coefficients

Sometimes, all the coefficients of a linear programming are imprecise. This problem can be formulated as:

$$\max \quad \tilde{c}x$$
$$\text{s.t.} \quad (\tilde{A}x)_i \leq \tilde{b}_i, \; \forall i,$$
$$x \geq 0 \tag{3.25}$$

For Equation (3.25), Carlsson and Korhonen [31] proposed a fully trade-off approach. They argued that Chanas's approach does not consider any continuous trade-off between grades of constraint violation. For example, if $\mu_1 = 1 - \theta_1$ and $\mu_i > 1 - \theta_1$, $i = 2,..., m$, then $\mu_c = 1 - \theta_1$. Thus, the better possible solution will come out through releasing the μ_i, which is greater than $1 - \theta_1$, to $1 - \theta_i$ by trade-off with $\mu_i \geq 1 - \theta_2$, $i = 2,..,m$. It is obvious θ_2 will be less than θ_1. Therefore, the better solution with a higher degree of satisfaction may be obtained with $\mu_D = \mu_0 = \mu_1 = \mu_2 = ...= \mu_m = 1 - \theta_2$.

Before going into the solution procedure, the requirement of the solution $z^* = z^*(c, -A, b)$ of the non-fuzzy version of Equation (3.25) is an increasing function of the parameters c, -A and b is needed. Also, it is assumed that the user can specify the intervals $[c^0, c^1)$, $[A^0, A^1)$ and $[b^0, b^1)$ for the possible values of the parameters. The lower bounds represent 'risk-free' values in the sense that a solution most certainly should be implementable. The upper bounds, on the other hand, represent parameter values which are most certainly unrealistic and impossible and the solution obtained by using these values is not implementable. When we move from 'risk-free' towards "impossible" parameter value, we move from solutions with a high grade to ones with a low grade on implementing from secure to optimistic solutions. Then, our task is to find an optimal compromise "in-between" as a function of the grades of imprecision in the parameter.

Carlsson and Korhonen proposed a relationship between a solution in Equation (3.25) and its parameters: the solution $z^* = z^*(c, -A, b)$ of Equation (3.25) is an increasing function of the parameter c, -A and b. Thus, we can reasonably assume that membership functions are monotonically decreasing functions of the parameters, c, -A, b. The monotonically decreasing functions may be linear, piecewise linear, hyperbolic exponential, etc. For the purpose of illustrating non-linear membership functions, we will discuss the exponential functions proposed by Carlsson and Korhonen. That is (see Figure 3.14):

$$\mu_c = a_c[1 - \exp\{-b_c(c - c^1)/(c^0 - c^1)\} \tag{3.26}$$

where $b_c > 0$ or $b_c < 0$, and $a_c = 1/[1-\exp(-b_c)]$ and $c \in [c^0, c^1]$. $\mu_c = 1$ when $c \le c^0$, $\mu_c = 0$ when $c > c^1$. b_c is specified by the decision maker. In the same manner, the membership function of A and b can be obtained. It is worth noting that for a maximization problem, the membership functions of the objective function should be non-decreasing in the sense of preference. However, let us keep the originality of Carlsson and Korhonen [31] here.

After full trade-off between c, -A and b, the solution will always exist at:

$$\mu = \mu_c = \mu_A = \mu_b. \tag{3.27}$$

Therefore, we can obtain the following equation:

$$c = g_c(\mu), \ A = G_A(\mu) \text{ and } b = g_b(\mu) \tag{3.28}$$

where $\mu \in [0, 1]$ and g_c, G_A and g_b are inverse functions of μ_c, μ_A and μ_b. Then Equation (3.25) becomes:

$$\text{max} \quad [g_c(\mu)]x$$
$$\text{s.t.} \quad [G_A(\mu)]x \le g_b(\mu) \text{ and } x \ge 0. \tag{3.29}$$

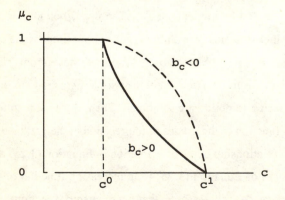

Figure 3.14 The membership function of fuzzy set c
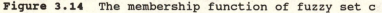

Obviously, Equation (3.29) is a nonlinear programming problem. However, it can be solved by any linear programming technique if μ is given. Thus, we can obtain a set of solutions corresponding to a set of μs and then plot these solution pairs (z^*, μ). After referring to this relationship, the decision maker can choose his/her preferred solution for implementation.

3.1.4.1 Example: A Production Scheduling Problem

Consider a production scheduling problem. A leather goods factory manufactures 3 types of handbags, which must pass through four work centers and the man-hours in each are: chinking [18, 22), paring [10, 40), stitching [96, 110) and finishing [96, 110). The input data are described as fuzzy parameters because they are estimates and cannot be stated precisely in nature. The unit profits of the three products and the technical coefficients are given in the following fuzzy linear programming model:

$$\begin{aligned}
\max \quad & [1, 1.5)x_1 + [1, 3)x_2 + [2, 2.2)x_3 \\
\text{s.t.} \quad & [3, 2)x_1 + [2, 0)x_2 + [3, 1.5)x_3 \leq [18, 22) \\
& [1, 0.5)x_1 + [2, 1)x_2 + [1, 0)x_3 \leq [10, 40) \\
& [9, 6)x_1 + [20, 18)x_2 + [7, 3)x_3 \leq [96, 110) \\
& [7, 6.5)x_1 + [20, 15)x_2 + [9, 8)x_3 \leq [96, 110) \\
& x_1, x_2, x_3 \geq 0
\end{aligned}$$

where the numerical data are based on Carlsson and Korhonen [31]. Suppose that the membership function of c, A, b (Figure 3.15) are:

$$\mu_{aij} = (a_{ij} - a_{ij}^1)/(a_{ij}^0 - a_{ij}^1)$$
$$\mu_{bi} = \{1 - \exp[-0.8(b_i-b_i^1)/(b_i^0-b_i^1)]\}/[1-\exp(-0.8)]$$
$$\mu_{ci} = \{1 - \exp[3(c_i-c_i^1)/(c_i^0-c_i^1)]\}/[1-\exp(3)]$$

where the constants (-0.8) and (3) determined by the decision maker indicate that a positive number implies convex exponential function and a negative number implies concave exponential function.

Next, through the appropriate transformation with the assumption of absolute trade-off between fuzzy numbers of a_{ij}, b_i, c_j, $\forall i,j$, we obtained:

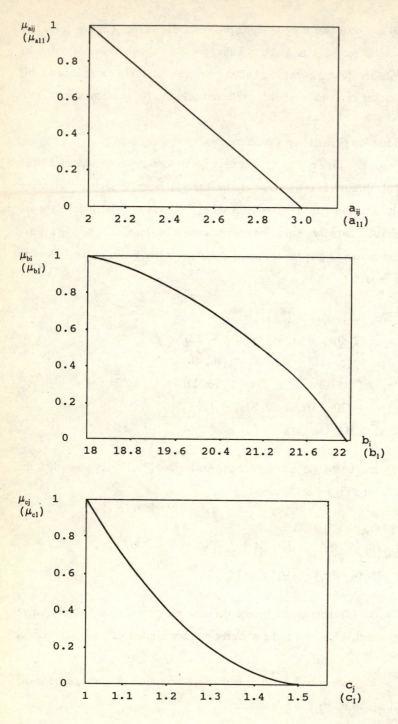

Figure 3.15   The membership functions of the fuzzy coefficients

$$a_{ij} = a_{ij}{}^1 + \mu(a_{ij}{}^0 - a_{ij}{}^1)$$

$$b_i = b_i - (1/0.8)\ln\{1 - \mu_{bi}[1 - \exp(-0.8)]\}(b_i{}^0 - b_i{}^1)$$

$$c_j = c_j + (1/3)\ln\{1 - \mu_{cj}[1 - \exp(3.0)]\}(c_j{}^0 - c_j{}^1).$$

Thus, the original fuzzy linear programming with absolute trade-off is then equivalent to the following:

max $[1.5-0.167\ln(1+19.1\mu)]x_1+[3-0.667\ln(1+19.1\mu)]x_2+[2.2-0.067\ln(1+19.1\mu)]x_3$

s.t. $(2+\mu)x_1+(2\mu)x_2+(1.5+1.5\mu)x_3 \le 22+5\ln(1-0.55\mu)$

$(0.5+0.5\mu)x_1+(1+\mu)x_2+(\mu)x_3 \le 40+37.5\ln(1-0.55\mu)$

$(6+3\mu)x_1+(18+2\mu)x_2+(3+4\mu)x_3 \le 110+17.5\ln(1-0.55\mu)$

$(6.5+0.5\mu)x_1+(15+5\mu)x_2+(8+\mu)x_3 \le 110+17.5\ln(1-0.55\mu)$

$\mu \in [0, 1]$ and $x_i \ge 0$, i = 1, 2, 3.

Obviously, the above nonlinear programming cannot be solved by any standard linear programming technique. However, it can be solved if μ is pre-determined. That is: for each specific value of μ, one can get an optimal solution for the original solution. Therefore, one may choose n number of experiments (n different μ value) in order to obtain n optimal solutions and then present these optimal solutions to the decision maker.

The solution procedure is described as follows:

Step 1. Let us start at $\mu = 0$ and take n = 10 experiment, and set $\delta = 0.1$.

Step 2. Solve the following linear programming:

max $1.5x_1 + 3x_2 + 2.2x_3$

s.t. $2x_1 + 1.5x_3 \le 22$

$0.5x_1 + x_2 \le 40$

$6x_1 + 18x_2 + 3x_3 \le 110$

$6.5x_1 + 15x_2 + 8x_3 \le 110$

$x_1, x_2, x_3 \ge 0.$

The optimal solution is: $x^* = (0, 0, 13.75)$ and $z^* = \$30.25$.

Step 3. Plot $(\mu, z^*) = (0, 30.25)$ in Figure 3.16. Then, set $\mu = \mu + 0.1$. If $\mu > 1$, and go to Step 4. Otherwise, go to Step 2 and solve the linear programming with $\mu = \mu + 0.1$.

Step 4. Figure 3.16 and Table 3.12 show the relationship between the optimal profits and the corresponding membership grades. According to this relationship, the decision maker can then get his optimal solution under a pre-determined imprecision allowable. For instance, when the decision maker considers 0.3 degree imprecision is acceptable, the optimal solution is: $z^* = 15.8$ and $x^* = (0, 2.53, 6.28)$.

Table 3.12 The optimal solution for the absolute trade-off model of a fuzzy linear programming

μ	Profit z^*	Decision x^*	Resources used b_1	b_2	b_3	b_4
0	30.25	(0, 0, 13.75)	19.9	0.0	39.8	110.0
0.1	28.35	(0, 0.17, 13.14)	21.7	1.5	47.7	109.1
0.2	26.02	(0, 0.73, 11.74)	21.4	3.1	58.0	108.0
0.3	23.76	(0, 1.22, 10.44)	21.1	4.7	66.5	106.8
0.4	21.60	(0, 1.64, 9.26)	20.8	6.0	73.4	105.7
0.5	19.55	(0, 1.99, 8.18)	20.4	7.1	78.7	104.4
0.6	17.62	(0, 2.29, 7.19)	20.0	8.0	82.8	103.1
0.7	15.80	(0, 2.53, 6.28)	19.6	8.7	85.5	101.4
0.8	14.24	(0, 0, 7.07)	19.1	5.7	43.8	62.2
0.9	13.08	(0, 0, 6.52)	18.8	5.9	43.0	58.0
1.0	12.00	(0, 0, 6)	18.0	6.0	42.0	54.0

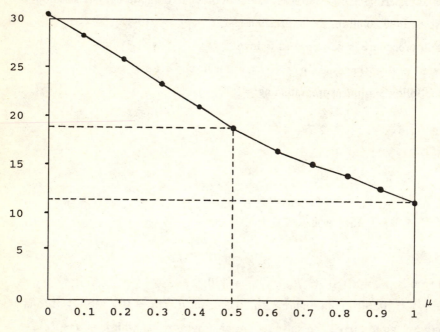

Figure 3.16 The relationship between optimal values and their corresponding membership grades

3.2 Interactive Fuzzy Linear Programming

In this section, an interactive fuzzy linear programming approach (Lai and Hwang [137] is investigated to improve the flexibility and robustness of linear programming (LP) techniques. This section provides a practical modelling method which is a symmetric integration of Zimmermann's, Werners's, Verdegay's and Chanas's fuzzy LP approaches and additionally provides a decision support system for solving a specific domain of practical LP problems. The interactive concept provides a learning process about the system, whereby the decision maker can learn to recognize good solutions, relative importance of factors in the system and then, design a high-productivity system, instead of optimizing a given system. This interactive fuzzy LP system provides integration-oriented, adaptation and learning features by considering all possibilities of a specific domain of LP problems which are integrated in logical order using an IF-THEN rule.

3.2.1 Introduction

In the early twentieth century, reducing complex real-world systems into precise mathematical models became the main trend in science and engineering. Since then, Operations Research (OR) has been applied to real-world decision making problems and has become one of the most important fields in science and engineering. However, practical situations are often not well-defined and thus cannot be described precisely. This imprecise nature is actually fuzziness rather than randomness. Therefore, the traditional OR approaches may not really be suitable for solving practical decision making problems. Since the fuzzy set theory was proposed by Zadeh [274] in 1965, we have been able to handle not only well-defined, precise data, but also vague or fuzzy data such as "profit is around 30 %," "cost is essentially less than $55," "lifetime is essentially larger than 68 hours," etc. The role of fuzzy sets in decision processes is best described in the original statements of Bellman and Zadeh [12]:

Much of the decision-making in the real world takes place in an environment in which the goals, the constraints and the consequences of possible actions are not known precisely. To deal quantitatively with imprecision, we usually employ the concepts and techniques of probability theory and, more particularly, the tools provided by decision theory, control theory and information theory. In so doing, we are tacitly accepting the premise that imprecision -- whatever its nature -- can be equated with randomness. This, in our view, is a questionable assumption...

Thus, decision processes are better described and solved by using fuzzy set theory, rather than precise approaches. However, the decision maker (DM) himself is always playing the most important role in using fuzzy set theory. Therefore, an interactive process between the DM and the decision process is necessary to solve our problems. That is actually a user-dependent fuzzy LP technique. Furthermore, a problem-oriented concept is also a vitally important concept in solving practical problems, as noted by Simon [224]:

> Perhaps what is most important in making rapid progress towards our goals is that we adopt a problem-oriented point of view. We must let the problem that we are trying to solve determine the methods we apply to it, instead of letting our techniques determine what problems we must be willing and able to tackle. That means we must be willing to apply our entire armory of techniques, both heuristic and algorithmic.

By use of fuzzy set theory, and user-dependent (interactive) and problem-oriented concepts, the flexibility and robustness of linear programming (LP) techniques are improved. An interactive fuzzy LP approach is then developed which is a symmetric integration of Zimmermann's, Werners's, Verdegay's and Chanas's approaches (as discussed in the previous sections) and additionally provides a decision support system for solving a specific domain of real-world LP systems.

Moreover, the interactive concept provides a learning process about the system and makes allowance for psychological convergence of the DM, whereby, (s)he learns to recognize good solutions, relative importance of factors in the system, and then design a high-productivity system [293], instead of optimizing a given system. Thus, we can improve some deficits of operations research methodology as indicated in Ackoff [1, 2].

This interactive fuzzy LP system provides integration-oriented, adaptation and learning features by considering all possibilities of a specific domain of LP problems which are integrated in logical order using an IF-THEN rule.

In the following sections, we will first discuss Zimmermann's, Werners's, Chanas's and Verdegay's approaches.

3.2.2 Discussion Of Zimmermann's, Werners's, Chanas's and Verdegay's Approaches

After a survey of various models of fuzzy linear programming techniques, we find that Zimmermann's, Werners's, Chanas's and Verdegay's approaches and concepts are the most practical ones among various techniques in the literature. However, each of them solves one kind of linear programming problems and decision making tasks.

In order to understand the differences and contributions of these approaches and concepts, it is necessary to discuss some major points among them in this section.

In the Zimmermann fuzzy linear programming model, goal b_0 and its maximum tolerance p_0 should be given initially. In real-world problems, it is unrealistic to initially ask the decision maker to give b_0 and p_0 without providing any information about them. Therefore, the membership function of the fuzzy objective is questionable and so the solution is also questionable. For example, if the b_0 given is too large, there will be no solution in Zimmerman's FLP model. At the same time, if the p_0 given is too large, then there will be no meaning for the membership function. Thus, the solution is questionable. Instead of asking the decision maker for b_0 and p_0 to establish the membership function of the fuzzy objective, Werners provided two possible extreme points z^0 and z^1 which are:

$$z^0 = \inf_{x \epsilon X} \quad (\max cx)$$

$$z^1 = \sup_{x \epsilon X} \quad (\max cx)$$

where $X = \{x|\ (Ax)_i \leq b_i, \forall i, \text{ and } x \geq 0\}$. The difference between Werners's and Zimmermann's membership function is depicted in Figure 3.17. In this figure, Zimmermann's b_0 and p_0 are assumed to be rational, i.e., $z^0 \leq b_0 \leq z^1$ and $b_0 - p_0 \geq z^0$. If so, the decision maker may consider Zimmermann's membership function as a more acceptable one most of the time.

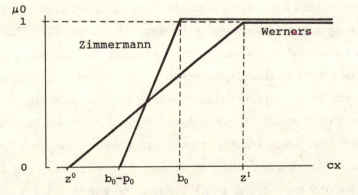

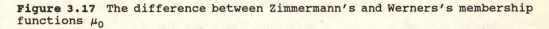

Figure 3.17 The difference between Zimmermann's and Werners's membership functions μ_0

While Tanaka et al. [236] and Zimmermann [296] connected fuzzy set theory and the max-min operator with linear programming, Verdegay [250] and Chanas [34] obtained the equivalent relation between parametric programming with parametric resources and fuzzy linear programming with fuzzy resources defined by assumed linear membership functions. Thus, for a symmetric fuzzy linear programming problem, we can first consider its sub-problem of nonsymmetric fuzzy linear programming with a crisp objective and fuzzy constraints, and then construct and solve its corresponding parametric programming problem. The solutions will be presented in a tableau which not only provides an optimal solution with respect to a parameter, but also provides the actual consumed resources. Therefore, the difficulty of providing b_0 and p_0 initially is overcome by presenting the solution table of a parametric programming problem. After referring to this solution table, the decision maker can precisely locate his subjective b_0 and p_0. The solution of Zimmermann's model is then reliable.

Chanas [34] suggested that the membership function of the fuzzy objective be constructed directly by the parametric optimal solution. Thus, μ_0 is a function of the parameter instead of the general function of x. However, in any real-world problems, the number of constraints is always rather large (say 50), and so are the decision variables. Therefore, Chanas's approach for formulating the membership function of the fuzzy objective is not practical.

3.2.3 Interactive Fuzzy Linear Programming - I

3.2.3.1 Problem Setting

The interactive fuzzy linear programming proposed here is a problem-oriented and user-dependent approach. It also considers a large variety of situations that the decision maker might meet in modelling and solving a linear programming problem. Interactive fuzzy linear programming provides an efficient and systematic approach to solving the indicated linear programming problem, and then shows the results (solutions and resources used) to the decision maker. The decision maker may be satisfied with this solution or he may want to modify some situations and/or change the original model. If satisfied, the problem, of course, has been solved. If not satisfied, an interactive process will then proceed. The problem-solving procedure is continued until the decision maker obtains a satisfying solution. It is worth noting that the approach proposed here is not only to solve a given LP problem, but also to design a high productivity system. A high productivity system was stated by Zeleny [293] as:

Modern production systems are increasingly characterized by novel high-productivity attributes: minimal buffers, close-to-zero inventories, just-in-time operations, minimal waste, full utilization of resources, flexibility of layouts and resource portfolios, and so on. ... Finally, the problem has shifted from "How to optimize a given system" to "How to design an optimal system". ... It is therefore important to design high-productivity systems with no excess, waste or buffers. Only then can the desirable "buffers" be added with some insight and competence. If such optimal design is impossible, then at least improved redistribution of the excess capacity should be attempted.

Interactive fuzzy programming methods have been studied since 1980. Typical works are Baptistella and Ollero [8], Fabian, Ciobanu and Stoica [84], Sakawa and Yano [220], Slowinski [227], Werners [257] and Zimmermann [BM64]. Zimmermann [BM64] described some general concepts and modelling methods of decision support systems and expert systems in a fuzzy environment. Others developed interactive approaches to solve multiple criteria decision making problems (crisp or fuzzy).

The approach proposed here is based on an interaction with the decision maker in order to find a satisfying solution for a linear programming problem. In a decision process using an LP model, resources available may be fuzzy, instead of a precisely given number as in the traditional (crisp) LP model. For example, machine hours, labor force, material needed and so on, are always imprecise, because of various potential suppliers and environments. Therefore, they should be considered as fuzzy resources, and then the LP problem should be solved by use of the fuzzy set theory. Thus under the fuzzy resource constraints, the objective function (e.g., profit function) to be maximized may be considered as either a crisp objective function or a fuzzy objective function symmetrically. Therefore, we can define a linear programming problem with crisp or fuzzy resource constraints, and a crisp or fuzzy objective as:

$$\overset{(\sim)}{\max} \quad z = cx$$
$$\text{s.t.} \quad (Ax)_i \leq \overset{(\sim)}{b_i}, \ i = 1, 2, \dots, m$$
$$x \geq 0 \tag{3.30}$$

where fuzzy resources $\tilde{b}_i$, $\forall i$, have the same forms of membership functions as Figure 3.8. We may also consider the following fuzzy inequality constraints:

$$\overset{(\sim)}{\max} \quad z = cx$$
$$\text{s.t.} \quad (Ax)_i \underset{(\sim)}{\leq} b_i, \quad i = 1, 2, \dots, m$$
$$x \geq 0 \tag{3.31}$$

where fuzzy inequality constraints have the membership functions of Figure 3.8. Even though

Equations (3.30) and (3.31) are different in some points of view, we can use the same approach to handle them under the pre-assumption of the membership functions of the fuzzy available resources and fuzzy inequality constraints.

According to Equations (3.30) and (3.31), we can derive the following five problems:

Problem 1. When the resources can be determined precisely, a traditional linear programming problem is considered as:

$$\max \quad z = cx$$
$$\text{s.t.} \quad (Ax)_i \leq b_i, \forall i$$
$$x \geq 0 \tag{3.32}$$

where c, A, and b_i, $\forall i$, are precisely given. The solution of Equation (3.32) is a unique optimal solution.

Problem 2. (Approach of Chanas [34] and Verdegay [250]) A decision maker might wish to make a post-optimization analysis. Thus, a parametric programming problem is formulated as:

$$\max \quad z = cx$$
$$\text{s.t.} \quad (Ax)_i \leq b_i + \theta p_i, \forall i$$
$$\theta \in [0, 1] \text{ and } x \geq 0 \tag{3.33}$$

where c, A, b_i and p_i, $\forall i$, are precisely given and θ is a parameter (a fraction of the maximum tolerances). p_i, $\forall i$, are maximum tolerances which are always positive. The solutions $z^*(\theta)$ of Equation (3.33) are functions of θ. That is, for each θ, we can obtain an optimal solution; with the increasing use of resources, the optimal solution behavior can be figured out. The results are displayed in Table 3.13 and then presented to the decision maker. The decision maker can then choose his satisfying solution for implementation.

On the other hand, the available resources may be fuzzy. Then the linear programming problem with fuzzy resources becomes:

$$\max \quad z = cx$$
$$\text{s.t.} \quad (Ax)_i \leq \tilde{b}_i, \forall i$$
$$x \geq 0. \tag{3.34}$$

Table 3.13 Solutions of a parametric programming problem

θ	$z^*(\theta)$	Resources Actually Used			
		b_1	b_2	$\cdots$	b_m
0.0					
0.1					
0.2					
.					
.					
.					
0.8					
0.9					
1.0					

It is possible to determine the maximum tolerances p_i, of the fuzzy resources b_i, $\forall i$. Then we can construct the membership functions μ_i assumed linear for each fuzzy constraints, as follows:

$$\mu_i(x) = \begin{cases} 1 & \text{if } (Ax)_i < b_i \\ 1 - [(Ax)_i - b_i]/p_i & \text{if } b_i \leq (Ax)_i \leq b_i + p_i \\ 0 & \text{if } (Ax)_i > b_i + p_i \end{cases} \tag{3.35}$$

Verdegay and Chanas proposed that Equations (3.34) and (3.35), however, are equivalent to Equation (3.33), a parametric linear programming where c, A, b_i and p_i, $\forall i$, are given, by use of the α-level cut concept, where $\alpha = 1 - \theta$.

Problem 3. (Werners's [257] approach) A decision maker may want to solve a fuzzy linear programming problem with a fuzzy objective and fuzzy constraints, while the goal b_0 is not given. That is:

$$\begin{aligned} &\widetilde{\max} \quad z = cx \\ &\text{s.t.} \quad (Ax)_i \leq \widetilde{b_i}, \forall i \\ &\qquad\quad x \geq 0 \end{aligned} \tag{3.36}$$

which is equivalent to:

$$\begin{aligned} &\widetilde{\max} \quad z = cx \\ &\text{s.t.} \quad (Ax)_i \leq b_i + \theta p_i, \forall i \\ &\qquad\quad \theta \in [0, 1] \text{ and } x \geq 0 \end{aligned} \tag{3.37}$$

where c, A, b_i and p_i, $\forall i$, are given, but the goal of the fuzzy objective is not given.

By using Werners's approach, we first obtain z^0 and z^1 for Equation (3.37). Then Werners's membership function μ_0 of the fuzzy objective is (see Figure 3.17):

$$\mu_0(x) = \begin{cases} 1 & \text{if } cx > z^1 \\ 1 - (z^1 - cx)/(z^1 - z^0) & \text{if } z^0 \leq cx \leq z^1 \\ 0 & \text{if } cx < z^0. \end{cases} \tag{3.38}$$

The membership functions μ_i, $\forall i$, of the fuzzy constraints are defined as Equation (3.35). But using the max-min operator, we have:

$$\text{max} \quad \alpha$$
$$\text{s.t.} \quad \mu_0(x) \text{ and } \mu_i(x) \geq \alpha$$
$$\alpha, \mu_0(x) \text{ and } \mu_i(x), \forall i, \alpha \in [0, 1] \text{ and } x \geq 0$$

where c, A, b_i and p_i, $\forall i$, are given, and $\alpha = \min(\mu_0, \mu_1, ..., \mu_m)$. Let $\alpha = 1 - \theta$, then the problem given becomes:

$$\text{min} \quad \theta$$
$$\text{s.t.} \quad cx \geq z^1 - \theta(z^1 - z^0)$$
$$(Ax)_i \leq b_i + \theta p_i, \quad \forall i$$
$$\theta \in [0, 1] \text{ and } x \geq 0 \tag{3.39}$$

where c, A, b_i and p_i, $\forall i$, are given and θ is a fraction of $(z^1 - z^0)$ for the first constraint and a fraction of the maximum tolerances for others. The solution is a unique optimal solution.

Problem 4. (Zimmermann's [296] approach) A decision maker may want to solve a fuzzy linear programming problem with a fuzzy objective and fuzzy constraints, when the goal b_0 of the fuzzy objective and its maximum tolerance are given. That is:

$$\widetilde{\text{max}} \quad cx$$
$$\text{s.t.} \quad (Ax)_i \leq \widetilde{b}_i, \forall i, \text{ and } x \geq 0 \tag{3.40}$$

where c, A, b_0, p_0, b_i and p_i, $\forall i$, are given. It is noted that the decision maker should refer to Table 3.13 for deciding b_0 and p_0. Zimmermann's membership function of the fuzzy objective μ_0 is defined as (see Figure 3.17):

$$\mu_0(x) = \begin{cases} 1 & \text{if } cx > b^0 \\ 1 - (b_0 - cx)/p_0 & \text{if } b_0 - p_0 \leq cx \leq b_0 \\ 0 & \text{if } cx < b_0 - p_0. \end{cases} \tag{3.41}$$

Let $\alpha = 1 - \theta$. Then Equation (3.40) becomes:

$$\begin{aligned}
\text{min} \quad & \theta \\
\text{s.t.} \quad & cx \geq b_0 - \theta p_0 \\
& (Ax)_i \leq b_i + \theta p_i, \forall i \\
& \theta \in [0, 1] \text{ and } x \geq 0
\end{aligned} \tag{3.42}$$

where c, A, b_0, p_0, b_i and p_i, $\forall i$, are given and θ is a fraction of the maximum tolerances. The optimal solution of the Equation (3.42) is unique.

Problem 5. A decision maker may want to solve a fuzzy linear programming problem with a fuzzy objective and fuzzy constraints, while only the goal b_0 of the fuzzy objective is given, but its tolerance p_0 is not given. That is:

$$\begin{aligned}
\widetilde{\text{max}} \quad & cx \\
\text{s.t.} \quad & (Ax)_i \leq \tilde{b}_i, \forall i \\
& x \geq 0
\end{aligned} \tag{3.43}$$

where c, A, b_0, b_i and p_i, $\forall i$, are given, but p_0 is not given. While p_0 is not given, we do know that p_0 should be in between 0 and $b_0 - z^0$ (z^0 as defined in the Problem 4 section). For each $p_0 \in [0, b_0 - z^0]$, we can obtain the membership function of the fuzzy objective as Equation (3.41). Since in a high-productivity system the objective value should be larger then z^0 at $\theta = 0$, there is no meaning to giving a positive grade of membership for those which are less than z^0. Figure 3.18 depicts the possible range of p_0.

The difference between Problem 5 and Problem 4 is that p_0 is not initially given in Problem 5. Therefore, we may assume a set of p_0s, where $p_0 \in [0, b_0 - z^0]$. Then, a problem with each given p_0 is a Problem 4. The solutions for this given set of p_0s are presented in Table 3.14. After referring to this table, the decision maker may choose a refined p_0. Then Problem 4 with the decision maker's refined p_0 is solved. This solution will be the final optimal solution for Equation (3.43).

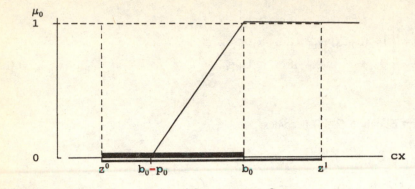

Figure 3.18   The reasonable range of p_0

Table 3.14   Optimal solutions of a symmetric fuzzy linear
programming for a given set of p_0s

p_0	θ	z^{**}	x^{**}	Resources Actually Used			
				b_1	b_2	$\cdots$	b_m
0							
$\cdot$							
$\cdot$							
$b_0 - z^0$							

3.2.3.2 The Algorithm of IFLP-I

Based on Lai and Hwang [137], the algorithm of interactive fuzzy linear programming -
I is (see Figure 3.19) then:

Step 1. Solve a traditional linear programming problem of Equation (3.32) by use of the simplex
method. The unique optimal solution with its corresponding consumed resources is presented to
the decision maker.

Step 2. Does this solution satisfy the decision maker? Consider the following cases:

```
IF                                    THEN
1) solution is satisfied      ->      print out results and stop
2) resource i, for some i     ->      reduce resources available
   are idle                           and go to Step 1
3) available resources are    ->      make a parametric analysis
   not precise and some               and go to Step 3
   tolerances are possible
```

Figure 3.19 Flow chart of IFLP-I (Lai and Hwang [137])

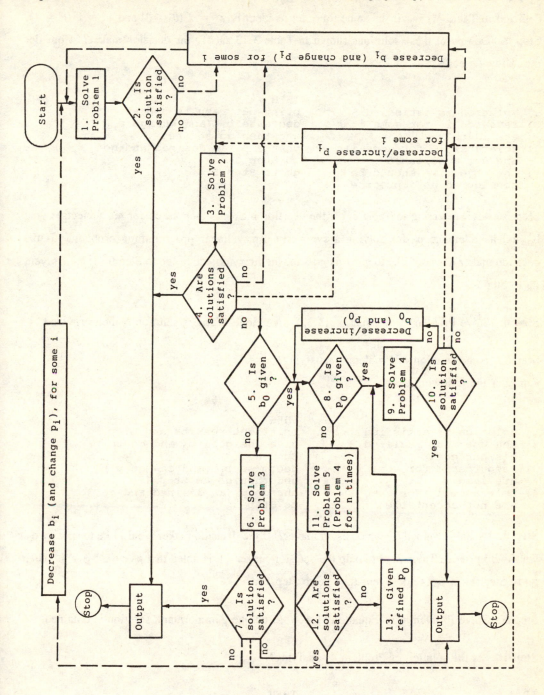

Step 3. Solve a parametric linear programming problem of the Equation (3.33). The results are depicted in Table 3.13. At the same time, let us identify $z^0 = z^*(\theta = 0)$ and $z^1 = z^*(\theta = 1)$.

Step 4. Do any of these solutions shown in Table 3.13 satisfy the decision maker? Consider the following cases:

```
IF                                    THEN
1) solution is satisfied      ->      print out results and stop
2) resource i, for some i     ->      decrease b_i (and change p_i)
   are idle                           and then go to Step 1
3) tolerance i, for some i    ->      change p_i as desired and go
   are not acceptable                 to Step 3
4) the objective should be    ->      go to Step 5
   considered as imprecise
```

Step 5. After referring to Table 3.13, the decision maker is then asked for his subjective goal b_0 and its tolerance p_0 for solving a symmetric fuzzy linear programming problem. If the decision maker does not like to give his goal for the fuzzy objective, go to Step 6. If b_0 is given, go to Step 8.

Step 6. Solve Problem 3 of Equation (3.39). A unique Werners solution is then provided.

Step 7. Is the solution of Problem 3 satisfying?
Consider the following cases:

```
IF                                    THEN
1) solution is satisfied      ->      print out results and stop
2) the user has realized      ->      give the goal b_0 and go to
   his/her goal                       Step 8.
3) resource i, for some i     ->      decrease b_i (and change p_i)
   are idle                           and then go to Step 1
4) tolerance i, for some i    ->      change p_i as desired and go
   are not acceptable                 to Step 3
```

Step 8. Is p_0 determined by the decision maker? If the decision maker would like to specify p_0, we should provide Table 3.14 to help the decision maker. It is noted that $p_0 \in [0, b_0 - z^0]$. Then go to Step 9. If p_0 is not given, then go to Step 11.

Step 9. Solve Problem 4 of Equation (3.42). A unique Zimmermann solution is obtained.

Step 10. Is the solution of Problem 4 satisfying?

```
IF                                    THEN
1) solution is satisfied      ->      print out results and stop
2) the user has realized      ->      give the goal b_0 (and p_0)
   better his/her goal                and go to Step 8
```

```
                              (and its tolerance)
3)  resource i, for some i    ->    decrease b_i (and change p_i)
    are idle                        and then go to Step 1
4)  tolerance i, for some i    ->    change p_i as desired and go
    are not acceptable               to Step 3
```

Step 11. Solve Problem 5. That is: call Step 9 to solve Problem 4 for a set of p_0s. Of course, $p_0 \in [0, b_0 - z^0]$. Then the solutions are depicted in Table 3.14.

Step 12. Are the solutions satisfying? If yes, print out the solution and terminate the solution procedure. Otherwise, go to Step 13.

Step 13. Ask the decision maker to specify the refined p_0, and then go to Step 9. It is rather reasonable to ask the decision maker p_0 at this step, because he has a good idea about it now.

For implementing the above interactive fuzzy linear programming, we need only two solution-finding techniques: simplex method and parametric method. Therefore, the interactive fuzzy linear programming approach proposed here can be easily programmed in a PC system for its simplicity.

In order to understand the interactive fuzzy linear programming, it is necessary to have a numeric example as follows.

3.2.3.3 Example: The Knox Production-Mix Selection Problem

Let us consider the above Knox Mix company problem by using our interactive fuzzy linear programming system to obtain a satisfying solution.

Step 1. Solve the above linear programming problem by use of the simplex method. The optimal solution is: $x^* = (50/7, 0, 55/7, 0) = (7.14, 0, 7.86, 0)$ and $z^* = \$695/7 = \$99.29$. The actual resources used are 15, 73.57 and 100 units for man-weeks, material Y and material Z, respectively.

Step 2. The decision maker feels that 73.57 units of material are actually enough to reach a \$99.25 profit. Thus, $120 - 73.57 = 46.43$ pounds of material Y that are really idle resources. The decision maker, therefore, would like to try to find the solution of a new linear programming with the resources changed from $(b_1, b_2, b_3) = (15, 120, 100)$ to $(15, 73.57, 100)$. Then, go to Step 1.

Step 1. Solve the crisp linear programming problem with $b_2 = 73.57$. The optimal solution is the same as previously indicated: $x^* = (7.14, 0, 7.86, 0)$, and $z^* = \$99.29$. The actual resources used are 15, 73.57, and 100. Now there are no idle resources. Instead of just optimizing a given LP system, the procedure has yielded a high-productivity system.

Step 2. If the decision maker is satisfied with this solution, print out the results and stop the solution procedure. Sometimes a decision maker may want to further analyze his problem. Then the following two cases may be considered.

Case 1.

Step 3-1. The decision maker may feel that a posterior analysis of the previous LP problem is necessary before implementing its solution.

Let us assume that the decision maker provides that the maximum tolerances for man-weeks, material Y and material Z are 3, 0, and 20 units, respectively. Then, the parametric programming is:

$$\max_{x \in X} z$$

where $X = \{x \mid g_1(x) \le 15 + 3\theta, g_2(x) \le 120, g_3(x) \le 100 + 20\theta, x \ge 0, \text{ and } \theta \in [0, 1]\}$ and θ is a parameter. The optimal solution is: $x^* = (7.14 + 1.43\theta, 0, 7.86 + 1.57\theta, 0)$, and $z^* = \$(99.29 + 19.86\theta)$. In order to more clearly display these results, Table 3.15 is provided to the decision maker.

Table 3.15 Solutions of the parametric programming problem

| θ | z^* | resources actually used | | |
		man-weeks	material Y	material Z
0.0	99.29	15.00	73.57	100.00
0.1	101.28	15.30	75.04	102.00
0.2	103.27	15.60	76.51	104.00
0.3	105.26	15.90	77.98	106.00
0.4	107.25	16.20	79.45	108.00
0.5	109.24	16.50	80.92	110.00
0.6	111.23	16.80	82.39	112.00
0.7	113.22	17.10	83.86	114.00
0.8	115.21	17.40	85.33	116.00
0.9	117.20	17.70	86.80	118.00
1.0	119.19	18.00	88.27	120.00

Step 4-1. After an overview of the above table, the decision maker may pick out a satisfying solution for implementation. If so, print out the results and terminate the solution procedures. Otherwise, the following steps may be considered:

(i) Reduce the amount of material Y from 120 to 73.57 and give its maximum tolerance. Then go to Step 1.

(ii) Decrease/increase the maximum tolerances p_1 and/or p_3, and then go to Step 3-1. (Just modify the results of the parametric process.)

The solution process will continue until the satisfying solution is reached.

Case 2.

Step 3-2. The decision maker may consider that the traditional linear programming is not enough to solve his problem because of the imprecise properties of the resources in nature. After a detailed analysis, he feels that the consumption of man-weeks, material Y and material Z should be "substantially less than or equal to" 15, 80 and 100 units, respectively, with the corresponding maximum tolerances of 5, 40 and 30 units. Thus the problem becomes a fuzzy linear programming problem with a crisp objective and fuzzy constraints which are defined by their membership function assumed to be linear. That is:

$$\max_{x \in X'} \; z$$

where $X' = \{x \mid g_1(x) \leq \tilde{15}, \; g_2(x) \leq \tilde{80}, \; g_3(x) \leq \tilde{100}, \text{ and } x \geq 0\}$ and the membership function μ_i for the ith fuzzy constraints, $\forall i$, are:

$$\mu_1(x) = \begin{cases} 1 & \text{if } g_1(x) < 15 \\ 1 - [g_1(x)-15]/5 & \text{if } 15 \leq g_1(x) \leq 20 \\ 0 & \text{if } g_1(x) > 20 \end{cases}$$

$$\mu_2(x) = \begin{cases} 1 & \text{if } g_2(x) < 80 \\ 1 - [g_2(x)-80]/40 & \text{if } 80 \leq g_2(x) \leq 120 \\ 0 & \text{if } g_2(x) > 120 \end{cases}$$

$$\mu_3(x) = \begin{cases} 1 & \text{if } g_3(x) < 100 \\ 1 - [g_3(x)-100]/30 & \text{if } 100 \leq g_3(x) \leq 130 \\ 0 & \text{if } g_3(x) > 130. \end{cases}$$

This nonsymmetric fuzzy linear programming is then equivalent to:

$$\max_{x \in X''} z$$

where $X'' = \{ x \mid g_1(x) \leq 15 + 5\theta, g_2(x) \leq 80 + 40\theta, g_3(x) \leq 100 + 30\theta, x \geq 0, \text{ and } \theta \in [0, 1] \}$ and θ is a parameter. Therefore, we can use the parametric programming approach to solve this problem and obtain its solutions as shown in Table 3.16.

Table 3.16 Solutions of a nonsymmetric fuzzy linear programming problem with a crisp objective and fuzzy constraints

θ	z^*	resources actually used man-weeks	material Y	material Z
0.0	99.29	15.00	73.57	100.00
0.1	102.36	15.50	76.22	103.00
0.2	105.43	16.00	78.75	106.00
0.3	108.50	16.50	81.50	109.00
0.4	111.57	17.00	84.15	112.00
0.5	114.64	17.50	86.80	115.00
0.6	117.71	18.00	89.45	118.00
0.7	120.78	18.50	92.10	121.00
0.8	123.85	19.00	94.75	124.00
0.9	126.92	19.50	97.40	127.00
1.0	130.00	20.00	100.00	130.00

Step 4-2. After referring to this solution table, the decision maker may choose a satisfying solution for implementation, and then terminate the solution procedure. Otherwise, three procedures (as those in the Algorithm) may be considered. Assume that a symmetric fuzzy linear programming with a fuzzy objective and fuzzy constraints is considered, and then go to Step 5.

Step 5. Present Table 3.16 to the decision maker and ask him to determine his subjective goal b_0. If b_0 is given, go to Step 8. Otherwise, go to the next step.

Step 6. Solve Problem 3 given in (3.39). First, according to Table 3.16, let us identify:

$$z^0 = z_*(\theta = 0) = 99.29$$

$$z^1 = z^*(\theta = 1) = 130.$$

The symmetric linear programming problem can then be formulated as follows:

$$\widetilde{\max}_{x \in X'} \; z$$

where the membership functions of the fuzzy constraints are described in Step 3-2 and the membership function μ_0 of the fuzzy objective is defined as:

$$\mu_0(x) = \begin{cases} 1 & \text{if } z > 130 \\ 1 - (130\text{-}z)/(130\text{-}99.29) & \text{if } 99.29 \leq z \leq 130 \\ 0 & \text{if } z < 99.29. \end{cases}$$

This problem is then equivalent to:

$$\min_{x \in X_\theta} \; \theta$$

where $X_\theta = \{ x \mid f(x) \geq 130 - 30.71\theta, \; g_1(x) \leq 15 + 5\theta, \; g_2(x) \leq 80 + 40\theta, \; g_3(x) \leq 100 + 30\theta, \; x \geq 0 \text{ and } \theta \, \epsilon \, [0, 1]\}$. The solution is: $x^{**} = (8.57, 0, 8.93, 0)$ and $z^{**} = \$114.65$ at $\theta = 0.5$, while the actually used resources are 17.5, 86.78 and 115.01 for man-weeks, material Y and material Z, respectively.

Step 7. The decision maker may be satisfied with this solution. Print out the results and stop the solution procedure. If he is not satisfied with this solution, three steps in Algorithm might be considered. Let us assume that after recalling Table 3.16, determining b_0 is considered and then go to Step 8.

Step 8. Presenting Table 3.16, ask the decision maker to determine p_0. If p_0 is given, go to Step 9. Otherwise, go to Step 11.

Step 9. Solve Problem 4. Now let us assume $b_0 = 111.57$ at $\theta = 0.4$ and $p_0 = 10$. It is noted that p_0 should be in between 0 and 12.28. Our problem becomes:

$$\text{find} \quad x$$
$$\text{such that} \quad f(x) \geq 1\widetilde{1}1.57 \text{ and } x \, \epsilon \, X'$$

where the membership functions of the fuzzy constraints are described in Step 3-2 and the membership function μ_0 of the fuzzy objective is defined as:

$$\mu_0(x) = \begin{cases} 1 & \text{if } z > 111.57 \\ 1 - (111.57\text{-}z)/10 & \text{if } 101.57 \le z \le 111.57 \\ 0 & \text{if } z > 101.57. \end{cases}$$

The problem is then equivalent to:

$$\min_{x \in X_\theta'} \quad \theta$$

where $X_\theta' = \{ x \mid f(x) \ge 111.57 - 10\theta, g_1(x) \le 15 + 5\theta, g_2(x) \le 80 + 40\theta, g_3(x) \le 100 + 30\theta, x \ge 0 \text{ and } \theta \in [0, 1]\}$. The solution is: $x^{**} = (8.01, 0, 8.50, 0)$ and $z^{**} = \$108.54$ at $\theta = 0.30$, while the resources actually used are 16.51, 81.57 and 109.03 for man-weeks, material Y and material Z, respectively.

Step 10. Consider four steps shown in Algorithm. Here, let us assume that the solution is satisfied. Then the solution procedure of our interactive fuzzy linear programming system is completed. After such detailed communication between the system and the decision maker, the solution will be quite good for implementation.

Step 11. Take a simulation by choosing a set of p_0s. Let us take 5 possible values of $p_0 \in [0, 12.28]$; 0, 3, 6, 9, 12.28. For each p_0 value, we call Step 9 to obtain the solution of Problem 4. The results are depicted in Table 3.17.

Step 12. After referring to Table 3.17, the decision maker may choose a satisfying solution and terminate the solution procedure. Otherwise, he is asked to specify a refined p_0. It is quite reasonable here to obtain a p_0 which can really satisfy the decision maker's need. Table 3.17 provides very detailed information of the optimal solution of a symmetric fuzzy linear programming problem with $b_0 = 111.57$ for each p_0 in the given set. Here, let us assume that the optimal solution should be in between $p_0 = 9$ and 12.28.

Step 13. The decision maker finally indicates that $p_0 = 10$ will be perfect. Then go to Step 9.

Finally, it is necessary to compare the solutions of Problem 3 and Problem 4. Table 3.18 and Figure 3.20 show the differences.

Table 3.17 The solutions of Problem 5

p_0	θ^*	z^{**}	x^{**}
0	0.400	111.570	(8.286, 0, 8.714, 0)
3	0.364	110.478	(8.184, 0, 8.638, 0)
6	0.335	109.562	(8.099, 0, 8.574, 0)
9	0.309	108.788	(8.027, 0, 8.520, 0)
12.28	0.286	108.057	(7.959, 0, 8.469, 0)

p_0	resources actually used man-weeks	material Y	material Z
0	17.000	84.144	111.998
3	16.822	83.202	110.932
6	16.673	82.415	110.037
9	16.547	81.749	109.281
12.28	16.428	81.120	108.567

Table 3.18 The comparison of the solutions of Problem 3 and Problem 4

Problem 3	Problem 4
$\theta = 0.5$	0.3
$z^{**} = 114.65$	108.54
$x^{**} = (8.57, 0, 8.93, 0)$	(8.01, 0, 8.50, 0)
resources used	
man-week: 17.5	16.51
material Y: 86.78	81.57
material z: 115.01	109.03

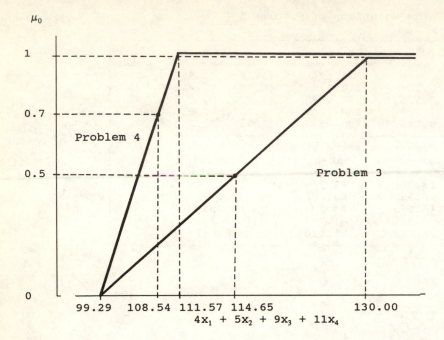

μ_0

1

0.7

Problem 4

0.5

Problem 3

0

99.29   108.54 111.57 114.65   130.00

$4x_1 + 5x_2 + 9x_3 + 11x_4$

Figure 3.20   Comparison of the solutions of Problem 3 and Problem 4

3.2.4 Interactive Fuzzy Linear Programming - II

To develop a decision support system to solve all possible linear programming problems, we first need to classify all possible situations. After a thorough literature survey, we classify all possible situations into 11 cases and 7 problems as shown in Table 3.19.

In this section, we propose a revised version of our previous interactive fuzzy linear programming. Interactive fuzzy linear programming - II (IFLP-II) integrates all possible situations of fuzzy linear programming problems, and further provides a decision support system for improving the flexibility and robustness of linear programming techniques (Lai and Hwang [141]).

We have developed an interactive fuzzy linear programming for solving Problem 1 through Problem 5, in the previous section. Thus we will only discuss Problems 6 and 7 in the following.

Table 3.19 Classification of fuzzy linear programming problems [141]

max/min cx s.t. Ax ≤ b x ≥ 0	fuzzy objective z	fuzzy coefficients c
	fuzzy resources b	A

Case	Problem	Approach		
1. Crisp LP	Prob. 1	Simplex method		
2. $\tilde{b}$	Prob. 2	Parametric programming Verdegay's or Chanas's		
3. $\tilde{z}$ & $\tilde{b}$	Prob. 3 Prob. 4 Prob. 5	Werners's Zimmermann's Lai and Hwang's	IFLP-I	
4. $\tilde{c}$	Prob. 2	Parametric programming		
5. $\tilde{A}$				IFLP-II
6. $\tilde{b}$ & $\tilde{c}$				
7. $\tilde{A}$ & $\tilde{b}$	Prob. 6	Carlsson and Korhonen's		
8. $\tilde{A}$ & $\tilde{c}$				
9. $\tilde{A}$, $\tilde{b}$ & $\tilde{c}$				
10. $\tilde{z}$ & $\tilde{A}$	Prob. 7	Lai and Hwang's		
11. $\tilde{z}$, $\tilde{A}$ & $\tilde{b}$				

Noted: Case 4 is essentially the dual problem of Case 2.

Problem 6. (Carlsson and Korhonen's approach [31]) The problem is a general one where A, b and c might be fuzzy. Carlsson and Korhonen have provided an efficient approach to solve the problem of fuzzy A, b and c (Case 9). Their approach can also be applied to solve the problems of fuzzy A (Case 5), fuzzy b and c (Case 6), fuzzy A and b (Case 7), and fuzzy A and c (Case 8).

Now, let us formulate the problems of Cases 5, 6, 7, 8 and 9 as:

$$\max \quad z = \overset{(\sim)}{c}x$$
$$\text{s.t.} \quad \overset{(\sim)}{A}x \le \overset{(\sim)}{b} \text{ and } x \ge 0 \tag{3.44}$$

where the membership functions of fuzzy coefficients A, and/or b, and/or c are assumed linear as follows (see Figure 3.21):

$$\mu_{aij}(a_{ij}) = \begin{cases} 1 & \text{if } a_{ij} < a_{ij}^0 \\ 1 - (a_{ij}-a_{ij}^0)/p_{ij} & \text{if } a_{ij}^0 \leq a_{ij} \leq a_{ij}^0+p_{ij} \\ 0 & \text{if } a_{ij} > a_{ij}^0+p_{ij} \end{cases}$$

$$\mu_{bi}(b_i) = \begin{cases} 1 & \text{if } b_i < b_i^0 \\ 1 - (b_i-b_i^0)/p_i & \text{if } b_i^0 \leq b_i \leq b_i^0+p_i \\ 0 & \text{if } b_i > b_i^0+p_i \end{cases}$$

$$\mu_{cj}(c_j) = \begin{cases} 1 & \text{if } c_j > c_j^0 \\ 1 - (c_j^0 - c_j)/p_j & \text{if } c_j^0-p_j \leq c_j \leq c_j^0 \\ 0 & \text{if } c_j < c_j^0-p_j. \end{cases}$$

It is noted that the membership functions of fuzzy c are for a maximization problem so that the larger the coefficients, the more we prefer. On the other hand, the less the resources are consumed, the better we feel, and so the technique coefficients. Carlsson and Korhonen pointed out that if the full trade-off between A, b and c is assumed, the optimal solution will always exist at $\mu = \mu_A = \mu_b = \mu_c$, where the membership functions are non-increasing (or non-decreasing) monotonic linear functions. Therefore, we obtain the following auxiliary problem:

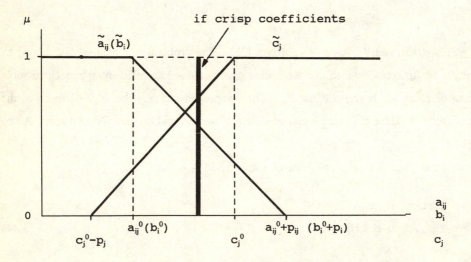

Figure 3.21 Membership functions of Fuzzy A, b and c

$$\max \quad [\mu^{-1}{}_c(\mu)]x$$
$$\text{s.t.} \quad [\mu^{-1}{}_A(\mu)]x \le \mu^{-1}{}_b(\mu) \text{ and } x \ge 0$$

or

$$\max \quad \Sigma_j \ [c_j{}^0 - (1-\mu)p_j]x_j$$
$$\text{s.t.} \quad \Sigma_j \ [a_{ij}{}^0 + (1-\mu)p_{ij}]x_j \le b_i{}^0 + (1-\mu)p_i, \ \forall i$$
$$x_j \ge 0, \ \forall j \tag{3.45}$$

where $c_j{}^0$, $a_{ij}{}^0$ and $b_i{}^0$ are given critical values, and p_j, p_{ij} and p_i, respectively, are given tolerances, for all i and j. Obviously, Equation (3.45) is a linear programming problem if μ is fixed first. Thus, for each given μ value, we can obtain the optimal solution of Equation (3.45). In practice, we may set $\mu = 0.0, 0.1, ..., 0.9, 1$. Here, μ is the overall degree of satisfaction from the critical points (target points or goals). Finally, Table 3.20 is provided to decision makers for further decisions.

Table 3.20 Solutions of Problem 6

μ	Decision z^*	x^*	Resources Used b_1	b_2	...	b_m
0.0						
0.1						
.						
.						
.						
0.9						
1.0						

Problem 7. (Lai and Hwang's approach [141]) First let us formulate Cases 10 and 11 as follows:

$$\max \quad \tilde{z} = cx$$
$$\text{s.t.} \quad \tilde{A}x \le {}^\backprime\tilde{b}{}^\backprime \text{ and } x \ge 0 \tag{3.46}$$

where the membership functions of $\tilde{a}_{ij}$ and/or $\tilde{b}_i$ are assumed linear as in Problem 6. Based on the concept of non-symmetrical models [BM64], the membership function of $\tilde{z}$ can be established by the analysis of fuzzy constraints. Here we consider that Werners's and Zimmermann's membership functions are suitable for this situation. Thus we can solve Equation (3.46) by the approaches provided by Carlsson and Korhonen (see Problem 6), and Werners (see Problem 3).

That is: we first solve the following problem:

$$\max \; \Sigma_j \, c_j x_j$$
$$\text{s.t.} \quad \Sigma_j \, [a_{ij}^0 + (1-\mu)p_{ij}]x_j \leq b_i^0 + (1-\mu)p_i, \; \forall i$$
$$x_j \geq 0, \; \forall j \tag{3.47}$$

where b_i^0 and a_{ij}^0 are given critical points, and p_i and p_{ij}, respectively, are given tolerances. As discussed in Problem 6, let $\mu = 0.0, 0.1, \ldots, 0.9, 1.0$. The solution table (as Table 3.20) and the corresponding piece-wise linear (assumed) function (see Figure 3.22) between z^* and μ are generated. Thus, Werners's membership function of the objective function (z) can be constructed by z^0 and z^1, where $z^0 = z^*(\mu = 0)$ and $z^1 = z^*(\mu = 1)$. The optimal solution z_w^* is at the intersection point of the membership function of the objective function and the assumed linear segment of the relation functions between objective values (z^*) and degree of satisfaction (μ) as shown in Figure 3.22.

On the other hand, the decision maker may like to provide his/her goal (b_0) and its tolerance (p_0), after referring to Table 3.20. Then Zimmermann's membership function is more meaningful. The solution (z_z^*) can be found by solving two simultaneous equations - Zimmermann's membership function and the assumed linear segment of the relation functions between objective values (z^*) and degree of satisfaction (μ) as shown in Figure 3.22.

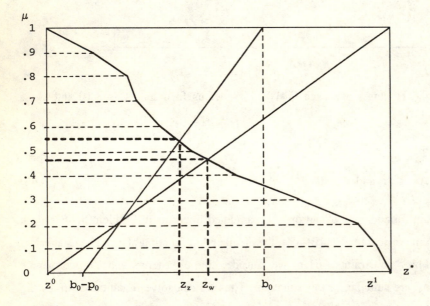

Figure 3.22 The solution of Problem 7

3.2.4.1 The Algorithm of IFLP-II

The IFLP-II is basically a systematic integration of Zimmermann's [296], Werners's [257], Verdegay's [250, 251], Chanas's [34], Lai and Hwang's [137, 141], Lai [139] and Carlsson and Korhonen's [31] fuzzy linear programming approaches, and additionally provdes a decision support system for solving a specific domain of practical linear programming problems.

The algorithm (see Figure 3.23) is shown in the following:

Step 1. Solve a traditional linear programming problem of Problem 1 by use of the simplex method. The unique optimal solution with its corresponding consumed resources is presented to the decision maker.

Step 2. Does this solution satisfy the decision maker? Consider the following cases:

```
IF                              THEN
1) solution is satisfied   ->   print out results and stop
2) resource i, for some i  ->   reduce resources available
   are idle                     and go to Step 1
3) available resources are ->   make a parametric analysis
   not precise and some         and go to Step 3
   tolerances are possible
```

Step 3. Solve a parametric linear programming problem of Problem 2. The results are depicted in Table 3.13. At the same time, let us identify $z^0 = z^*(\theta = 0)$ and $z^1 = z^*(\theta = 1)$.

Step 4. Do any of the solutions shown in Table 3.13 satisfy the decision maker? Consider the following cases:

```
IF                              THEN
1) solution is satisfied   ->   print out results and stop
2) resource i, for some i  ->   decrease b_i (and change p_i)
   are idle                     and then go to Step 1
3) tolerance i, for some i ->   change p_i as desired and go
   are not acceptable           to Step 3
4) the objective should be ->   go to Step 5
   considered as imprecise
5) coefficients of the     ->   go to Step 14
   objective are imprecise
6) objective function and  ->   go to Step 17
   technique coefficients
   (and/or resources available)
   are imprecise
7) technique coefficients  ->   go to Step 15
   (and/or resources available
   and/or coefficients of
   objective function) are
   imprecise
```

Figure 3.23 Flow chart of IFLP-II (Lai and Hwang [141])

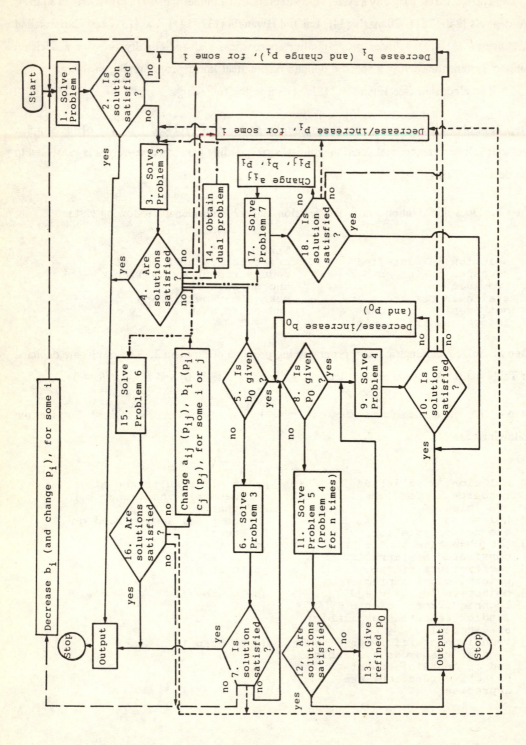

Step 5. After referring to Table 3.13, the decision maker is then asked for his subjective goal b_0 and its tolerance p_0 for solving a symmetric fuzzy linear programming problem. If the decision maker does not like to give his goal for the fuzzy objective, go to Step 6. If b_0 is given, go to Step 8.

Step 6. Solve Problem 3. A unique Werners solution is then provided.

Step 7. Is the solution of Problem 3 satisfying? Consider the following cases:

```
IF                               THEN
1) solution is satisfied    ->   print out results and stop
2) the user has realized    ->   give the goal b₀ and go to
   his/her goal                  Step 8.
3) resource i, for some i   ->   decrease bᵢ (and change pᵢ)
   are idle                      and then go to Step 1
4) tolerance i, for some i ·->   change pᵢ as desired and go
   are not acceptable            to Step 3
```

Step 8. Is p_0 determined by the decision maker? If the decision maker would like to specify p_0, we should provide Table 3.13 to help. It is noted that $p_0 \in [0, b_0 - z^0]$. Then go to Step 9. If p_0 is not given, then go to Step 11.

Step 9. Solve Problem 4. A unique Zimmermann solution is obtained.

Step 10. Is the solution of Problem 4 satisfying?

```
IF                               THEN
1) solution is satisfied    ->   print out results and stop

2) the user has better      ->   give the goal b₀ (and p₀)
   realized his/her goal         and go to Step 8
   (and its tolerance)
3) resource i, for some i   ->   decrease bᵢ (and change pᵢ)
   are idle                      and then go to Step 1
4) tolerance i, for some i  ->   change pᵢ as desired and go
   are not acceptable            to Step 3
```

Step 11. Solve Problem 5. That is: call Step 9 to solve Problem 4 for a set of p_0s. Of course, $p_0 \in [0, b_0 - z^0]$. Then the solutions are depicted in Table 3.14.

Step 12. Are the solutions satisfying? If yes, print out the solution and terminate the solution procedure. Otherwise, go to Step 13.

Step 13. Ask the decision maker to specify the refined p_0, and then go to Step 9.

Step 14. Obtain the dual problem of the primal problem and then go to Step 3. The dual problem now is a linear programming problem with fuzzy resources.

Step 15. Solve Problem 6. A solution table (Table 3.20) is then presented to the decision maker.

Step 16. Does this solution table satisfy the decision maker? We can consider the following cases:

```
IF                                THEN
1) solution is satisfied    ->    print out results and stop
2) resource i, for some i   ->    decrease b_i (and change p_i)
   are idle, and all the           and then go to Step 1
   imprecise coefficients
   are fixed
3) tolerance of resource i  ->    change p_i as desired and go
   for some i are not              to Step 3
   acceptable and all imprecise
   coefficients,except
   resources, are fixed
4) some of the technique    ->    change a_ij (p_ij), b_i (p_i),
   and/or objective               and/or c_j (p_j), for some i
   coefficients, and/or           and/or j and go to Step 15
   resources should be
   revised
```

Step 17. Solve Problem 7. A solution table (Table 3.20 or Figure 3.22) is presented. The decision maker is then asked to choose his/her preferred membership function for the objective function, either Werners's or Zimmermann's membership function. After that, a unique optimal solution is obtained.

Step 18. Does the optimal solution or any solution in the solution table satisfy the decision maker? We can consider the following cases:

```
IF                                THEN
1) solution is satisfied    ->    print out results and stop
2) resource i, for some i   ->    decrease b_i (and change p_i)
   are idle, and all the           and then go to Step 1
   imprecise coefficients
   are fixed
3) tolerance of resources   ->    change p_i as desired and go
   i, for some i, are not         to Step 3
   acceptable and all imprecise
   coefficients,except
   resources, are fixed
4) some of the technique    ->    change a_ij (p_ij) and/or b_i
   coefficients, and/or           (p_i), for some i and/or j
   resources should be            and go to Step 17
   revised
```

3.3 Some Extensions Of Fuzzy Linear Programming Problems

Beyond the above discussed problems and approaches, we would like to introduce some extensions of fuzzy linear programming problems in the following four sub-sections. These extensions could be applied to possibilistic linear programming, multiple objective linear programming (MOLP), fuzzy multiple objective linear programming (FMOLP) and possibilistic multiple objective linear programming (PMOLP) as in the following chapter.

3.3.1 Membership Functions

In this chapter, we have introduced some important fuzzy linear programming (FLP) models based on linear, exponential and triangular membership functions, and max-min operators. However, there are still other forms of membership functions in the literature such as piecewise linear, hyperbolic, trapezoid, some specific power and exponential functions, and so on. Some of the previous fuzzy linear programming models may be used with these types of membership functions in appropriate situations. It cannot be over-emphasized that the shapes of the membership functions must depend on the meanings of the particular problems. In this chapter, we basically use continuous monotonically decreasing functions for "fuzzy $\leq$" constraints, continuous monotonically increasing functions for "fuzzy $\geq$" constraints, and triangular functions for "fuzzy $=$" constraints. For fuzzy parameters, we also use similar membership functions to model our problems. In this monograph, we will use triangular and trapezoid membership functions to model possibilistic linear programming (PLP) problems. But, this does not mean that PLP have to use triangular and trapezoid membership functions and FLP have to use continuous monotonically functions. The types of membership functions are dependent on our problems and their meanings. In this sub-section, we will discuss piecewise linear membership function as follows. Other considerations of membership functions as discussed in Chapter 3 can also be applied in FLP problems.

Let us discuss Zimmermann's FLP model (Section 3.1.2) with piecewise membership functions. Since the piecewise linear function can be used to approach many non-linear functions, it is noteworthy to be especially introduced here. Furthermore, one way to obtain the membership function is to ask the decision maker to scale his/her goal(s). For example, suppose that the analyst provides a set of possible goals, say {45, 50, 55, 60, 65, 70}, to the decision maker and asks the decision maker to provide his/her degree of satisfaction for each of the numbers. Then we can obtain his/her subjective membership function of the goal as Figure 3.24, for instance.

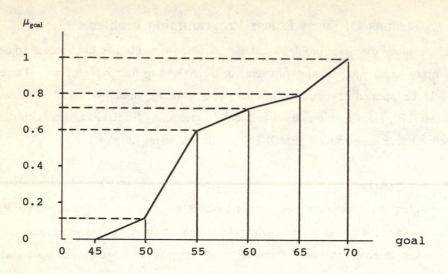

Figure 3.24 A piecewise linear membership function

Hannan, Nakamura, and Inuiguchi, Ichihashi and Kume have provided their models to solve FLP problems with piecewise linear membership functions. Here, let us discuss Nakamura's [174] approach.

According to the fact that a fuzzy logical function which contains only $\wedge$ (conjunction) and $\vee$ (disjunction) can be reduced to polynomial form with respect to $\wedge$ whose monomials are in $\vee$, Nakamura proposed piecewise linear membership functions which could act as approximations of convex and/or concave curves. Suppose A to be a convex fuzzy set in Y with its membership function expressed in a function form which combines only a finite number of linear functions, $\sigma_k^j(y)$, with $\wedge$ and $\vee$ operators. Then, the membership function can be represented as:

$$\mu_A(y) = [\{ \bigwedge_{j=1}^{q} v_j(y)\} \wedge 1] \vee 0 , \quad y \in Y \tag{3.48}$$

where $v_i(y) = \bigvee_{k=1}^{r(j)} \sigma_k^j(y).$

Now, let us consider three examples, as shown in Figure 3.25:

a) $\mu_A(y) = [\{ v_1(y) \wedge v_2(y) \wedge v_3(y) \wedge v_4(y) \wedge v_5(y)\} \wedge 1] \vee 0$

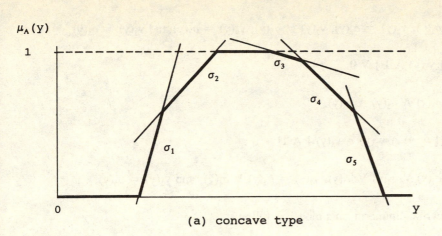

(a) concave type

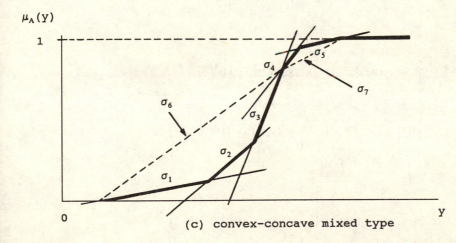

(b) convex type

(c) convex-concave mixed type

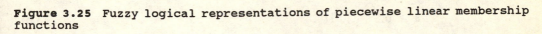

Figure 3.25 Fuzzy logical representations of piecewise linear membership functions

where $v_1(y) = \sigma_1(y)$, $v_2(y) = \sigma_2(y)$, $v_3(y) = \sigma_3(y)$, $v_4(y) = \sigma_4(y)$, and $v_5(y) = \sigma_5(y)$.

(b) $\mu_A(y) = [\ v_1(y) \wedge 1\ ] \vee 0$

where $v_1(y) = \sigma_1(y) \wedge \sigma_2(y) \wedge \sigma_3(y)$.

(c) $\mu_A(y) = [\{\ v_1(y) \wedge v_2(y) \wedge v_3(y)\ \} \wedge 1] \vee 0$

where $v_1(y) = \sigma_1(y) \vee \sigma_2(y) \vee \sigma_3(y)$, $v_2(y) = \sigma_4(y) \vee \sigma_6(y)$, and $v_3(y) = \sigma_5(y)$.

Let us rewrite Zimmermann's model as follows:

$$y_0 = cx \leq z^0, \quad y_0 \in Y_0$$
$$y_i = (Ax)_i \leq b_i, \quad y_i \in Y_i, \ i = 1, 2, \ldots, m$$
$$x \geq 0. \tag{3.49}$$

Suppose $m+1$ convex fuzzy subset A_i in Y_i with the piecewise linear membership functions as Equation (3.49). Then, Equation (3.13), becomes:

$$\max \ \alpha$$
$$\text{s.t.} \quad \alpha \leq f_i((Bx)_i) = \mu_{Ai}(y_i), \ i = 0, 1, \ldots, m$$
$$x \geq 0, \ \alpha \in [0,1] \tag{3.50}$$

or

$$\max \ \alpha$$

$$\text{s.t.} \quad \alpha \leq \mu_{A1}(y_1) = [\{\ \overset{g_1}{\underset{j=1}{\wedge}}\ v_{1j}(y_1)\ \} \wedge 1] \vee 0$$

$$\alpha \leq \mu_{A2}(y_2) = [\{\ \overset{g_2}{\underset{j=1}{\wedge}}\ v_{2j}(y_2)\ \} \wedge 1] \vee 0$$

$$\vdots$$

$$\alpha \leq \mu_{Am+1}(y_m) = [\{\ \overset{g_{m+1}}{\underset{j=1}{\wedge}}\ v_{m+1,j}(y_{m+1})\ \} \wedge 1] \vee 0$$

$$x \geq 0, \ \alpha \in [0,1] \tag{3.51}$$

where $v_{ij}(y_1) = \overset{r_i(j)}{\underset{k=1}{V}} \sigma^j_{ik}(y_i)$, $i = 1, 2, ..., m+1$, and and $\mu_{Ai}((Bx)_i)$ may be in the form of

Figure 3.26. If $r_i(j) = 1$ for all i and j, the Equation (3.51) becomes:

$$\max \quad \alpha$$
$$\text{s.t.} \quad \alpha \leq \sigma^1_{1,1}((Bx)_1)$$
$$\alpha \leq \sigma^2_{1,1}((Bx)_1)$$
$$\vdots$$
$$\alpha \leq \sigma^{\bar{r}_m+1}_{m+1,\,\bar{r}_m+1}((Bx)_{m+1})$$
$$x \geq 0 \text{ and } \alpha \in [0, 1]. \tag{3.52}$$

It is obvious that Equation (3.52) can be solved by linear programming as Equation (3.13). In many situations, the convex form and/or the concave-convex mixed form of membership function may occur. For example, $\mu_{Ai}(y_i)$ is convex; there then exists at least one $v_{ij}(y_i) = V_{k=1,\bar{r}_i(j)} \sigma^j_{ik}(y_1)$ with $r_i(j) > 1$. Thus we should consider the effective region, $\sigma^j_{ik}*((Bx)_i)$, with the assumption that $[d \, \sigma^j_{ik}((Bx)_i)]/[d((Bx)_i)] \neq 0$, all i and j. The solution procedure is described as follows (see Figure 3.27):

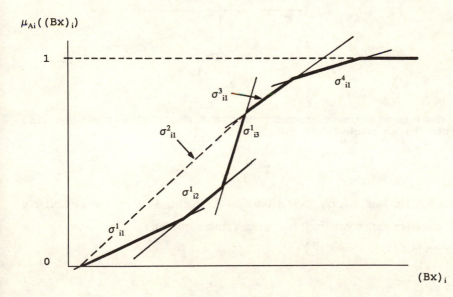

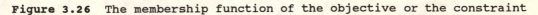

Figure 3.26 The membership function of the objective or the constraint

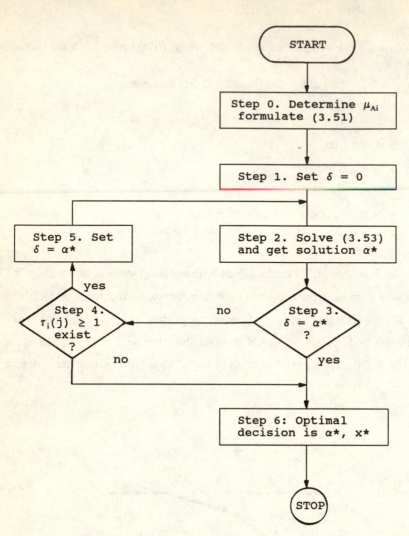

Figure 3.27 The computational procedure for the fuzzy linear programming with piecewise linear membership function

Step 0. Express $\mu_{Ai}((Bx)_i)$ as a fuzzy logical function as Figure 3.25 for the membership functions of the objective and constraints, and then, formate the fuzzy linear programming problem as Equation (3.51).

Step 1. Set $\delta = 0$.

Step 2. According to the value of δ, the effective regions of $\sigma^j_{ik}{}^*{}_{(\delta)}((Bx)i)$ are determined. Then we solve the following subproblem:

$$\max \quad \alpha$$
$$\text{s.t.} \quad \alpha \leq \sigma^{j}_{ik}{}^{*}{}_{(\delta)}((Bx)i), \quad \forall i,j$$
$$x \geq 0 \text{ and } \alpha \in [0, 1]. \tag{3.53}$$

Step 3. If $\alpha^* = \delta$, then go to Step 4, otherwise go to Step 6.

Step 4. If there is no $r_i(j) > 1$, Equation (3.53) will be equivalent to Equation (3.51), and then go to step 6. Otherwise, go to Step 5.

Step 5. Set $\delta = \alpha^*$, then go to Step 2.

Step 6. α^* is the optimal value of the original problem and the corresponding x^* is the optimal decision. STOP.

Finally, assume that α^0 is the optimal solution of Equation (3.51) and α^* is the optimal solution of Equation (3.53). Then for $\alpha \in [0, \alpha^0]$ we have $\alpha^0 \geq \alpha^* \geq \alpha$ and $\alpha^0 = \alpha$ iff $\alpha^* = \alpha$.

3.3.1.1 Example: A Truck Fleet Problem

Consider a truck fleet problem [296] and assume the fuzzy goal and constraints are modelled by piecewise membership functions.

This problem is to find an optimal decision on the size and structure of the truck fleet of a company at a minimal cost. The following requirements should be satisfied.

Four differently sized trucks x_1, x_2, x_3 and x_4 are considered. The objective is to minimize cost and the constraints are to supply all customers (who have a strong seasonally fluctuating demand). That is: certain quantities have to be moved (the quantity constraint) and a minimum number of customers per day have to be contacted (routing constraint). For other reasons, it is required that at least six of the smallest trucks be included in the fleet. The manager wants to use quantitative analysis and agrees to the following suggested linear programming:

$$\min \quad y_1 = 41,400x_1 + 44,300x_2 + 48,100x_3 + 49,100x_4 \quad \text{(cost)}$$
$$\text{s.t.} \quad y_2 = 0.84x_1 + 1.44x_2 + 2.16x_3 + 2.40x_4 \geq 170 \quad \text{(quantity)}$$
$$y_3 = 16x_1 + 16x_2 + 16x_3 + 16x_4 \geq 1300 \quad \text{(customer)}$$
$$y_4 = x_1 \geq 6 \quad \text{(minimum level)}$$

$$x_1, x_2, x_3, x_4 \geq 0 \qquad\qquad\qquad\qquad \text{(non-negative)}$$

where $41,400, $44,300, $48,100, $49,100 are unit costs, and 0.84, 1.44, 2.16, 2.4 are the unit loads (in tons) per truck for trucks x_1, x_2, x_3 and x_4 in the truck fleet, respectively. 170 is the least total quantity needed to be moved (in tons). The number of customers contacted by each truck is 6 and the least total number of customers needed to be contacted is 1300.

The solution of this crisp LP problem is $x^* = (6, 17.85, 0, 58.64)$ with the minimum cost = $3,918,850.

Since the manager feels that (s)he is forced into giving precise constraints in spite of the fact (s)he would rather have some given intervals, this leads to fuzzy linear programming problems. Therefore, the following assumptions are agreed to by the manager:

a) Total cost should be between $3,700,000 and $4,200,000. Furthermore, the manager also provides degrees of his/her satisfaction levels as follows:

$$1.0/3,700,000, \qquad 0.7/3,950,000,$$
$$0.25/4,075,000, \qquad 0.0/4,200,000.$$

b) The manager feels much better if there is some leeway on the constraints. After more detailed study, the manager provided the following data for three constraints, respectively.

$$1.0/180, \quad 0.9/176, \quad 0.5/175, \quad 0.1/172, \quad 0.0/170$$
$$1.0/1400, \quad 0.7/1350, \quad 0.4/1325, \quad 0.9/1300$$
$$1.0/12, \quad 0.9/11, \quad 0.6/9, \quad 0.25/8, \quad 0.1/7, \quad 0.0/6.$$

c) The linear approximations between available points are acceptable. Piecewise linear membership functions are then assumed (see Figure 3.28).

d) There are no inter-dependencies between the constraints.

Finally we obtain the following fuzzy linear programming problem:

$$\text{find} \quad x$$
$$\text{such that} \quad y_1 = (Bx)_1 \leq \tilde{z}^0$$
$$y_i = (Bx)_i \geq \tilde{b}_i \, , \, i = 2,3,4$$
$$x_1 , x_2, x_3, x_4 \geq 0.$$

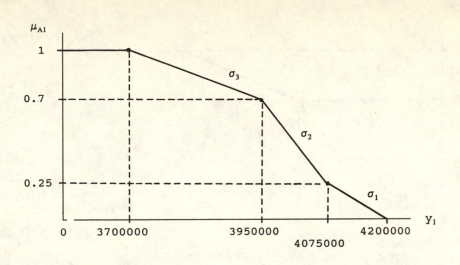

(a) A_1: $\sigma_1 = 8.40 - 0.000002y_1,$ $\sigma_2 = 14.92 - 0.0000036y_1$

$\sigma_3 = 5.44 - 0.0000012y_1$

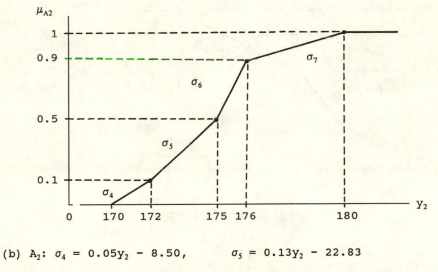

(b) A_2: $\sigma_4 = 0.05y_2 - 8.50,$ $\sigma_5 = 0.13y_2 - 22.83$

$\sigma_6 = 0.40y_2 - 68.30,$ $\sigma_7 = 0.025y_2 - 3.50$

Figure 3.28 Piecewise linear membership functions for the objective and constraints of the truck fleet problem

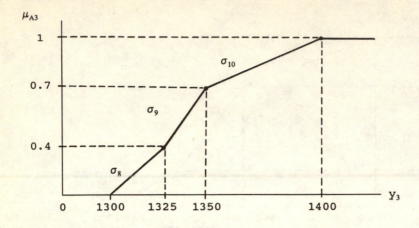

(c) A_3: $\sigma_8 = 0.016y_3 - 20.80,$ $\sigma_9 = 0.04y_3 - 15.82$

$\sigma_{10} = 0.006y_3 - 7.40$

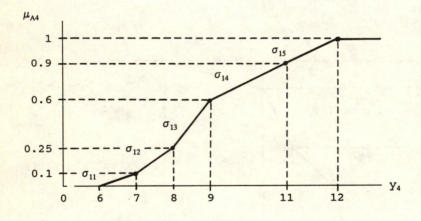

(d) A_4: $\sigma_{11} = 0.10y_4 - 0.60,$ $\sigma_{12} = 0.15y_4 - 0.95$

$\sigma_{13} = 0.35y_4 - 2.55,$ $\sigma_{14} = 0.15y_4 - 0.75$

$\sigma_{15} = 0.10y_4 - 0.20$

Figure 3.28 (Continous)

Let A_i be convex fuzzy set in $Y_i = \{ y_i \mid y_i \geq \tilde{b}_i \ (\tilde{b}_i = \tilde{z}^0) \}$ for $i = 1, 2, 3, 4$ and their membership functions (see Figure 3.28) $\mu_{Ai}(y_i)$ be given as follows:

Step 0. $\mu_{A1}(y_1) = [\{ \sigma_1 \vee \sigma_2 \} \wedge \sigma_3 \wedge 1] \vee 0$

$\mu_{A2}(y_2) = [\{ \sigma_4 \vee \sigma_5 \vee \sigma_6 \} \wedge \sigma_7 \wedge 1] \vee 0$

$\mu_{A3}(y_3) = [\sigma_8 \wedge \sigma_9 \wedge \sigma_{10} \wedge 1] \vee 0$

$\mu_{A4}(y_4) = [\{ \sigma_{11} \vee \sigma_{12} \vee \sigma_{13} \} \wedge \sigma_{15} \wedge \sigma_{16} \wedge 1] \vee 0.$

Step 1. Let $\delta = 0$.

Step 2. Solve the following subproblem:

$$\max \quad \alpha$$
$$\text{s.t.} \quad \alpha \leq \sigma_1 \wedge \sigma_3 = \sigma_1$$
$$\alpha \leq \sigma_4 \wedge \sigma_7 = \sigma_4$$
$$\alpha \leq \sigma_8 \wedge \sigma_9 \wedge \sigma_{10}$$
$$\alpha \leq \sigma_{11} \wedge \sigma_{14} \wedge \sigma_{15} \wedge \sigma_{16} = \sigma_{11}$$
$$\alpha \in [0, 1]$$

which is equivalent to:

$$\max \quad \alpha$$
$$\text{s.t.} \quad \alpha \leq 8.40 - 0.00002 y_1$$
$$\alpha \leq 0.05 y_2 - 8.50$$
$$\alpha \leq 0.016 y_3 - 20.80$$
$$\alpha \leq 0.04 y_3 - 15.82$$
$$\alpha \leq 0.006 y_3 - 7.40$$
$$\alpha \leq 0.10 y_4 - 0.60$$
$$\alpha \in [0, 1]$$

or

$$\max \quad \alpha$$
$$\text{s.t.} \quad \alpha \leq -0.0828 x_1 - 0.0886 x_2 - 0.0962 x_3 - 0.0982 x_4 + 8.400$$
$$\alpha \leq -0.040 x_1 + 0.070 x_2 + 0.108 x_3 + 0.120 x_4 - 8.500$$
$$\alpha \leq 0.256(x_1 + x_2 + x_3 + x_4) - 20.800$$
$$\alpha \leq 0.640(x_1 + x_2 + x_3 + x_4) - 15.820$$

$$\alpha \leq 0.096(x_1+x_2+x_3+x_4)\text{-}7.400$$

$$x_1, x_2, x_3, x_4 \geq 0 \text{ and } \alpha \epsilon [0, 1]$$

whose solution is: $x^* = (16.667, 0, 0, 70.323)$ with $\alpha^* = 0.445$.

Step 3. $\alpha^* = 0.445 > \alpha = 0$. Go to Step 4.

Step 4. $r_2(j) = r_4(j) = 3$. Go to Step 5.

Step 5. Set $\delta = 0.445$. Go to Step 2.

Iteration 2.

Step 2. Solve the following subproblem:

$$\max \quad \alpha$$
$$\text{s.t.} \quad \alpha \leq \sigma_2 \wedge \sigma_3$$
$$\alpha \leq \sigma_5 \wedge \sigma_7 = \sigma_5$$
$$\alpha \leq \sigma_9 \wedge \sigma_{10}$$
$$\alpha \leq \sigma_{13} \wedge \sigma_{14} \wedge \sigma_{15}$$
$$\alpha \epsilon [0, 1]$$

or

$$\max \quad \alpha$$
$$\text{s.t.} \quad \alpha \leq 14.920\text{-}0.1490x_1\text{-}0.1595x_2\text{-}0.1730x_3\text{-}0.1768x_4$$
$$\alpha \leq 5.440\text{-}0.0497x_1\text{-}0.0532x_2\text{-}0.0577x_3\text{-}0.0589x_4$$
$$\alpha \leq 0.1094x_1+0.1872x_2+0.2808x_3+0.3120x_4\text{-}22.830$$
$$\alpha \leq 0.640(x_1+x_2+x_3+x_4)\text{-}15.820$$
$$\alpha \leq 0.096(x_1+x_2+x_3+x_4)\text{-}7.400$$
$$\alpha \leq 0.350x_1\text{-}2.550, \ \alpha \leq 0.150x_1\text{-}0.750$$
$$\alpha \leq 0.100x_1\text{-}0.200$$
$$x_1, x_2, x_3, x_4 \geq 0 \text{ and } \alpha \epsilon [0, 1]$$

whose solution is: $x^* = (8.932, 5.651, 0, 66.503)$ with $\alpha^* = 0.576$.

Step 3. $\alpha^* = 0.576 > \alpha = 0.445$. Go to Step 4.

Step 4. $r_4(j) = 2$. Go to Step 5.

Step 5. Set $\alpha = 0.576$. Go to Step 2.

Iteration 3.

Step 2. Solve the following subproblem:

$$\max \quad \alpha$$
$$\text{s.t.} \quad \alpha \leq \sigma_2 \wedge \sigma_3$$
$$\alpha \leq \sigma_6 \wedge \sigma_7$$
$$\alpha \leq \sigma_9 \wedge \sigma_{10}$$
$$\alpha \leq \sigma_{13} \wedge \sigma_{14} \wedge \sigma_{15}$$
$$\alpha \epsilon [0, 1]$$

or

$$\max \quad \alpha$$
$$\text{s.t.} \quad \alpha \leq 14.920 - 0.1490x_1 - 0.1595x_2 - 0.1732x_3 - 0.1768x_4$$
$$\alpha \leq 5.440 - 0.0497x_1 - 0.0532x_2 - 0.0577x_3 - 0.0589x_4$$
$$\alpha \leq 0.336x_1 + 0.576x_2 + 0.864x_3 + 0.960x_4 - 68.300$$
$$\alpha \leq 0.210x_1 + 0.360x_2 + 0.540x_3 + 0.600x_4 - 3.500$$
$$\alpha \leq 0.640(x_1 + x_2 + x_3 + x_4) - 15.820$$
$$\alpha \leq 0.096(x_1 + x_2 + x_3 + x_4) - 7.400$$
$$\alpha \leq 0.350x_1 - 2.550, \quad \alpha \leq 0.150x_1 - 0.750$$
$$\alpha \leq 0.100x_1 - 0.200$$
$$x_1, x_2, x_3, x_4 \geq 0, \text{ and } \alpha \epsilon [0, 1]$$

whose solution is : $x^* = (9.229, 14.715, 0, 59.747)$ with $\alpha^* = 0.634$.

Step 3. $\alpha^* = 0.634 > \alpha = 0.576$. Go to Step 4.

Step 4. $r_i(j) = 1$, all i. STOP. The optimal solution for the original problem is: $x^* = (9.229, 14.715, 0, 59.747)$ with $\alpha^* = 0.634$.

Test. Solve the following sub-problem:

$$\max \quad \alpha$$
$$\text{s.t.} \quad \alpha \leq \sigma_2 \wedge \sigma_3, \quad \alpha \leq \sigma_6 \wedge \sigma_7$$
$$\alpha \leq \sigma_9 \wedge \sigma_{10}, \quad \alpha \leq \sigma_{14} \wedge \sigma_{15}$$

$$\alpha \in [0, 1]$$

whose solution is: $x^* = (9.229, 14.715, 0, 59.747)$ with $\alpha^* = 0.634 = \delta$ and the minimum cost $= 3,970,430.8$, so the optimal solution is reached.

The differences between the two solutions are provided as follows:

	Original/Crisp System	Fuzzy System
Decision		
$x_1^* =$	6.00	9.23
$x_2^* =$	17.85	14.72
$x_3^* =$	0.00	0.00
$x_4^* =$	58.64	59.75
$z^* =$	3,918,850	3,988,257
Constraints:		
1-st	171.50	172.35
2-nd	1,320.00	1,339.20
3-rd	6.00	9.23

3.3.2 Operators

In this chapter, all fuzzy linear programming approaches are based on the max-min operator. However, the max-min operator is a non-compensatory operator. But in some practical situations, a compensatory operator might be required to aggregate fuzzy constraints/objective(s). For the FLP problem, Zimmermann and Zysno [299] suggested that an acceptable compromise between empirical fit and computational efficiency seemed to be the convex combination of the min operator and the max operator. Thus the "compensatory and" operator was introduced as:

$$\mu_C(x) = \Gamma \min_i \mu_i(x) + (1 - \Gamma) \max_i \mu_i, \quad \Gamma \in [0, 1]. \tag{3.54}$$

If we use this operator, the original FLP (Section 3.1.2) will become the following mix integer programming problem [304]:

$$\max_{x \epsilon X} \{\Gamma \min_{i=0}^{m} \mu_i(x) + (1 - \Gamma) \max_{i=0}^{m} \mu_i\} \tag{3.55}$$

which is equivalent to:

$$\max \quad \Gamma\alpha_1 + (1-\Gamma)\alpha_2$$
$$\text{s.t.} \quad \mu_i(x) \geq \alpha_1, i = 0, 1, 2, ..., m$$
$$\mu_i(x) \geq \alpha_2, \text{ for at least one } i \epsilon \{0, ..., m\}$$
$$x \epsilon X$$

or

$$\max \quad \Gamma\alpha_1 + (1-\Gamma)\alpha_2$$
$$\text{s.t.} \quad \mu_i(x) \geq \alpha_1, \quad i = 0, 1, ..., m$$
$$\mu_i(x) \geq \alpha_2 + M_{yi}, i = 0, 1, ..., m$$
$$\Sigma_i y_i \leq m$$
$$x \epsilon X \text{ and } y_i \epsilon \{0, 1\}, \forall i \tag{3.56}$$

where M_{yi} are very large real numbers.

Other practical operators are "fuzzy and" and "fuzzy or" proposed by Werners [259]. "Fuzzy and" is defined as:

$$\mu_{and} = \Gamma \min_i \mu_i(x) + (1-\Gamma)[\Sigma_i \mu_i(x)]/(m+1) \tag{3.57}$$

with $\Gamma \epsilon [0, 1]$. And, "fuzzy or" is defined as:

$$\mu_{or} = \Gamma \max_i \mu_i(x) + (1-\Gamma)[\Sigma_i \mu_i(x)]/(m+1) \tag{3.58}$$

with $\Gamma \epsilon [0, 1]$.

These two operators (connectives) are not inductive and associative, but they are commutative, idempotent, strictly monotonic increasing in each component, continuous, and compensatory. By use of the "fuzzy and" operator, the FLP model (in Section 3.1.2) will become:

$$\max_{x \epsilon X} \{\Gamma \min_i \mu_i(x) + (1-\Gamma)[\Sigma_i \mu_i(x)]/(m+1)\} \tag{3.59}$$

which is equivalent to:

$$\max \quad \alpha + (1\text{-}\Gamma)[\ \Sigma_i\ \alpha_i]/(m+1)$$
$$\text{s.t.} \quad \alpha + \alpha_i \leq \mu_i,\ i = 0,\ 1,\ ...,\ m$$
$$x\ \epsilon\ X,\ \text{and}\ \alpha,\ \alpha_i,\ \mu_i,\ \forall i,\ \epsilon\ [0,\ 1]. \tag{3.60}$$

Finally, we would like to mention that some other operators discussed in Chapter 3 can also be applied to FLP problems, but some of them will cause computational inefficiencies.

3.3.3 Sensitivity Analysis and Dual Theory

Speaking of linear programming, it seems important to discuss sensitivity analysis and dual theory. For a fuzzy linear programming, we could study its sensitivity analysis and dual theory after obtaining its auxiliary crisp linear programming. Thus usual sensitivity analysis and dual approaches can be applied for fuzzy linear programming (or after some modifications if necessary).

Sensitivity analysis and dual theory of fuzzy linear programming problems have been discussed in the literature of Hamacher, Leberling and Zimmermann [112], Rodder and Zimmermann [215], Verdegay [250], Kabbara [122], Llena [153], Tanaka, Ichihashi and Asai [242], and Ostermark [199]. However, for the space limitation of this monograph, we are not going to discuss these topics. Interested people may refer to the references.

3.3.4 Fuzzy Non-Linear Programming

In this chapter, we have discussed several important techniques to solve linear programming problems in a fuzzy environment. These techniques can also be applied on some non-linear programming problems. Like fuzzy linear programming, a fuzzy non-linear programming problem should be transferred into an equivalent crisp non-linear programming problem. Then we can solve it by conventional solution techniques or packages of non-linear programming, instead of linear programming.

As an example, let us discuss Zimmermann's symmetric model for a symmetric fuzzy mathematical programming problem (see Section 3.1.2). Other models and fuzzy mathematical programming problems can be applied in a similar way. Consider the following problems:

$$\widetilde{\min} \quad g_0(x)$$
$$\text{s.t.} \quad g_i(x) \geq \widetilde{b}_i,\ i = 1,\ 2,\ ...,\ m$$
$$x \geq 0 \tag{5.61}$$

which, according to Zimmermann's concept, is equivalent to:

find   x

such that $g_0(x) \leq \tilde{b}_0$

          $g_i(x) \geq \tilde{b}_i$, $\forall i$, and $x \geq 0$ (3.62)

where g_0 and g_i, may be nonlinear functions. Assume that membership functions are linear as in Figure 3.6. We obtain the following equivalent problem:

max α

s.t. $g_0(x) \leq b_0 + (1-\alpha)p_0$

      $g_i(x) \geq b_i - (1-\alpha)p_i$

      $\alpha \in [0, 1]$ and $x \geq 0$ (3.63)

where b_0, b_i, p_0 and p_i are given goals and their tolerances, respectively. This crisp non-linear programming problem can then be solved by any non-linear programming solution package, for example KSV-SUMT or GREG packages.

3.3.4.1 Example: A Fuzzy Machining Economics Problem

Let consider the following fuzzy machining economics (ME) problem [248]. A crisp cost model for machining a D_w diameter, I_w length workpiece on a lathe, i.e. a single-pass turning operation, can be written as:

min $C_{pr} = C_b/N_b + C_m t_h + C_m[t_i + t_m + t_m(t_c + C_e/C_m)/t_s]$

s.t. $v_{min} \leq v \leq v_{max}$ (the restriction of cutting speed)

     $f_{min} \leq f \leq f_{max}$ (the allowable feed rates)

     $F_c \leq F_{c,max}$ (the maximum cutting force)

     $P_m \leq P_{max}$ (the necessary machining power)

     $R_a \leq R_{a,max}$ (the limited surface-roughness level)

     $1 \leq a/f \leq 20$ (depth of cut and feed rate)

     v and $f \geq 0$ (non-negative)

where the parameters and decision variables are explained in the following:

C_b/N_b: unit set up and preparation cost for each workpiece

C_m: total machine and operator rate

t_h: handling time for each workpiece

t_i: idle time for each workpiece

t_m: machining time for each workpiece

t_c: tool change time

t_s: tool life

C_e: cost of each tool cutting edge

$C_m t_m t_c/t_s$: unit tool change cost for each workpiece

$C_m t_m t_c/(t_s C_m)$: unit cost of a sharp tool edge for each workpiece

$t_m = I_w/[1000fv/(\pi D_w)]$

$t_s = C'/(v^{\alpha T}f^{\beta T})$

$$C_{pr} = (C_b/N_b + C_m t_h + C_m t_i) + \frac{C_m \pi D_w I_w}{1000fv} + (C_m t_c + C_e) \frac{v^{\alpha T-1}f^{\beta T-1} \pi D_w I_w}{1000C'}$$

Doubtlessly, defining a machining optimization problem, including the construction of an objective equation and constraints, is a complicated decision making process. Depending on the tool material, the workpiece material, and the type of turning process, constraint resources have to be decided before the entire non-linear programming model can be constructed. Determining the machining limitations and specifications can involve many human factors . Therefore, the ME model can be represented by fuzzy non-linear programming to make the model more flexible and adaptable to the human decision process. Furthermore, the fuzzy ME model emphasizes the qualitative viewpoint of the decision maker, and thus is more human-oriented.

The fuzzy machining economics model is:

find $\quad$ f and v

such that $\quad K_0 + K_1 C_m v^{-1} f^{-1} + K_2 v^{\alpha T-1} f^{\beta T-1} \leq \tilde{b}_0$

$-v \leq -\pi D_w \tilde{n}_{s,min}/1000, \ v \leq \pi D_w \tilde{n}_{s,max}/1000$

$-f \leq -\tilde{f}_{min}, \ f \leq \tilde{f}_{max}$

$K_F f^{\beta F} a^{\Gamma F} \leq \tilde{F}_{c,max}$

$F_c v/(6120\delta_m) \leq \tilde{P}_{max}$

$$K_R v^{\alpha R} f^{\beta R} a^{\Gamma R} \leq \tilde{R}_{a,max}$$

$$-af^{-1} \leq -\tilde{1}, \; af^{-1} \leq \tilde{20}$$

$$f \text{ and } v \geq 0$$

where $K_0 = C_b/N_b + C_m t_h + C_m t_i$, $K_1 = \pi D_w I_w/1000$, $K_2 = K_1(C_m t_c + C_e)/C'$. If the goal b_0, and all the tolerances of the fuzzy goal and constraints are given, an equivalent crisp model will be obtained. Let us consider the following numerical data for the fuzzy ME problem:

```
Tool:   ISO SNMA 120XXX-P20
Holder: ISO PBSNR 2525
Material: SAE 1045 CD
Machine: ENGINE LATHE
Feed speed f: 0.05-2.5 mm/rev
Speed n_s: 20-1699 rpm
at 80% efficiency

C_m: 0.25/min          t_h: 1.35 min/pc
C_e: 0.50/edge         t_i: 0.2min/pass
C_b: 7.2/batch         t_c: 1.0 min/edge
D_w: 100 mm            I_w: 250 mm
e:   55mm              N_b: 25 parts/batch
P:   0.1               a : 3 mm
F_c:  250v^{-0.12}f^{0.75}a ≤ 170 kp
P_c:  F_cf/6120 ≤ 6.0 kw
R_a:  2.43x10^4 v^{-1.52}f ≤ 1.5 μm

Tolerances for the constraints are 1, 4, 200, 0, 1,
10, 4, 1, 0 and 0, respectively.
```

The equivalent crisp non-linear programming problem is:

$$\max \quad \alpha$$

s.t.
$$0.6755 + 19.635v^{-1}f^{-1} + 1.178e^{-6}v^{2.33}f^{0.11} \leq b_0 + (1-\alpha)1$$

$$-1000v/314 \leq -20 + (1-\alpha)4$$

$$1000v/314 \leq 1600 + (1-\alpha)200$$

$$-f \leq -0.05 + (1-\alpha)0$$

$$f \leq 2.50 + (1-\alpha)1$$

$$250v^{-0.12}f^{0.75}(3) \leq 170 + (1-\alpha)10$$

$$0.041v^{0.88}f^{0.75}(3) \leq 6.0 + (1-\alpha)4$$

$$2.43x10^4v^{-1.52}f \leq 1.5 + (1-\alpha)1$$

$$-3f^{-1} \leq -1 + (1-\alpha)0$$

$$3f^{-1} \leq 10 + (1-\alpha)0$$

f and v ≥ 0.

Three aspiration costs, $b_0 = 1.4$, 1.6 and 1.7, are used for comparison. The solutions for these aspirations are provided in the following:

Aspiration (Goal)	1.4	1.6	1.7
cost	1.4	1.6	1.65
α (satisfaction)	73.2%	93.3%	100%
v (m/min)	328.48	322.47	316.68
f (mm/r)	0.247	0.215	0.207

3.3.5 Fuzzy Integer Programming

Consider the following fuzzy integer programming problem:

$$\text{max} \quad cx$$
$$\text{s.t.} \quad (Ax)_i \lesssim b_i, \forall i$$
$$x \geq 0 \text{ and } x \text{ integer} \tag{3.64}$$

where the tolerances p_i, $\forall i$, are given for b_i. Assume that the membership functions of the fuzzy constraints are linear. To solve the above fuzzy constraints may be equivalent to the following auxiliary crisp constraints:

$$(Ax)_i \leq b_i + T_i \text{ and } 0 \leq T_i \leq p_i w_i \tag{3.65}$$

where w_i, $\forall i$, are binary variables attached to the constraints. To solve Equation (3.64), Fabian and Stoica [83] proposed the following auxiliary crisp model:

$$\text{max} \quad cx + \Sigma_i r_i[b_i-(Ax)_i]w_i$$
$$\text{s.t.} \quad (Ax)_i - p_i w_i \leq b_i, w_i \in \{0, 1\}, \forall i$$
$$x \geq 0 \text{ and } x \text{ integer} \tag{3.66}$$

where r_i, $\forall i$, are penalty coefficients. Obviously, Equation (3.66) becomes a non-linear programming problem.

Example. Consider the following integer linear programming problem:

$$\max \quad 2x_1 + 5x_2$$
$$\text{s.t.} \quad 2x_1 - x_2 \lesseqgtr 9$$
$$2x_1 + 8x_2 \lesseqgtr 31$$
$$x_1, x_2 \geq 0 \text{ and } x_1, x_2 \text{ integer.}$$

Assume that $r_1 = r_2 = 1$, $p_1 = 3$ and $p_2 = 4$. Then we have:

$$\max \quad 2x_1 + 5x_2 + (9-2x_1+x_2)w_1 + (31-2x_1-8x_2)w_2$$
$$\text{s.t.} \quad 2x_1 - x_2 -3w_1 \leq 9$$
$$2x_1 + 8x_2 - 4w_2 \leq 31$$
$$w_1, w_2 \in \{0, 1\}, x_1, x_2 \geq 0 \text{ and } x_1, x_2 \text{ integer.}$$

In Figure 3.29, the optimal solution of the crisp model ($p_1 = p_2 = 0$) is $x = (3, 3)$ and the objective value is 21. The fuzzy model has the solution $x = (4, 3)$ with the objective value 22. For this result the second constraint is violated by 1 which means 3.3% from the right-hand side.

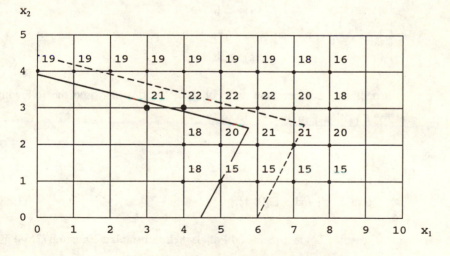

Figure 3.29 Solution of the numerical example

3.3.5.1 Fuzzy 0-1 Linear Programming

If the goal of the objective in Equation (3.64) is given as b_0, then we will have the following symmetrical model:

find $\quad$ x

such that $\quad cx \gtrsim b_0$

$\qquad (Ax)_i \lesssim b_i, \forall i$

$\qquad x \geq 0$ and x integer.

Now, if x_j, $j = 1, ..., n$, are either 0 or 1, the above problem will become the following 0-1 linear programming problem:

find $\quad (x_1, x_2, ..., x_n)$

such that $\quad (Ax)_i \lesssim b_i, \forall i$, and $x_j = 0$ or 1, $\forall j$ $\qquad\qquad$ (3.67)

where the tolerance p_i for b_i is given, $\forall i$, and the membership functions of the fuzzy constraints are assumed to be:

$$\mu_i(x) = \begin{cases} 1 & \text{if } (Ax)_i < b_i \\ 1 - [(Ax)_i - b_i]/p_i & \text{if } b_i \leq (Ax)_i \leq b_i + p_i \\ 0 & \text{if } (Ax)_i > b_i + p_i. \end{cases}$$

Based on the max-min operator, Zimmermann and Pollatschek [303] proposed the following auxiliary crisp model for Equation (3.67):

max $\quad \alpha$

s.t. $\quad \alpha \leq \mu_i(x) = 1 - [(Ax)_i - b_i]/p_i, \forall i$

$\qquad x_j = 0$ or 1, $\forall j$ and $\alpha \in [0, 1].$ $\qquad\qquad$ (3.68)

For computational reasons, Zimmermann and Pollatschek reformulate Equation (3.68) as:

$$\max \{0, \min[1, 1 + \max_x \min_i (b_i' - (Ax)_i')]\} \qquad\qquad (3.69)$$

where $b_i' = b_i/p_i$ and $(Ax)_i' = (Ax)_i/p_i$. Then, for $\alpha' = \max_x \min_i [b_i' - (Ax)_i']$, Equation (3.69) will be:

$$\text{max} \quad \{0, \min[1, 1+\alpha']\}$$
$$\text{s.t.} \quad x_j = 0 \text{ or } 1, \forall j. \tag{3.70}$$

Since $\alpha' = \max_x \min_i [b_i'-(Ax)_i']$, the following basic inequality holds:

$$\alpha' \leq \min_i \max_x [b_i'-(Ax)_i']. \tag{3.71}$$

The right-hand side of Equation (3.71) then serves as an upper bound for α'. This equation is important for the suggested branch and bound algorithm which is imbedded into a LIFO (last-in-first-out) scanning procedure.

To use the branch and bound algorithm, let us first discuss the following definition.

Let N be the index set of variables. $N = \{1, 2, ..., n\}$. J is an ordered subset of N such that each element of J has either a positive or a negative sign and may be underlined or not. The sign of the elements in J indicates the values of the variables belonging to the indices in J. For example, $J = \{2, -1\}$ means $x_2 = 1$ and $x_1 = 0$. The set N - J contains the indices of the variables (free variables) to which numerical values have not yet been assigned. J is called a node in a tree of the branch and bound algorithm and contains all relevant information about that subset of variables which are no longer free variables at this stage.

Branching is done by computing values based on Equation (3.69) or (3.71), respectively, for not completed nodes. If an increase of the bound can be achieved by adding indices of free variables to J, one moves from node J_t to node J_{t+1}. At first $J_t = \emptyset$. If a branch of the tree has been terminated, a new node for further explorations is selected (backtracking). Which of the nodes is selected is uniquely defined by the order of the elements of J, considering also whether they are underlined or not. The right-most element in J not underlined is underlined and all elements to the right of it are excluded from J.

Now, let us define:

$$BL(J_t) = \min_i \max_{j \in N-Jt} \{b_i' - \Sigma_{j \in N-Jt} a_{ij}'x_j - \Sigma_{j \in J} ta_{ij}'\} \tag{3.72}$$

as the upper bound of α' for a complete J_t, where $a_{ij}' = a_{ij}/p_i$ and $A = [a_{ij}]$. This upper bound can be easily calculated by:

$$BL(J_t) = \min_i \{(b_i' - \Sigma_{j \in Jt} a_{ij}') - \Sigma_{j \in N-Jt} \min(0, a_{ij}')]. \tag{3.73}$$

If $BL(J_t)$ is smaller than $L = \min_i [b_i' - \Sigma_j a_{ij}' x_j]$ of a known solution x (the so-called potential solution with L = the potential value), then all J_t of this branch of the tree will be inferior to the potential solution and one has to backtrack. On the other hand, if $BL(J_t) > L$, an appropriate variable for branching has to be selected. For a good selection, we need to calculate the following two bounds for each free variable:

$$BL(J_t \cup \{+j\}) \text{ and } BL(J_t \cup \{-j\}). \tag{3.74}$$

If for any variable both bounds are smaller than L, then backtracking will be started for the same reason as before. If only one bound is smaller than L, the variable will be selected. For instance, if $BL(J_t \cup \{+j\}) < L$, then J_t is augmented by -j. If this condition holds for several variables, they are all included into J_t. If both bounds are larger than L for any free variable, the branching variable is chosen heuristically by using any of the following three rules:

(1) Select $j \in N - J_t$ for which $BL(J_t \cup \{+j\}) - BL(J_t \cup \{-j\})$ is minimal.
(2) Select $j \in N - J_t$ for which $\max [BL(J_t \cup \{+j\}), BL(J_t \cup \{-j\})]$ is minimal.
(3) Select $j \in N - J_t$ for which $\min [BL(J_t \cup \{+j\}), BL(J_t \cup \{-j\})]$ is minimal.

By repeating the above procedures, we can obtain a solution $\alpha^* = 1 + L$ where L is initially set to -1.

It is noted that Rules (1) and (2) are intended to reduce as much as possible the chances that the alternative branch will have to be explored later, and that Rule (3) is intended to reduce the chances of completions (termination) in the selected branch. Zimmermann and Pollatschek have found that Rule (1) is the most effective, while Rule (3) is the least effective. Thus Rule (1) is primarily used. If Rule (1) does not lead to a unique decision, Rule (2) will be used. And, if a variable still has not been uniquely determined, Rule (3) will then be applied. In the case of none of the rules leading to a unique solution, the minimal j is chosen.

Example. Consider the following simplified problem:

$$\text{find} \quad x_1, x_2, x_3 \in \{0, 1\}$$
$$\text{s.t.} \quad g_1(x) = -10x_1 - 20x_2 - 20x_3 \leqq -45$$
$$g_2(x) = x_1 + x_2 + x_3 \leqq 2.5$$
$$g_3(x) = 2x_1 + x_2 + 3x_3 \leqq 5$$
$$g_4(x) = 0.5x_1 + 3x_2 + x_3 \leqq 3$$

where the tolerances $(p_1, p_2, p_3, p_4) = (10, 0.75, 1.5, 2)$. The membership functions of the fuzzy constrains are:

$$\mu_1(x) = \begin{cases} 1 & \text{if } g_1(x) < -45 \\ 1 - [(g_1(x)+45)/10] & \text{if } -45 \leq g_1(x) \leq -35 \\ 0 & \text{if } g_1(x) > -35 \end{cases}$$

$$\mu_2(x) = \begin{cases} 1 & \text{if } g_2(x) < 2.5 \\ 1 - [(g_2(x)-2.5)/0.75] & \text{if } 2.5 \leq g_2(x) \leq 3.25 \\ 0 & \text{if } g_2(x) > 3.25 \end{cases}$$

$$\mu_3(x) = \begin{cases} 1 & \text{if } g_3(x) < 5 \\ 1 - [(g_3(x)-5)/1.5] & \text{if } 5 \leq g_3(x) \leq 6.5 \\ 0 & \text{if } g_3(x) > 6.5 \end{cases}$$

$$\mu_4(x) = \begin{cases} 1 & \text{if } g_4(x) < 3 \\ 1 - [(g_4(x)-3)/2] & \text{if } 3 \leq g_4(x) \leq 5 \\ 0 & \text{if } g_4(x) > 5. \end{cases}$$

The algorithm starts with $L = -1$, $t = 0$ and $J_0 = \emptyset$. Thus the general upper bound of α' is:

$$BL(J_0) = \min_i [b_i' - \Sigma_{j=1,2,3} \min(0, a_{ij}')] = \min [1/2, 10/3, 10/3, 3/2] = 1/2.$$

Since $BL(J_0) > L$, one (or several) appropriate variable has to be selected for branching. Therefore, we must consider the following bounds:

$$BL(\{+1\}) = \min_i [b_i' - a_{i1}' - \Sigma_{j=2,3} \min(0, a_{ij}')]$$
$$= \min [1/2, 2, 2, 5/4] = 1/2 \ (> L)$$

$$BL(\{-1\}) = \min_i [b_i' - \Sigma_{j=2,3} \min(0, a_{ij}')]$$
$$= \min [-1/2, 10/3, 10/3, 3/2] = -1/2 \ (> L)$$

$$BL(\{+2\}) = \min_i [b_i' - a_{i2}' - \Sigma_{j=1,3} \min(0, a_{ij}')]$$
$$= \min [1/2, 2, 8/3, 0] = 0 \ (> L)$$

$$BL(\{-2\}) = \min_i [b_i' - \Sigma_{j=1,3} \min(0, a_{ij}')]$$

$$= \min\ [-3/2,\ 10/3,\ 10/3,\ 3/2] = -3/2\ (< L)$$

$$BL(\{+3\}) = \min_i\ [b_i' - a_{i3}' - \Sigma_{j=1,2}\ \min(0,\ a_{ij}')]$$
$$= \min\ [1/2,\ 2,\ 4/3,\ 1] = 1/2\ (> L)$$

$$BL(\{-3\}) = \min_i\ [b_i' - \Sigma_{j=1,2}\ \min(0,\ a_{ij}')]$$
$$= \min\ [-3/2,\ 10/3,\ 10/3,\ 3/2] = -3/2.\ (< L)$$

Since $BL(\{-2\}) < L < BL(\{+2\})$ and $BL(\{-3\}) < L < BL(\{+3\})$, J_0 is augmented by $+2$ and $+3$. $J_1 = \{+2,\ +3\}$. Now, the only one free variable is x_1. We calculate:

$$BL(J_1U\{+1\}) = \min_i\ [b_i' - \Sigma_{j=1,2,3}\ a_{ij}']$$
$$= \min\ [1/2,\ -2/3,\ -2/3,\ -3/4] = 13/4$$

$$BL(J_1U\{-1\}) = \min_i\ [b_i' - \Sigma_{j=2,3}\ a_{ij}']$$
$$= \min\ [-1/2,\ 2/3,\ 2/3,\ -1/2] = -1/2$$

and choose -1 to complete the node J_1. For the potential solution $(x_1,\ x_2,\ x_3) = (0,\ 1,\ 1)$, we obtain:

$$L = \min_i\ [b_i' - \Sigma_{j=1,2,3}\ a_{ij}'x_j]$$
$$= \min\ [-1/2,\ 2/3,\ 2/3,\ -1/2] = -1/2.$$

But from $BL(\{-2\}) = BL(\{-3\}) = -3/2$, it follows for any J_t including $\{-2\}$ or $\{-3\}$ that $BL(J_t)$ $\leq -3/2 \leq -1/2 = L$. Hence, $\min[1,\ 1+\alpha'] = 1 + \alpha' \leq 1/2$ and $\alpha^* = \max\ \{0,\ \min[1,\ 1+\alpha]\}$ $= 0$. It obviously results that all other branches are of no interest and $x = (0,\ 1,\ 1)$ is the optimal solution with $\alpha^* = 1 + L = 1 -1/2 = 1/2$.

3.3.5.1a Example: A Fuzzy Location Problem

For the network location problems, the p-center and p-median types of problems are most concerned. The p-center problem is that p new facilities are located with respect to m locations so as to minimize a maximum of weighted distance between each existing location and its closest center. On the other hand, the objective function of the p-median problem is to minimize a sum of weighted distances between each existing facility and its closest median. Both problems have been transferred by Krarup and Pruzan into a mixed integer programming problem as shown in Table 3.21.

Table 3.21   Krarup and Pruzan's plant location model

---

min $z = \sum\limits_{i \in I} \sum\limits_{j \in J} (c_i + t_{ij}) u_{ij} + \sum\limits_{i \in I} f_i y_i$ (cost)

s.t. $\sum\limits_{i \in I} u_{ij} \geq d_j,\ j \in J$ (demand)

       $k_i y_i - \sum\limits_{j \in J} u_{ij} \geq 0,\ i \in I$ (balance)

       $u_{ij} \geq 0,\ \forall i, j,$ and $y_i = 0$ or 1

where

   $i \in I = \{1, \ldots, m\}$: potential locations for new facilities

   $j \in J = \{1, \ldots, n\}$: locations to be served by the new facilities

   c_i: the operating cost per unit for facilities i

   t_{ij}: the cost of transporting a unit from i to j

   f_i: the fixed cost of opening the facility i

   u_{ij}: the units to be transported from i to j

   d_j: the demand by j

   k_i: a constant greater than the maximum production potential of i
        (usually, k_i is replaced by the sum of d_j's)

   $y_i = 1$ if a facility is located at i; = 0 otherwise.

Discrete locations models can also be formulated as 0-1 integer programming problems. The location problem can then be stated as the optimal location of a number of facilities at some points of a set of candidate points S so as to serve the entire membership of S.

Let $S = \{1, \ldots, i, \ldots, n\}$ be the set of n potential locations and $K = \{\{d_{ij}, j \in J\}, i \in I\}$ be a class of subsets of S where $d_{ij} \in K$ locates a facility at i to serve the rest of the points in the subset (points in d_{ij}). Also, let

$$y_{ij} = \begin{cases} 1 & \text{if } d_{ij} \text{ is in the cover} \\ 0 & \text{otherwise.} \end{cases}$$

Then, Darzentas [50] proposed the following model:

$$\min \quad z = \sum_{i \in I} \sum_{j \in J} a_{ij} y_{ij}$$

$$\text{s.t.} \quad \sum_{\{d_{ij}, j \in J\}} y_{ij} \geq 1, \text{ i} \in \text{I, and } y_{ij} = 0 \text{ or } 1. \tag{3.75}$$

where a_{ij}, $\forall i,j$, represent the cost of serving the group of points of the subset d_{ij} by a facility located at i. The objective of the above problem is to minimize the cost of subsets of S (members of K) necessary to cover the members of S, and thus obtain the minimum cost needed, according to certain criteria quantified by a_{ij}. The constraints of Equation (3.75) guarantee that every i ϵ S will be covered by at least one d_{ij} ϵ K, since for every i there is a constraint forcing at least one d_{ij} to be in the cover ($y_{ij} = 1$).

Now, consider the fuzzy version of Equation (3.75) as:

$$\text{find} \quad y_{ij}, \forall i,j$$

$$\text{s.t.} \quad \sum_{i \in I} \sum_{j \in J} a_{ij} y_{ij} \leq z_0$$

$$\sum_{\{d_{ij}, j \in J\}} y_{ij} \geq 1, \text{ i} \in \text{I, and } y_{ij} = 0 \text{ or } 1 \tag{3.76}$$

where the goal z_0 and its tolerance p_0 are given. Furthermore, the membership functions of the fuzzy constraints are assumed to be:

$$\mu_i(y_{ij}) = \begin{cases} 1 & \text{if } \sum a_{ij} y_{ij} \geq 1 \\ 1 - (1 - \sum a_{ij} y_{ij})/p_i & \text{if } 1-p_i \leq \sum a_{ij} y_{ij} \leq 1 \\ 0 & \text{if } \sum a_{ij} y_{ij} \leq 1-p_i \end{cases}$$

where p_i, $\forall i$, are the tolerances of the fuzzy constraints. With the above assumptions, we can then solve Equation (3.76) by using Zimmermann and Pollatschek's approach.

To illustrate the location problem, let us consider the numerical example of the road network shown in Figure 3.30 [50] where the points 1, 2, 3, and 4 represent villages. The population of these villages and the distances between the villages are also given. Our problem is to optimally locate three facilities in order to serve (cover) each village by only one facility.

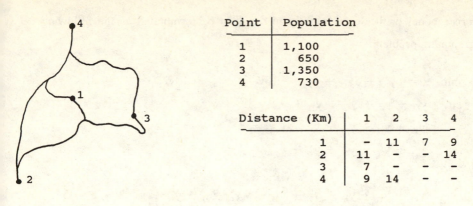

Point	Population
1	1,100
2	650
3	1,350
4	730

Distance (Km)	1	2	3	4
1	–	11	7	9
2	11	–	–	14
3	7	–	–	–
4	9	14	–	–

Figure 3.30 The numerical road network problem

First, let us construct the class K of subsets of the set S = {1, 2, 3, 4}, and assume that each member of K should not have more than 2,000 total population and the distance between villages should not be more than 15 Km. Table 3.22 gives the members of K for this example and all other required information for the model of Equation (3.75).

Table 3.22 Numerical input data for the road network problem

(1) Members of K	(2) d_j	(3) Total population	(4) Total distance	(5) y_{ij}	(6) $a_{ij}=(4)/(3)$	(7) $a_{ij} \times 1000$
1	1	1,100	0	y_{11}	0	0
2	1,2	1,750	11	y_{12}	0.006	6
3	1,4	1,830	9	y_{13}	0.005	5
4	2	650	0	y_{24}	0	0
5	2,1	1,750	11	y_{25}	0.006	6
6	2,4	1,380	14	y_{26}	0.010	10
7	3	1,350	0	y_{37}	0	0
8	4	730	0	y_{48}	0	0
9	4,1	1,830	9	y_{49}	0.005	5
10	4,2	1,380	14	$y_{4\,10}$	0.010	10

The crisp model of the location problem can then be formulated as the following 0-1 integer programming problem:

$$\min \quad z = 6y_{12} + 5y_{13} + 6y_{25} + 10y_{26} + 5y_{49} + 10y_{4\,10}$$

$$\text{s.t.} \quad y_{11} + y_{12} + y_{13} + y_{25} + y_{49} = 1$$

$$y_{12} + y_{24} + y_{25} + y_{26} + y_{4\,10} = 1$$

$$y_{13} + y_{26} + y_{48} + y_{49} + y_{4\,10} = 1$$

$$y_{37} = 1$$

$$y_{11} + y_{12} + y_{13} + y_{24} + y_{25} + y_{26} + y_{37} + y_{48} + y_{49} + y_{4\,10} = 3$$

$$y_{ij} = 0 \text{ or } 1, \; \forall i,j.$$

Here, based on the enumeration algorithm, we can search for a solution from y_{11} to $y_{4\,10}$ as in the constraints of the above problem. When $y_{ij} = 1$, then row i and column j are removed, the number of $y_{ij} \neq 0$ in the j column are recorded. In this example, $y_{37} = 1$ because Village 3 can only be served by itself.

The enumeration gives the following feasible cover: (1) y_{12} and y_{48}, (2) y_{13} and y_{24}, and (3) y_{25} and y_{48}. The locations of the facilities and the villages they serve (cover) are given in the following:

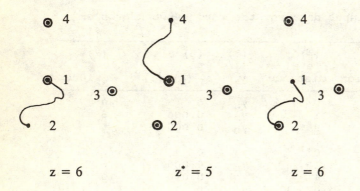

$$z = 6 \qquad\qquad z^* = 5 \qquad\qquad z = 6$$

The optimal solution is the cover of y_{13} and y_{24} and has the minimal cost of 5.

The fuzzy version of Equation (3.76) for this numerical example can also be formulated and solved if the goal (z_0) and its tolerance (p^0), and the membership functions of the fuzzy constraints are given (including the tolerances p_i).

Consider the following fuzzy version of the road network problem:

find y_{ij}, $\forall i,j$

s.t. $g_1(y) = 6y_{12}+5y_{13}+6y_{25}+10y_{26}+5y_{49}+10y_{4\,10} \leq 5$

        $g_2(y) = -(y_{11} + y_{12} + y_{13} + y_{25} + y_{49}) \leq -1$

        $g_3(y) = -(y_{12} + y_{24} + y_{25} + y_{26} + y_{4\,10}) \leq -1$

        $g_4(x) = -(y_{13} + y_{26} + y_{48} + y_{49} + y_{4\,10}) \leq -1$

        $y_{37} = 1$

        $y_{11} + y_{12} + y_{13} + y_{24} + y_{25} + y_{26} + y_{37} + y_{48} + y_{49} + y_{4\,10} = 3$

        $y_{ij} = 0$ or 1, $\forall i,j$

where the tolerances for the fuzzy constraints are all equal to 1 ($p_1 = p_2 = p_3 = p_4 = 1$). For this particular example, y_{37} should reserve one facility. Also, let us rename the variables y_{11} - $y_{4\,10}$ as x_1 - x_{10}, respectively. Then, the above problem can be formulated as:

find x_j, $\forall j$ (Note: $x_7 = 1$)

s.t. $g_1(x) = 6x_2+5x_3+6x_5+10x_6+5x_9+10x_{10} \leq 5$

        $g_2(x) = -(x_1 + x_2 + x_3 + x_5 + x_9) \leq -1$

        $g_3(x) = -(x_2 + x_4 + x_5 + x_6 + x_{10}) \leq -1$

        $g_4(x) = -(x_3 + x_6 + x_8 + x_9 + x_{10}) \leq -1$

        $x_j = 0$ or 1, $\forall j$

where $x_1 + x_2 + x_3 + x_4 + x_5 + x_6 + x_8 + x_9 + x_{10} = 2$. With the assumption of linear membership functions, we then use Zimmermann and Pollatschek's approach to solve this fuzzy 0-1 integer programming problem.

Let $L = -1$, $t = 0$ and $J_0 = \emptyset$. Then the general upper bound α' is:

$$BL(J_0) = \min_i [b_i' - \Sigma_j \min(0, a_{ij}')] = \min [5, -1, -1, -1] = -1.$$

Since $BL(J_0) \geq L$, one or several appropriate variables have to be selected for branching. Therefore, we must consider the following bounds:

$$BL(\{+1\}) = \min_i [b_i' - a_{i1}' - \Sigma_{j=2->10} \min (0, a_{ij}')] = \min [5, 4, 5, 5] = 4$$

$$BL(\{-1\}) = \min_i [b_i' - \Sigma_{j=2->10} \min (0, a_{ij}')] = \min [5, 3, 5, 5] = 3$$

$BL(\{+2\}) = -1,$ $BL(\{-2\}) = 3$

$BL(\{+3\}) = 0,$ $BL(\{-3\}) = 3$

$BL(\{+4\}) = 4,$ $BL(\{-4\}) = 3$

$$BL(\{+5\}) = -1, \qquad BL(\{-5\}) = 3$$
$$BL(\{+6\}) = -5, \qquad BL(\{-6\}) = 3$$
$$BL(\{+8\}) = 4, \qquad BL(\{-8\}) = 3$$
$$BL(\{+9\}) = 0, \qquad BL(\{-9\}) = 3$$
$$BL(\{+10\}) = -5, \qquad BL(\{-10\}) = 3.$$

If $BL(\{\}) \leq -1 = L$, then $\{\}$ will be inferior to the potential solution. Thus, the remaining possible candidates for J_1 are x_1, x_3, x_4, x_8 and x_9. However, J_1, Vt, can only contain two elements (or variables) because of $x_1 + x_2 + x_3 + x_4 + x_5 + x_6 + x_8 + x_9 + x_{10} = 2$. Thus, we have the following ten possible situations:

$$J_1 = \{1, 3\} \quad => \quad L = \min [5, 0, -1, -1] = -1$$
$$J_1 = \{1, 4\} \quad => \quad L = \min [5, 0, 0, -1] = -1$$
$$J_1 = \{1, 8\} \quad => \quad L = \min [5, 0, -1, 0] = -1$$
$$J_1 = \{1, 9\} \quad => \quad L = \min [0, 0, -1, 0] -1$$
$$J_1 = \{3, 4\} \quad => \quad L = \min [0, 0, 0, 0] = 0 \qquad *$$
$$J_1 = \{3, 8\} \quad => \quad L = \min [0, 0, -1, 1] = -1$$
$$J_1 = \{3, 9\} \quad => \quad L = \min [-5, 1, -1, 1] = -5$$
$$J_1 = \{4, 8\} \quad => \quad L = \min [5, -1, 0, 0] = -1$$
$$J_1 = \{4, 9\} \quad => \quad L = \min [0, 0, 0, 0] = 0 \qquad *$$
$$J_1 = \{8, 9\} \quad => \quad L = \min [0, 0, -1, 1] = -1.$$

Obviously, $\alpha^* = 1 + L = 1 + 0 = 1$ at $J_1 = \{3, 4\}$ and $\{4, 9\}$, while $\alpha^* = 1 - 1 = 0$ at the other situations. Thus the optimal solutions are: $y_{13} = y_{24} = y_{37} = 1$ and $y_{24} = y_{37} = y_{49} = 1$ where $z^* = 5$. The locations of the facilities and the villages they serve (cover) are given in the following:

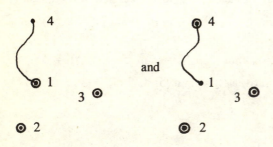

4 POSSIBILISTIC PROGRAMMING

Stochastic programming has been used since the late 1950s for decision models where input data (coefficients in LP problems) have been given probability distributions. The pioneer works were done by Dantzig [47], Beale [9], Tintner [247], Simon [223], Charnes, Cooper and Symonds [39], and Charnes and Cooper [40]. Since then, a number of stochastic programming models have been formulated in inventory theory, system maintenance, micro-economics, and banking and finance. The most recent summaries of the development of stochastic programming methods are Stancu-Minasian [BM48] and Wets [260].

Even after nearly four decades of development of both theorem and applications, stochastic programming approaches are still limited in solving practical problems. We believe that the main problems of applying stochastic programming models are: (1) lack of computational efficiency, and (2) inflexible probabilistic doctrines which might not be able to model the real imprecise meaning of decision makers. In practice, the common probability distributions may not be correct in the decision maker's mind. And the meanings of randomness for some imprecise/uncertain situations may not be correct, either. Zadeh [284] stated:

"The pioneering work of Wiener and Shannon on the statistical theory of communication has lead to a universal acceptance of the belief that information is intrinsically statistical in nature and, as such, must be dealt with by the methods provided by probability theory.

Unquestionably, the statistical point of view has contributed deep insights into the fundamental processes involved in the coding, transmission and reception of data, and played a key role in the development of modern communication, detection and telemetering systems. In recent years, however, a number of other important applications have come to the fore in which the major issues center not on the transmission of information but on its meaning. In such applications, what matters, is the ability to answer questions relating to information that is stored in a database - as in nature language processing, knowledge representation, speech recognition, robotics, medical diagnosis, analysis of rare events, decision-making under uncertainty, picture analysis, information retrieval and related areas.

... our main concern is with the meanings of information - rather than with its measure - the proper framework for information analysis is possibilistic rather than probabilistic in nature, thus implying that what is needed for such an analysis is not probability theory but an analogous - and yet different - theory which might be called the theory of possibility."

Since Zadeh [284], there has been much research on the possibility theory. Possibilistic decision making models have provided an important aspect in handling practical decision making problems. In this chapter we will discuss some important applications of the possibility theory to linear programming problems with imprecise coefficients. We call linear programming problems with imprecise coefficients restricted by possibilistic distributions Possibilistic Linear Programming (PLP) problems. Unlike stochastic linear programming, PLP models provide computational efficiency and flexible doctrines, as you will see. Yazenin [272] and Buckley [27] have given some comments on the comparison of fuzzy/possibilistic and stochastic programming.

It is noted that the possibility measure of an event might be interpreted as the possibility degree of its occurrence under a possibility distribution, $\pi(\cdot)$, (an analogous to a probability distribution) in this study.

By considering all possible cases in a PLP problem, we proposed a classification of PLP problems as shown in Table 4.1. In the next section we will discuss all PLP models. Some other extensions of PLP will also be discussed in Section 4.2.

Table 4.1 The classification of possibilistic LP problems (Lai [139])

max/min	cx		Imprecise coefficient	
s.t.	Ax ≤ b		c	
	x ≥ 0	Imprecise b	A	

Case	Problem	Approach
1. $\tilde{b}$ & $\tilde{A}$	Prob. 1	Ramik & Rimanek, Tanaka, Ichihashi & Asai, or Dubois
2. $\tilde{c}$	Prob. 2	Lai & Hwang, Rommelfanger, Hanuscheck & Wolf, or Delgado, Verdegay & Vila
3. $\tilde{b}$		(A dual problem of Problem 2)
4. $\tilde{c}$, $\tilde{b}$ & $\tilde{A}$	Prob. 3	Lai & Hwang, Buckley, Negi or Fuller
5. $\tilde{c}$ & $\tilde{A}$	Prob. 4	Lai & Hwang
6. $\tilde{c}$ & $\tilde{b}$		
7. $\tilde{A}$		

4.1 Possibilistic Linear Programming Models

Consider the following LP problems:

$$\max \quad \tilde{c}x$$
$$\text{s.t.} \quad \tilde{A}x \leq \tilde{b}, \text{ and } x \geq 0$$

where c, A and b may be imprecise with possibilistic distributions. Thus we have eight possible cases. Except for the crisp LP problem, seven possible problems are listed in Table 4.1 (or in Figure 1.4). Among those problems, let us first discuss the problem of imprecise resources only.

4.1.1 Linear Programming with Imprecise Resources and Technological Coefficients

Soyster [232] had developed an inexact programming that maximized an objective function, cx, and was subject to $x \in X = \{x \mid x_1 K_{i1} + x_2 K_{i2} + ... + x_n K_{in} \subset B_i, \forall i,$ and $x \geq 0\}$. K_{ij} and B_i, $\forall i,j$, are crisp sets. Thus, if one can figure out all K_{ij} and B_i (as some kinds of functions with parameters), then the inexact linear programming may be solved by standard linear programming or some existing nonlinear programming.

Consider the following PLP problem:

$$\max \quad cx$$
$$\text{s.t.} \quad \tilde{A}x \leq \tilde{b}, \text{ and } x \geq 0 \tag{4.1}$$

where $\tilde{A}$ and $\tilde{b}$ are (L-R) fuzzy numbers with possibilistic distributions (π). To solve Equation (4.1), Soyster's concept of inclusion treats the inequality constraints as a set-inclusion relationship. It is quite reasonable because the left and right hand sides are considered as fuzzy numbers which are fuzzy sets.

In this sub-section we will discuss three approaches proposed by Ramik and Rimanek [210], Tanaka, Ichihashi and Asai [240] and Dubois [72].

4.1.1.1 Ramik and Rimanek's Approach

Reconsider Equation (4.1) as:

$$\max \quad cx$$
$$\text{s.t.} \quad \tilde{a}_{i1}x_1 + \tilde{a}_{i2}x_2 + ... + \tilde{a}_{in}x_n \leq \tilde{b}_i, \quad i = 1, ..., m$$
$$x \geq 0 \tag{4.2}$$

where $\tilde{a}_{i1}$, $\tilde{a}_{i2}$, ... $\tilde{a}_{in}$, $\tilde{b}_i$, all i, are L-R fuzzy numbers here. Although there are many different expressions for the L-R fuzzy number (trapezoid fuzzy numbers in this particular case), let us use the tetrad expression, $\tilde{a} = (m, n, \alpha, \beta)$ or $\tilde{a} = (m+\alpha|m, n|n+\beta)$ where m is the left main value and n is the right main value with complete membership; α is the left spread and β is the right spread (see Figure 4.1). By use of the set-inclusion concept, Ramik and Rimanek obtained the following problem:

$$\max \quad cx$$
$$\text{s.t.} \quad \tilde{a}_{i1}x_1 \oplus \tilde{a}_{i2}x_2 \oplus \ .... \ \oplus \tilde{a}_{in}x_n \trianglelefteq \tilde{b}_i, \ \forall i$$
$$x \geq 0 \tag{4.3}$$

where $\tilde{a}_{i1}x_1 \oplus \tilde{a}_{i2}x_2 \oplus \ .... \ \oplus \tilde{a}_{in}x_n = (\Sigma_j m_{ij}x_j, \ \Sigma_j n_{ij}x_j, \ \Sigma_j \alpha_{ij}x_j, \ \Sigma_j \beta_{ij}x_j)$. To solve this possibilistic linear programming problem, the fuzzy inequality $\trianglelefteq$ should be defined.

To define this fuzzy inequality, we may use any ranking approach discussed in Chapter 2. However, let us only introduce Ramik and Rimanek's approach here.

Consider two supposed fuzzy numbers $\tilde{a}_{ij} = (m_{ij}, n_{ij}, \alpha_{ij}, \beta_{ij})$ and $\tilde{b}_i = (p_i, q_i, \tau_i, \delta_i)$ (see Figure 4.1). Ramik and Rimanek asserted:

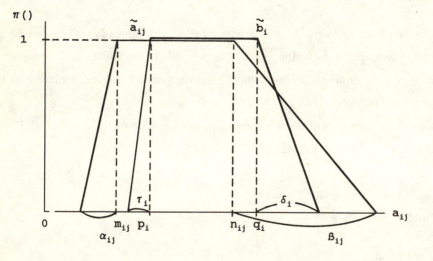

Figure 4.1 The representation of fuzzy inequality of two L-R fuzzy numbers a_{ij} and b_i

$$\tilde{a}_{ij} \leq \tilde{b}_i \text{ iff } m_{ij} \leq p_i, m_{ij} - \alpha_{ij} \leq p_i - \tau_i, n_{ij} \leq q_i, \text{ and } n_{ij} + \beta_{ij} \leq q_i + \delta_i. \quad (4.4)$$

Excluding the case of $x_1 = x_2 = \ldots = x_n = 0$, we obtain the following crisp linear programming problem from the facts of Equations (4.3) and (4.4):

$$\max \quad cx$$
$$\text{s.t.} \quad \Sigma_j m_{ij} x_j \leq p_i$$
$$\Sigma_j (m_{ij} - \alpha_{ij}) x_j \leq p_i - \tau_i$$
$$\Sigma_j n_{ij} x_j \leq q_i$$
$$\Sigma_j (n_{ij} + \beta_{ij}) x_j \leq q_i + \delta_i, \forall i$$
$$x \geq 0. \quad (4.5)$$

The optimal solution of Equation (4.5) can be considered as the solution of the possibilistic linear programming problem of Equation (4.2).

4.1.1.1a Example: A Profit Apportionment Problem

Consider the problem of profit apportionment in concern with mutual ownership [200]. The accounting problems of consolidated reporting can be divided into three broad categories:

a) elimination of reciprocal items, such as internal receivables and debts between the entities of the concern;

b) consolidation of nonreciprocal items, that is, aggregation of the resources and debts of the concern with respect to the outside world;

c) profit apportionment between concern majority and minority.

Ostermark assumed that points a) and b) have been solved according to some accepted consolidation procedures. The profit apportionment problem for a concern with mutual ownership is focused here.

Suppose that the concern structure of mutual ownership is shown as Figure 4.2. The figures along the arrows in Figure 4.2 represented the relative shares of the common stock owned directly by the numbers of the concern.

The separate net income was as follows:

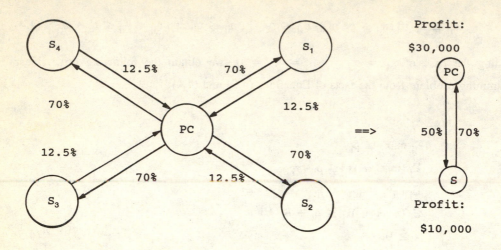

Figure 4.2 Concern structure (PC = Parent Company, S = Subsidiary)

```
        net income in PC:  $30,000
                     S₁:     1,000
                     S₂:     2,000
                     S₃:     3,000
                     S₄:     4,000
        ─────────────────────────────
        consolidated net profit  $40,000
```

net income in PC: $30,000

S_1: 1,000

S_2: 2,000

S_3: 3,000

S_4: 4,000

consolidated net profit $40,000

As indicated in the figures, each subsidiary owns 12.5% of the common stock in the parent company, whereas the parent company owns 70% of the common stock in each subsidiary. For each subsidiary, there is an external minority holding 30% of the shares. Since each subsidiary S_i (i = 1, ..., 4) owns 12.5% of the parent company, this fraction is affected by the mutual ownership relation of the concern. We can then obtain the following information:

a) Mutual holding matrix:

$$
\begin{array}{c}
\text{(owners)}
\end{array}
\quad
\begin{array}{c}
\text{(owned)} \\
\begin{array}{cc}
 \text{PC} & \text{S} \\
\end{array} \\
\begin{array}{c|cc}
\text{PC} & 0 & 0.7 \\
\text{S} & 4\text{x}.125 & 0 \\
\end{array}
\end{array}
\quad \Longrightarrow \quad
Q =
\begin{bmatrix}
0 & 0.7 \\
0.5 & 0
\end{bmatrix}
$$

b) Direct holdings:

$$
\begin{array}{c}
\text{(owners)}
\end{array}
\begin{array}{c}
 \\
\text{CI} \\
\text{MI}
\end{array}
\begin{array}{|cc|}
\hline
0.5 & 0 \\
0 & 0.3 \\
\hline
\end{array}
\qquad ==> \qquad R =
\begin{bmatrix}
0.5 & 0 \\
0 & 0.3
\end{bmatrix}
$$

with column headers (owned) PC   S

where CI = Concern Interest and MI = Minority Interest.

c) External profits:

$$
\begin{array}{c}
\text{(owners)}
\end{array}
\begin{array}{c}
 \\
\text{PC} \\
\text{S}
\end{array}
\begin{array}{|cc|}
\hline
30000 & 0 \\
0 & 10000 \\
\hline
\end{array}
\qquad ==> \qquad B =
\begin{bmatrix}
30000 & 0 \\
0 & 10000
\end{bmatrix}
$$

with column headers (owned) PC   S

First, Weil's basic formula for profit apportionment in concerns can be stated as:

$$Be = (I - Q)R^{-1}Xe$$

where I = identity matrix, X = profit apportionment matrix (unknown), e = a column vector with 1 in each position, and Q, R and B as indicated above. Obviously, we can easily obtain the solution: $(Xe)^T = [28460, 11540]$. That is: of the consolidated profit, 71.15% is apportioned to the majority interest and 28.85% to the minority interest.

Wallin introduced Zimmermann's fuzzy linear programming concept into the profit apportionment problem. The following model was then obtained:

$$
\begin{aligned}
\max \quad & \alpha_K + \alpha_M \\
\text{s.t.} \quad & R(I-Q)^{-1}Be + C\alpha \le Xe \\
& \alpha_K \text{ and } \alpha_M \in [0, 1]
\end{aligned}
$$

where the direction of the inequality in the first constraint depends on the form of the trade-off between majority and minority, and

$$C = \begin{pmatrix} c_K & 0 \\ 0 & c_M \end{pmatrix} \quad \text{and} \quad \alpha = [\alpha_K, \alpha_M]^T.$$

The objective function in this linear programming contains two scalars, one for the concern majority (α_K) and one for the external minority (α_M). The matrices R, Q, B and C are treated as constants: the future ownership relations (matrix R and Q), the expected adjusted net profits of the concern members (B) and profit aspirations (C) are assumed to be known. Thus, with α_K and α_M confined in the unit interval, the solution to the problem can be interpreted as a profit apportionment program, where majority and minority interests are simultaneously recognized.

Now, let us assume that $A = R(I-Q)^{-1}$, $w = Be$ and $b = Xe$. From the numeric data we obtain:

$$A = \begin{pmatrix} 0.7692 & 0.5385 \\ 0.2308 & 0.4616 \end{pmatrix}$$

and assume that:

$$C = \begin{pmatrix} -10000 & 0 \\ 0 & 3000 \end{pmatrix} \qquad b^T = [\ 25000,\ 12000].$$

The matrix C gives the trade-off intervals for the majority and minority interests, while the vector b contains the starting point for the trading. The following resulting problem is then obtained:

$$\max \quad \alpha_K + \alpha_M$$
$$\text{s.t.} \quad 0.7692w_1 + 0.5385w_2 - 10000\alpha_K \geq 25000$$
$$0.2308w_1 + 0.4616w_2 + 3000\alpha_M \leq 12000$$
$$\alpha_K \text{ and } \alpha_M \in [0, 1] \text{ and } w_1 \text{ and } w_2 \geq 0.$$

The sign of the elements in C, together with the direction of the inequalities, reveal that the problem formulation is based on the majority's (or concern management's) point of view. The first element in vector b, 25000, represents the minimum acceptable level of majority's profit share, whereas 12000 is the maximum tolerable level of minority's profit share. From majority's point of view, the best conceivable solution would be one yielding $\alpha_K = \alpha_M = 1$; since this would increase majority's profit share to the maximum value on the trade-off interval: 25000 +

10000 = 35000; while simultaneously minimizing minority's profit share: 12000 - 3000 = 9000. It is noted that the optimal shares of the majority and minority can be obtained directly by $K = 0.7692w_1 + 0.5385w_2$ and $M = 0.2308w_1 + 0.4616w_2$.

The management of the concern is uncertain of the maximum feasible profit share of the minority (b_M) and of the trade-off interval against the majority's profit share. Drawing upon data from the economic situation of the concern and recognizing issues of business policy and company law, management agrees upon the following estimation: from majority's point of view, the best feasible ceiling for minority's profit share is in the interval $b_M \in [12000, 20000]$, with left spread = 2000 and right spread = 3000. The best feasible trade-off against majority's interest is achieved by a trade-off interval between 3000 and 8000, with left spread = 1000 and right spread = 2000. Finally, we obtain the following possibilistic linear programming problem:

$$\text{max} \quad \alpha_K + \alpha_M$$
$$\text{s.t.} \quad 0.7692w_1 + 0.5385w_2 - 10000\alpha_K \geq 25000$$
$$(0.2308,0.2308,0,0)w_1 + (0.4616,0.4616,0,0)w_2 + (3000,8000,1000,2000)\alpha_M$$
$$\leq (12000,20000,2000,3000)$$
$$\alpha_K \text{ and } \alpha_M \in [0, 1] \text{ and } w_1 \text{ and } w_2 \geq 0.$$

By using Equation (4.5), the above possibilistic linear programming can then be transferred as follows:

$$\text{max} \quad \alpha_K + \alpha_M$$
$$\text{s.t.} \quad 0.7692w_1 + 0.5385w_2 - 10000\alpha_K \geq 25000$$
$$0.2308w_1 + 0.4616w_2 + 3000\alpha_M \leq 12000$$
$$0.2308w_1 + 0.4616w_2 + 2000\alpha_M \leq 10000$$
$$0.2308w_1 + 0.4616w_2 + 8000\alpha_M \leq 20000$$
$$0.2308w_1 + 0.4616w_2 + 6000\alpha_M \leq 17000$$
$$\alpha_K \text{ and } \alpha_M \in [0, 1] \text{ and } w_1 \text{ and } w_2 \geq 0.$$

The optimal solution is $\alpha_K^* = 0.1662$ and $\alpha_M^* = 1$. According to this optimal program, the parent company should only show profit. Thus the optimal profit apportionment to the problem is: majority's profit share = $0.7692w_1 = 26662$ and minority's profit share = $0.2308w_1 = 8000$.

4.1.1.2 Tanaka, Ichihashi and Asai's Approach

Tanaka, Ichihashi and Asai [240] assumed that the fuzzy coefficient ($\tilde{a}_{ij}$ or $\tilde{b}_i$) is a symmetric triangular possibilistic distribution as follows (see Figure 4.3 also):

$$\pi_{aij}(a_{ij}) = \begin{cases} 1 - 2(a_{ij}-a_{ij}^l)/(a_{ij}^u-a_{ij}^l) & \text{if } a_{ij}^l \leq a_{ij} \leq a_{ij}^l+(a_{ij}^u-a_{ij}^l)/2 \\ 1 - 2(a_{ij}^u-a_{ij})/(a_{ij}^u-a_{ij}^l) & \text{if } a_{ij}^l+(a_{ij}^u-a_{ij}^l)/2 \leq a_{ij} \leq a_{ij}^u \\ 0 & \text{otherwise} \end{cases}$$

or

$$\pi_{aij}(a_{ij}) = \begin{cases} 1 - |2a_{ij}-(a_{ij}^u+a_{ij}^l)| /(a_{ij}^u-a_{ij}^l) & \text{if } a_{ij}^l \leq a_{ij} \leq a_{ij}^u \\ 0 & \text{otherwise} \end{cases} \tag{4.6}$$

where a_{ij}^u and a_{ij}^l denote the upper and lower limits of the 0-level set a_{ij}, respectively. Let $\tilde{a}_{ij} = (a_{ij}^l, a_{ij}^u)$ denote this type of symmetric triangular possibilistic distribution.

Consider the following fuzzy linear function:

$$y_i = \tilde{a}_{i1}x_1 + \tilde{a}_{i2}x_2 + \ldots + \tilde{a}_{in}x_n = \tilde{a}_i x.$$

The possibilistic distribution of y_i can be obtained as:

$$\pi_{yi}(y_i) = \begin{cases} 1 & \text{if } x = 0 \text{ and } y = 0 \\ 1 - |2y_i-(a_i^u+a_i^l)x| /(a_i^u-a_i^l)x & \text{if } x \neq 0 \\ 0 & \text{if } x = 0 \text{ and } y \neq 0 \end{cases} \tag{4.7}$$

where $a_i^u = [a_{i1}^u, \ldots, a_{in}^u]$ and $a_i^l = [a_{i1}^l, \ldots, a_{in}^l]$.

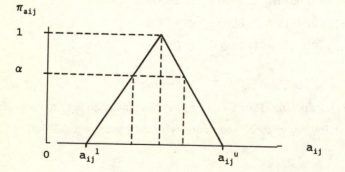

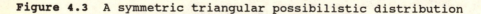

Figure 4.3 A symmetric triangular possibilistic distribution

From Equation (4.7), the α-level set of the fuzzy set $\tilde{y}_i$ $(\pi_{yi}(y_i) \geq \alpha)$ is the following interval:

$$R_\alpha(y_i) = [\{(\alpha-1)(a_i{}^u-a_i{}^l)x+(a_i{}^u+a_i{}^l)x\}/2, \{(1-\alpha)(a_i{}^u-a_i{}^l)x+(a_i{}^u+a_i{}^l)x\}/2].$$

Since $x \geq 0$, we obtain:

$$R_\alpha(y_i) = [\{(\alpha/2)a_i{}^u+(1-\alpha/2)a_i{}^l\}x, \{(1-\alpha/2)a_i{}^u+(\alpha/2)a_i{}^l\}x]. \tag{4.8}$$

Thus the possibility of y_i is easily calculated in the case of a linear function.

Based on the extension principle, Tanaka et al. defined "b_i is greater than y_i" as "$b_i \geq y_i$ $<=>$ max $[b_i, y_i] = b_i$." Thus they considered the problem of Equation (4.1) as:

$$\begin{aligned} \text{max} \quad & cx \\ \text{s.t.} \quad & \tilde{y}_i = \tilde{a}_i x \leq_\alpha \tilde{b}_i, \forall i \end{aligned} \tag{4.9}$$

where α is specified by the decision maker a priori and $\tilde{y}_i = \tilde{a}_i x \leq^\alpha \tilde{b}_i$ if $(y_i)_h{}^u \leq_\alpha (b_i)_h{}^u$ and $(y_i)_h{}^l \leq (b_i)_h{}^l$ hold for h ϵ [α, 1] as in Figure 4.4. The upper bound of $(b_i)_\alpha$ should be greater than or equal to the upper bound of $(y_i)_\alpha$, and the lower bound of $(b_i)_\alpha$ should be less than or equal to the lower bound of $(y_i)_\alpha$. Here, level α corresponds to the degree of optimism of the decision maker.

By using the above results, Equation (4.1) might be equivalent to the following auxiliary problem:

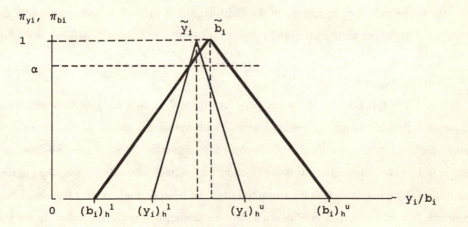

Figure 4.4 Order relation between fuzzy sets $\tilde{y}_i \leq_\alpha \tilde{b}_i$

$$\text{max} \quad cx$$

$$\text{s.t.} \quad \{(\alpha/2)a_i^u + (1-\alpha/2)a_i^l\}x \leq (\alpha/2)b_i^u + (1-\alpha/2)b_i^l, \forall i$$

$$\{(1-\alpha/2)a_i^u + (\alpha/2)a_i^l\}x \leq (1-\alpha/2)b_i^u + (\alpha/2)b_i^l, \forall i$$

$$x \geq 0 \qquad\qquad\qquad\qquad\qquad\qquad\qquad (4.10)$$

where α is given a priori by the decision maker.

Example. Consider the following simplified possibilistic linear programming problem:

$$\text{max} \quad 25x_1 + 18x_2$$

$$\text{s.t.} \quad (12, 18)x_1 + (32, 36)x_2 \leq_\alpha (750, 850)$$

$$(19, 21)x_1 + (7, 13)x_2 \leq_\alpha (380, 480)$$

$$x_1 \text{ and } x_2 \geq 0.$$

Assume that $\alpha = 0.4$ is given. The equivalent crisp linear programming is then:

$$\text{max} \quad 15x_1 + 18x_2$$

$$\text{s.t.} \quad 16.8x_1 + 35.2x_2 \leq 830$$

$$13.2x_1 + 32.8x_2 \leq 770$$

$$20.6x_1 + 11.8x_2 \leq 460$$

$$19.4x_1 + 8.2x_2 \leq 400$$

$$x_1 \text{ and } x_2 \geq 0.$$

The optimal solution is: $x^* = (12.14, 17.78)$ and the optimal value is about 444. In this problem, the upper values of b_1 and b_2 on the right-hand side provide restrictions, while there are leeways for the lower values of b_1 and b_2.

4.1.1.3 Dubois's Approach

Dubois [72] expressed the fuzzy numbers (in Figure 4.1) as fuzzy intervals which carry at once several possible ranges of values together with associated levels of confidence. The interval of $[m_{ij}, n_{ij}]$ with possibility $= 1$ is the most uncertain interval and the interval of $[m_{ij}+\alpha_{ij}, n_{ij}+\beta_{ij}]$ with possibility $= 0$ is the boundary of certain intervals. Thus, the degree of the possibility can be interpreted as the compliment of the level of confidence. By use of the α-level concept, $(a_{ij})_\alpha$ has a level of confidence $1 - \alpha$: the smaller α, the more certain we are about x taking its values in $(a_{ij})_\alpha$. Dubois [72] stated:

"The use of fuzzy intervals does not presuppose the existence of statistical data (although statistical interpretations do exist). They are manipulated by means of lattice operations, and do not rely upon an additive structure. Fuzzy intervals are a nature model of soft restrictions of parameter values, as specified by experts. The calculation with fuzzy intervals is a direct multiple-valued extension of sensitivity analysis; hence we may expect results which are less precise than what a probabilistic analysis would yield. But this loss in precision is balanced by a gain in computational efficiency: the required amount of calculation for fuzzy optimization problem-solving is often significantly less than that required for the same problem under a stochastic approach."

Let us consider a more general fuzzy constraint as follows (equality constraint):

$$f(x, \tilde{a}) = g(x, \tilde{b}) \tag{4.11}$$

where $f(x, \tilde{a}) = \sum_{i=1}^{n} \tilde{a}_i x_i + \tilde{a}_{n+1}$ and $g(x, \tilde{b}) = \sum_{i=1}^{n} \tilde{b}_i x_i + \tilde{b}_{n+1}$.

The respective ranges $\tilde{A}$ and $\tilde{B}$ of $\tilde{a}$ and $\tilde{b}$ are supposed to be separable, i.e., are Cartesian products of fuzzy intervals $\tilde{A}_i$ and $\tilde{B}_i$, $i = 1, 2, ..., n+1$, defined by:

$$\pi_A = \min_{i=1}^{n+1} \pi_{Ai} \text{ and } \pi_B = \min_{i=1}^{n+1} \pi_{Bi}. \tag{4.12}$$

This assumption means that the variables a_i (and b_i) are not linked. The fuzzy intervals A_i and B_i are represented by means of the L-R representation as in Section 4.1.1:

$$\tilde{A}_i = (m_i, n_i, \alpha_i, \beta_i) \text{ and } \tilde{B}_i = (p_i, q_i, \Gamma_i, \delta_i).$$

Now, $g(x,\tilde{B})$ can be expressed by means of fuzzy addition and scale multiplication:

$$g(x,\tilde{B}) = \tilde{B}_1 x_1 \oplus ... \oplus \tilde{B}_{n+1} x_{n+1} = \tilde{B}x = (px, qx, \Gamma x, \delta x).$$

where $x_{n+1} = 1$. Similarly, $f(x, \tilde{A})$ can be also expressed in the same way.

For the fuzzy constraints of Equation (4.11), Dubois provided two cases of "soft" and "hard" equivalent constraints. First, let us discuss the "soft constraints."

(a) Soft constraints can be expressed by the α-weak feasibility (α-WF) of x. That is:

$$WF(x) = Poss(f(x,\tilde{a}) = g(x,\tilde{b})) = \pi(\tilde{A}x, \tilde{B}x) \geq \alpha \tag{4.13}$$

where α is a required possibility threshold. Furthermore, Dubois obtained the following results (see Figure 4.5):

$$\text{x is } \alpha\text{-weak feasible} \quad \text{iff} \quad \pi_{[Bx,\infty)}^{-1}(\alpha) \leq \pi_{(-\infty,Ax]}^{-1}(\alpha) \text{ and}$$

$$\pi_{[Ax,\infty)}^{-1}(\alpha) \leq \pi_{(-\infty,Bx]}^{-1}(\alpha) \tag{4.14}$$

where $\pi_{[Ax,\infty)}^{-1}(\alpha) = \Sigma_i \ x_i(m_i - \alpha_i L_i(.))$.

Similar expressions hold for $\pi_{[Bx,\infty)}^{-1}(\alpha)$, $\pi_{(-\infty,Ax]}^{-1}(\alpha)$ and $\pi_{(-\infty,Bx]}^{-1}(\alpha)$. Thus, the weak feasibility constraints WF(x) $\geq \alpha$ are equivalent to two linear inequality constraints which express Equation (4.14):

$$\sum_{i=1}^{n+1} (p_i - \Gamma_i L'^{-1}_i(\alpha))x_i \leq \sum_{i=1}^{n+1} (n_i - \beta_i R_i^{-1}(\alpha))x_i \tag{4.15}$$

$$\sum_{i=1}^{n+1} (m_i - \alpha_i L_i^{-1}(\alpha))x_i \leq \sum_{i=1}^{n+1} (q_i - \delta_i R'^{-1}_i(\alpha))x_i. \tag{4.16}$$

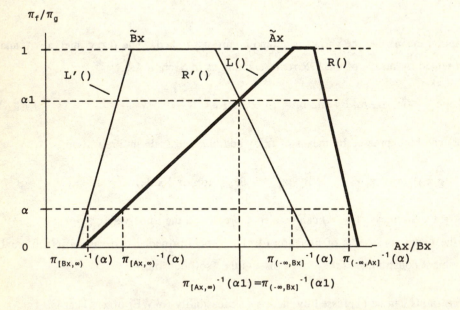

Figure 4.5 Soft constraint using possibility index

However, we may have the following possibility linear programming problem:

$$\min \quad cx$$
$$\text{s.t.} \quad \tilde{A}x = \tilde{b} \text{ and } x \geq 0 \tag{4.17}$$

where $\tilde{b} = (p, \infty, \Gamma, \infty)$ and $\tilde{A}x$ are the same as previously described. Equation (4.15) is only meaningful and Equation (4.17) is equivalent to:

$$\min \quad cx$$
$$\text{s.t.} \quad p - \Gamma^{-1}L(\alpha) \leq \Sigma_i (n_i + \beta_i R_i^{-1}(\alpha))x_i.$$
$$x \geq 0. \tag{4.18}$$

It is noted that in PLP problem $\tilde{b}$ is often (p, q, Γ, δ), instead of $(p, \infty, \Gamma, \infty)$. Thus Equations (4.15) and (4.16) are often considered at the same time. The original equality is then substituted by two inequality constraints.

(b) A hard constraint is considered as a fuzzy extension of the set-inclusive constraint $\tilde{A}x \subset \tilde{B}x$. x is said to be α-hard feasible iff:

$$SF(x) = Nes(g(x,\tilde{b}) = f(x,\tilde{a}) \mid \tilde{a}) = N_{f(x,A)}(g(x,\tilde{B})) \geq \alpha$$

that is (see Figure 4.6):

$$\pi_{[Bx,\infty)}^{-1}(\alpha) \leq \pi_{]Ax,\infty)}^{-1}(1-\alpha) \text{ and } \pi_{(-\infty,Ax[}^{-1}(1-\alpha) \leq \pi_{(-\infty,Bx]}^{-1}(\alpha) \tag{4.19}$$

Finally, Dubois obtained the following linear constraints:

$$\sum_{i=1}^{n+1} [p_i - \Gamma_i L'_i^{-1}(\alpha)]x_i \leq \sum_{i=1}^{n+1} [m_i - \alpha_i L_i^{-1}(1-\alpha)]x_i \tag{4.20}$$

$$\sum_{i=1}^{n+1} [n_i - \beta_i R_i^{-1}(1-\alpha)]x_i \leq \sum_{i=1}^{n+1} [q_i - \delta_i R'_i^{-1}(\alpha)]x_i. \tag{4.21}$$

When $\alpha_i = \beta_i = \Gamma_i = \delta_i = 0$, Soyster's set-inclusive constraints of the form $x^T\tilde{A} \subset x^T\tilde{B}$ are clearly recovered.

Consider the following problem:

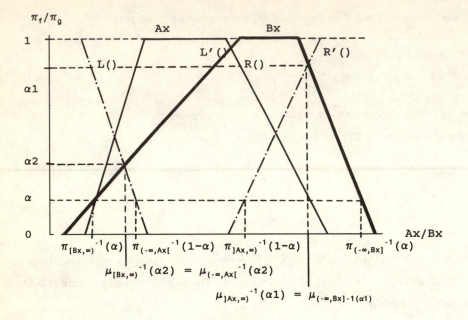

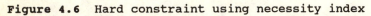

Figure 4.6 Hard constraint using necessity index

$$\min \quad cx$$
$$\text{s.t.} \quad \tilde{A}x \leq \tilde{b} \text{ and } x \geq 0 \tag{4.22}$$

where $\tilde{b} = (p, \infty, \Gamma, \infty)$ and $\tilde{A}x$ is the same as previously described. By using hard constraints, we obtain:

$$\min \quad cx$$

$$\text{s.t.} \quad p - \Gamma^{-1}L(\alpha) \leq \sum_{i=1}^{n+1} (m_i + \alpha_i L_i^{-1}(1-\alpha))x_i$$

$$x \geq 0. \tag{4.23}$$

Obviously, if the decision maker provides α value, the problems of Equations (4.18) and (4.23) will become crisp linear programming problems.

4.1.2 Linear Programming with Imprecise Objective Coefficients

In practice, the objective coefficients are always imprecise, such as the unit costs/profits of new products or new projects, the lending and borrowing interest rates and cash flows. Assume that this imprecise nature can be modelled by the current possibility theory and that the objective coefficients have their individual possibility distributions. Thus, we cannot directly solve this kind of possibilistic linear programming problem by use of any existing linear programming technique. Auxiliary programming should be first obtained.

This problem has been intensively discussed since 1987. In the following, we would like to present the three approaches of Lai and Hwang [138], Rommelfanger, Hanuscheck and Wolf [216], and Delgado, Verdegay and Vila [56].

4.1.2.1 Lai and Hwang's Approach

In this section, we will first discuss a linear programming problem with imprecise coefficients which can be stated as (Lai and Hwang [138]):

$$\max \quad \Sigma_i \tilde{c}_i x_i$$
$$\text{s.t.} \quad x \in X = \{x \mid Ax \le b \text{ and } x \ge 0\} \tag{4.24}$$

where $\tilde{c}_i = (c_i^m, c_i^p, c_i^o)$, all i, are imprecise unit profits and have triangular possibility distributions as in Figure 4.7. c_i^m is the most possible value, c_i^p is the most pessimistic value and c_i^o is the most optimistic value. When normalized, $\pi_i(c_i^m) = 1$ and $\pi_i(c_i^p) = \pi_i(c_i^o) = 0$. Now, let us reconsider Equation (4.24) as:

$$\max_{x \in X} \quad \Sigma_i (c_i^m x_i, c_i^p x_i, c_i^o x_i)$$

or

$$\max_{x \in X} \quad (c^m x, c^p x, c^o x) \tag{4.25}$$

where $c^m = (c_1^m, c_2^m, \dots c_n^m)$, $c^p = (c_1^p, c_2^p, \dots c_n^p)$ and $c^o = (c_1^o, c_2^o, \dots c_n^o)$. The objective function is actually an imprecise objective function with a triangular possibility distribution $(c^m x, c^p x, c^o x)$.

To solve Equation (4.25), Tanaka, Ichihashi and Asai [240] provided the weighted average of the upper and lower limits and then substituted this average into the problem. This results in the following problem:

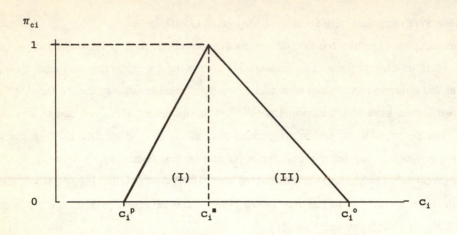

Figure 4.7 The triangular possibility distribution of c_i

$$\max_{x \in X} \quad \sum_i [w_i c_i^p + (1-w_i)c_i^o]x_i.$$

$$(4.26)$$

We have provided a most likely solution for a multiple objective linear programming problem with imprecise objective coefficients. This most likely solution is found by solving the following auxiliary linear programming problem of Equation (4.25):

$$\max_{x \in X} \quad [(4c^m + c^p + c^o)/6]x.$$

$$(4.27)$$

If the minimal acceptable possibility, α, is given (here, we do not imply the α-level set which is a crisp set), we will obtain the following problem:

$$\max_{x \in X} \quad [(4c^m + c^p_\alpha + c^o_\alpha)/6]x$$

$$(4.28)$$

where c^p_α and c^o_α are the most pessimistic and the most optimistic value of the acceptable events, respectively. It is noted that the possibilities of the acceptable events (the events with the possibility $\geq \alpha$) are kept, because we do not like to lose any valuable information at this stage. Keeping the possibility is important, especially when an interactive procedure is used. In regard to the solution procedure, the acceptable events with their possibilities are correctly presented by these three critical points. Therefore, the difference between Equations (4.26) and (4.27) is that Equation (4.27) includes the most possible value. Thus, we will have different solutions for different possibilistic distributions, as in Figure 4.8. Equation (4.26) will provide the same solution for the situation shown in Figure 4.8, where $\tilde{c}x$ should be preferred to $\tilde{c}'x$ if the minimal acceptable degree is α and the possibility distributions are considered. Of course, we might also like to change the weights among c^m, c^p_{α} and c^o_{α}, respectively.

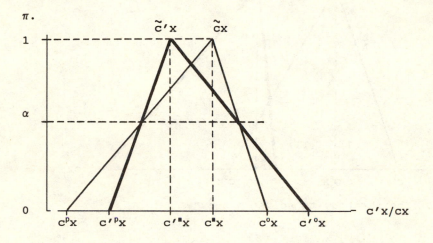

Figure 4.8 $\tilde{c}x$ is preferred to $\tilde{c}'x$ if the minimal accepted grade of possibility is α

On the other hand, we may reconsider the fuzzy objective function with a triangular distribution. This fuzzy objective is fully defined by three corner points $(c^mx, 1)$, $(c^px, 0)$ and $(c^ox, 0)$, geometrically. Thus, maximizing the fuzzy objective can be obtained by pushing these three critical points in the direction of the right-hand side. Fortunately, the vertical coordinates of the critical points are fixed at either 1 or 0. The only considerations then are the three horizontal coordinates. Therefore, our problem is to solve:

$$\max_{x \in X} \quad (c^mx, c^px, c^ox)$$

where (c^mx, c^px, c^ox) is the vector of three objective functions, c^mx, c^px and c^ox. In order to keep the triangular shape (normal and convex), of the possibility distribution, it is necessary to make a small change. Instead of maximizing these three objectives simultaneously, we are going to maximize c^mx, minimize $[c^mx - c^px]$ and maximize $[(c^ox - c^mx]$, where the last two objective functions are actually relative measures from c^mx, the first objective function (see Figure 4.9).

The three new objectives also guarantee the previous argument of pushing the triangular possibility distribution in the direction of the right-hand side.

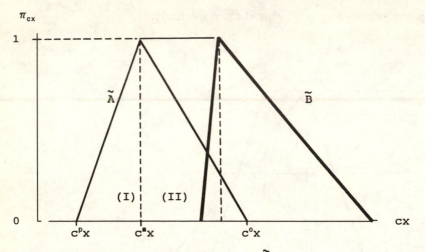

Figure 4.9 The strategy to solve "max $\tilde{c}x$"

This leads us to the following auxiliary problem of Equation (4.25):

$$\min \quad z_1 = (c^m - c^p)x$$
$$\max \quad z_2 = c^m x$$
$$\max \quad z_3 = (c^o - c^m)x$$
$$\text{s.t.} \quad x \in X. \tag{4.29}$$

The crisp MOLP of Equation (4.29) is equivalent to maximizing the most possible value of the imprecise profit (at the point of possibility degree $= 1$). At the same time, we have minimized the inferior side of the possibility distribution. This means minimizing the region (I) which, in our opinion, is equivalent to "the risk of obtaining lower profit." And, we have also maximized the region (II) of the possibility distribution, which is equivalent to "the possibility of obtaining higher profit." As in Figure 4.9, we would prefer the possibility distribution of $\tilde{B}$ to that of $\tilde{A}$. In portfolio theory, maximizing mean return, minimizing variance and maximizing skewness are often considered [BM9], while the considered return has a probability distribution. Clearly, our strategy is essentially analogous to that used in portfolio theory.

To solve Equation (4.29) we may use any MOLP technique (Hwang and Masud [BM19]) such as utility theory, goal programming, fuzzy programming or interactive approaches. However, in this study we suggest using Zimmermann's fuzzy programming method [274] with our normalization process.

First we should get the following Positive Ideal Solutions (PIS) and Negative Ideal Solutions (NIS) of the objective functions (Hwang and Yoon [BM20] and Lai and Hwang [143]):

$$z_1^{PIS} = \min_{x \in X} (c^m - c^p)x, \qquad z_1^{NIS} = \max_{x \in X} (c^m - c^p)x$$

$$z_2^{PIS} = \max_{x \in X} c^m x, \qquad z_2^{NIS} = \min_{x \in X} c^m x$$

$$z_3^{PIS} = \max_{x \in X} (c^o - c^m)x, \qquad z_3^{NIS} = \min_{x \in X} (c^o - c^m)x. \qquad (4.30)$$

The linear membership function of these objective functions can now be computed (see Figure 4.10) as:

$$\mu_{z1} = \begin{cases} 1 & \text{if } z_1 < z_1^{PIS} \\ (z_1^{NIS} - z_1)/(z_1^{NIS} - z_1^{PIS}) & \text{if } z_1^{PIS} \le z_1 \le z_1^{NIS} \\ 0 & \text{if } z_1 > z_1^{NIS} \end{cases} \qquad (4.31)$$

$$\mu_{z2} = \begin{cases} 1 & \text{if } z_2 > z_2^{PIS} \\ (z_2 - z_2^{NIS})/(z_2^{PIS} - z_2^{NIS}) & \text{if } z_2^{PIS} \le z_2 \le z_2^{NIS} \\ 0 & \text{if } z_2 < z_2^{NIS} \end{cases} \qquad (4.32)$$

and μ_{z3} is similar to μ_{z2}. This normalization has also been applied by Seo and Sakawa [BM45]

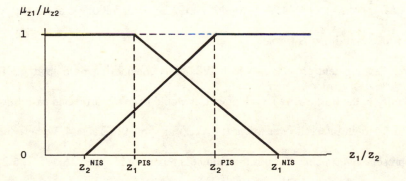

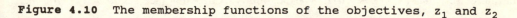

Figure 4.10   The membership functions of the objectives, z_1 and z_2

Finally, we solve the following Zimmermann's equivalent single-objective linear programming model:

$$\max \quad \alpha$$
$$\text{s.t.} \quad \mu_{zi}(x) \geq \alpha, \text{ i=1,2,3, and x } \epsilon \text{ X.} \tag{4.33}$$

The optimal solution of Equation (4.33) will provide a satisfying solution under the previous strategy of minimizing the risk of lower profit, and maximizing the most possible value and the possibility of higher profit.

For minimization linear programming (minimize the total cost), we will have the following equivalent problem:

$$\max \quad (c^m - c^p)x$$
$$\min \quad c^m x$$
$$\min \quad (c^o - c^m)x$$
$$\text{s.t.} \quad x \epsilon X \tag{4.34}$$

which can be solved by using a similar approach (as previously described).

Now, let us look at the following simplified investment problem from Gould, Eppen and Schmidt [BM13] with some modifications.

4.1.2.1a Example: A Winston-Salem Development Management Problem

Winston-Salem Development Management (WSDM) is trying to complete its investment plans for the next 3 years. Currently, WSDM has $2 million available for investment. At 6-month intervals over the next 3 years, WSDM expects the income stream from previous investments (in millions), as shown in Table 4.2. In this particular case, $300,000 at the end of the third year will not play any important role.

There are three development projects in which WSDM is considering participating. If WSDM participated fully in the Foster City Development, it would have the projected cash flow stream (in millions) at 6-month intervals over the next 3 years as shown in Table 4.2 where the negative numbers represent investments, and positive numbers represent income. (5.5, 5.1, 6.2) is an imprecise number and has a triangular possibility distribution (see Figure 4.11) with the

Table 4.2 The multiple project and multiple period investment problem (in million dollars)

period (year)	0	0.5	1	1.5	2	2.5	3
Income Stream	2.0	0.5	0.4	0.38	0.36	0.34	0.3
Project 1	-3.0	-1.0	-1.8	0.4	1.8	1.8	(5.5,5.1,6.2)
Project 2	-2.0	-0.5	1.5	1.5	1.5	0.2	(-1,-1.2,.85)
Project 3	-2.0	-2.0	-1.8	1.0	1.0	1.0	(6.0,5.0,6.5)

Note.
1) Total outstanding principle ≤ $2.0 million.
2) At most, $2.0 million can be borrowed at one time.
3) Invest surplus funds at 4% per half year.
4) The borrowing interest rate is 6% per half year.

most possible value = 5.5, the most pessimistic value = 5.1 and the most optimistic value = 6.2. A second project involves taking over the operation of some old, lower-middle income housing on the condition that certain initial repairs to it be made and that it be demolished at the end of 3 years. The third project is to invest in the Disney-Universe Hotel. The cash flow streams for these two projects, if fully participated in, are also shown in Table 4.2. WSDM can borrow money for half-year intervals at 6% interest per half-year. At most, $2 million can be borrowed at one time; that is, the total outstanding principle can never exceed $2 million. WSDM can invest surplus funds at 4% per half-year.

Consider the problem of maximizing WSDM's net worth at the end of 3 years. Disregard taxes, and assume that if WSDM participates in a project at less than 100%, all the cash flows of that project are reduced proportionately. Then this problem can be formulated as a linear programming model. In order to do this, let:

F = fractional participation in Foster City

M = fractional participation in Lower-Middle

D = fractional participation in Disney-Universe

B_i = amount borrowed in period i, i = 1, ..., 6

L_i = amount loaned in period i, i = 1, ..., 6

Z = net worth after the six periods

(without considering $300,000)

The linear programming model can be formulated as (in millions):

max $Z = (5.5,5.1,6.2)F+(-1.0,-1.2,-0.85)M+(6.0,5.0,6.5)D-1.06B_6+1.04L_6$

s.t. $3F+2M+2D-B_1+L_1 \leq 2$

$1F+0.5M+2D+1.06B_1-1.04L_1-B_2+L_2 \leq 0.5$

$1.8F-1.5M+1.8D+1.06B_2-1.04L_2-B_3+L_3 \leq 0.4$

$-0.4F-1.5M-1D+1.06B_3-1.04L_3-B_4+L_4 \leq 0.38$

$-1.8F-1.5M-1D+1.06B_4-1.04L_4-B_5+L_5 \leq 0.36$

$-1.8F-0.2M-1D+1.06B_5-1.04L_5-B_6+L_6 \leq 0.34$

$B_i \epsilon [0, 2], i = 1, 2, 3, 4, 5, 6,$

$F, M \text{ and } D \epsilon [0, 1].$ (4.35)

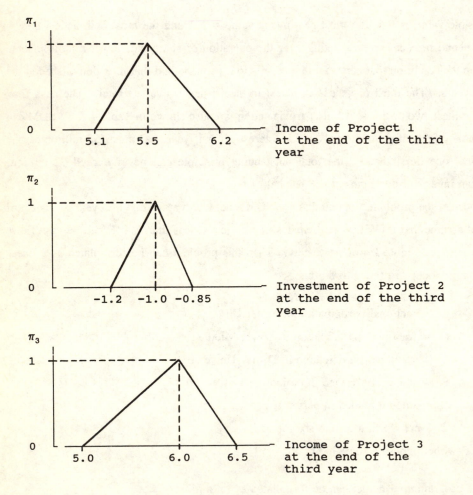

Figure 4.11 The possibilistic distributions of the cash flows of three projects at the end of the third year

It is noted that the most pessimistic and optimistic values should be reversed when the coefficients are negative for the maximization problem. Equation (4.35) can then be solved by the following auxiliary problem:

$$\begin{aligned} \min \quad & 0.4F-0.2M+D \\ \max \quad & 5.5F-1.0M+6.0D-1.06B_6+1.04L_6 \\ \max \quad & 0.7F-0.15M+0.5D \\ \text{s.t.} \quad & x \in X \end{aligned} \qquad (4.36)$$

where X is defined as the constraint of Equation (4.35).

First, we obtain the PIS and NIS of Equation (4.36) which are (0.2, 9.00, 0.58) and (0.84, 3.12, 0.15), respectively. Then, according to Equations (4.31), (4.32) and (4.33), the solution is obtained and shown in Table 4.3. Furthermore, Figure 4.12 shows the possibilistic distribution of the imprecise profit obtained. The upper and lower bounds of the profits of the problem (8.72, 7.91), and the most likely profit (8.20) are obtained.

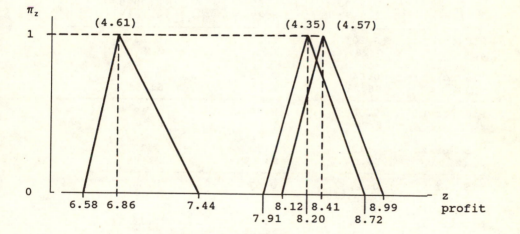

Figure 4.12   The possibilistic distributions of the optimal profits of Equations (4.35), (4.57) and (4.61)

It is worthy noting that the possibilistic linear programming problem (PLP) studied here is different from some fuzzy linear programming (FLP) problems in meaning. The FLP is based on the preferred concept to establish membership functions. On the other hand, the PLP is based on the degree of the occurrence of an event to obtain possibilistic distributions. In other

Table 4.3 Solution of Equations (4.35), (4.57) and (4.61)

Variable	Solution of Equation (4.35)	Solution of Equation (4.57) at $\beta = 0.5$	Solution of Equation (4.61) at $\beta = 0.5$
α	0.8634	0.8861	0.9149
F	0.7642	0.7254	0.7935
M	0.0914	0.0365	0.1650
D	0.0000	0.0000	0.0000
B_1	0.4755	0.2491	0.7106
L_1	0.0000	0.0000	0.0000
B_2	2.0000	2.0000	1.1292
L_2	1.1860	1.4924	0.0000
B_3	1.7251	1.4188	2.0000
L_3	0.0000	0.0000	0.0000
B_4	0.0000	0.0000	1.1851
L_4	0.2276	0.7736	0.0000
B_5	0.0000	0.0000	0.0000
L_5	2.1094	2.5280	0.7667
B_6	0.0000	2.0000	2.0000
L_6	3.9277	6.2983	4.5818
PIS for z_1	0.20	0.23	0.23
z_1^*	0.287	0.291	0.287
NIS for z_1	0.84	0.85	0.84
$\mu_{z1}(z_1^*)$	86.3%	90.3%	91.5%
PIS for z_2	9.00	9.09	7.19
z_2^*	8.20	8.41	6.86
NIS for z_2	3.12	3.13	3.12
$\mu_{z2}(z_2^*)$	83.0%	88.6%	91.9%
PIS for z_3	0.58	0.63	0.64
z_3^*	0.521	0.577	0.521
NIS for z_3	0.15	0.16	0.16
$\mu_{z3}(z_3^*)$	86.3%	88.6%	87.7%

The optimal profit is imprecise and has a triangular possibility distribution of $(z_2^*, z_2^*-z_1^*, z_2^*+z_3^*)$.
That is -- Equation (4.35): (8.20, 7.91, 8.72)
Equation (4.57): (8.41, 8.12, 8.99)
Equation (4.61): (6.86, 6.58, 7.44).

words, the FLP is an analogous of utility approaches, and the PLP is an analogous of probability methods. However, after transferring the possibilistic linear programming problem to an auxiliary crisp MOLP problem, we then use Zimmermann's fuzzy programming (preference-based membership functions of the objective functions) to obtain the compromise solution. Without using the β-level set concept, we can easily solve the imprecise objective function and obtain the whole possibilistic distribution of the satisfying objective value. Furthermore, we can solve rather general, imprecise problems through an interactive process with the decision maker. Three critical values (the most possible, the pessimistic and the optimistic values) are only necessary for each imprecise object. It is very important that the satisfying objective value should be imprecise since the unit profits/costs are imprecise and this nature is always existing. Furthermore, our solution has the nature of minimizing the possibility of lower profit, maximizing the most possible profit and maximizing the possibility of higher profit. As a matter of fact, with little effort, we really obtain a lot of information about the solution. Thus we do believe that our possibilistic linear programming technique is more efficient to solve the problems with imprecise nature in the real world.

4.1.2.2 Rommelfanger, Hanuscheck and Wolf's Approach

The imprecise objective coefficients of the linear programming problem discussed here, is assumed as follows:

$$\max \quad \tilde{c}x$$
$$\text{s.t.} \quad x \in X = \{x \mid Ax \leq b \text{ and } x \geq 0\} \tag{4.37}$$

where $\tilde{c} = [c^L, c^U]$ is an interval-valued coefficient vector. Analogously to the theory of Multiple Objective Linear Programming (MOLP), the complete solution of the linear optimization problem with interval-valued coefficients of Equation (4.37) is defined as:

$$E = \{x \in X \mid \text{there is no } x' \in X \text{ such that } cx' \geq cx \text{ for all } c \in [c^L, c^U] \text{ and}$$
$$cx' > cx \text{ for at least one } c \in [c^L, c^U]\}.$$

To solve Equation (4.37), let us first define:

$$z_{min}(x) = c^L x \text{ and } z_{max}(x) = c^U x$$

which are two extreme cases. Rommelfanger, Hanuscheck and Wolf [216] reduced the infinite objective functions of Equation (4.37) to search a solution of the following MOLP problem:

$$\max_{x \in X} \ [z_{min}(x), z_{max}(x)].$$

(4.38)

Rommelfanger et al. pointed out that this reduction of the objective function is also connected to a reduction of the set of feasible compromise solutions since the complete solution of Equation (4.38) is a mostly proper subset of E. By this procedure, the set of compromise solution, however, is not restricted to the corner points of simplex X.

To solve Equation (4.38), we can then use any MODM technique. Here, we use Zimmermann's [274] approach and first find:

$$z_{min}^* = z_{min}(x_{min}^*) = \max_{x \in X} z_{min}(x)$$

$$z_{max}^* = z_{max}(x_{max}^*) = \max_{x \in X} z_{max}(x), \text{ and}$$

$$z_{min}' = z_{min}(x_{max}^*) \text{ and } z_{max}' = z_{max}(x_{min}^*).$$

(4.39)

Then, we can establish the membership functions of the objectives of Equation (4.38). These are:

$$\mu_k(x) = \begin{cases} 1 & \text{if } z_k(x) > z_k^* \\ (z_k(x)-z_k')/(z_k^*-z_k') & \text{if } z_k' \le z_k(x) \le z_k^* \\ 0 & \text{if } z_k(x) < z_k' \end{cases}$$

where k = min or max.

Finally, we compute:

$$\max \ \alpha$$
$$\text{s.t.} \quad \mu_k(x) \ge \alpha, \ k = \min \text{ and } \max$$
$$x \in X.$$

(4.40)

Now, if the decision maker is capable of expressing his ideals about the imprecise objective coefficients in the form of fuzzy sets:

$$\tilde{c} = \{(c, \pi(c)) \mid c \in [c^L, c^U]\}$$

then the problem arises to take into account the additional information contained in possibility distributions $\pi(c)$. To handle this situation, let us first assume that the imprecise coefficients have convex possibility distributions. Then, we may take a finite set of degrees of possibility ($\{\beta_1, \beta_2, ..., \beta_r\}$) and consider r objectives with corresponding β_i-level, $i = 1, 2, ..., r$, specific interval-valued parameters. Thus, our problem becomes:

$$\max_{x \in X} \quad (c_{\beta 1}x, c_{\beta 2}x, ..., c_{\beta r}x)$$

where $c_{\beta i}$, $\forall i$, are interval-valued as shown in Figure 4.12. By use of the concept of Equation (4.38) for each objective function, we obtain the following auxiliary problem:

$$\max \quad \alpha$$
$$\text{s.t.} \quad \mu_{i,k}(x) \geq \alpha, \quad k = \min, \max \text{ and } i = 1, 2, ... r$$
$$x \in X. \tag{4.41}$$

where

$$\mu_{i,k}(x) = \begin{cases} 1 & \text{if } z_{i,k}(x) > z_{i,k}^* \\ (z_{i,k}(x)-z_{i,k}')/(z_{i,k}^*-z_{i,k}') & \text{if } z_{i,k}' \leq z_{i,k}(x) \leq z_{i,k}^* \\ 0 & \text{if } z_{i,k}(x) < z_{i,k}' \end{cases}$$

and

$$z_{i,\min}^* = z_{i,\min}(x_{i,\min}^*) = \max_{x \in X} z_{i,\min}(x)$$

$$z_{i,\max}^* = z_{i,\max}(x_{i,\max}^*) = \max_{x \in X} z_{i,\max}(x)$$

$$z_{i,\min}' = z_{i,\min}(x_{i,\max}^*) \text{ and } z_{i,\max}' = z_{i,\max}(x_{i,\min}^*), \forall i.$$

Example. Consider the following linear programming with imprecise objective coefficients:

$$\max \quad c_1 x_1 + c_2 x_2$$
$$\text{s.t.} \quad x \in X = \{x \mid x_1 + 4x_2 \leq 100, x_1 + 3x_2 \leq 76$$
$$x_1 + 2x_2 \leq 53, 3x_1 + 5x_2 \leq 138$$
$$3x_1 + 4x_2 \leq 120, 7x_1 + 8x_2 \leq 260$$
$$x_1 + x_2 \leq 36, 3x_1 + 2x_2 \leq 103$$
$$2x_1 + x_2 \leq 68 \text{ and } x_1, x_2 \geq 0\}$$

where c_1 and c_2 are imprecise and have their corresponding possibility distributions as shown in Figure 4.13.

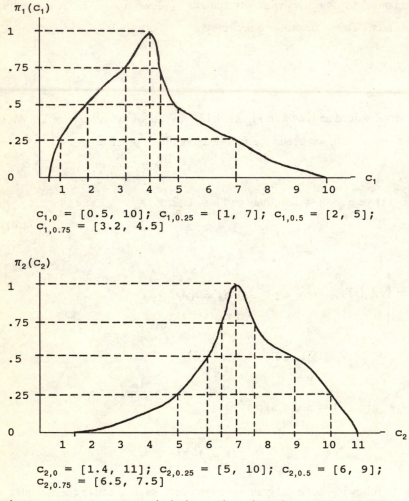

$$c_{1,0} = [0.5, 10]; \quad c_{1,0.25} = [1, 7]; \quad c_{1,0.5} = [2, 5];$$
$$c_{1,0.75} = [3.2, 4.5]$$

$$c_{2,0} = [1.4, 11]; \quad c_{2,0.25} = [5, 10]; \quad c_{2,0.5} = [6, 9];$$
$$c_{2,0.75} = [6.5, 7.5]$$

Figure 4.13 The possibility distributions of c_1 and c_2

To solve this problem, let $\beta_1 = 0.0$, $\beta_2 = 0.25$, $\beta_3 = 0.5$ and $\beta_4 = 0.75$. After computing Equation (4.39) for each β_i, we then solve the following problem:

$$\max \quad \alpha$$

$$\text{s.t.} \quad \mu_{1,\min}(x) = (0.5x_1 + 1.4x_2 - 25.2)/10.5 \geq \alpha$$

$$\mu_{1,\max}(x) = (10x_1 + 11x_2 - 223)/145 \geq \alpha$$

$$\mu_{2,min}(x) = (x_1+5x_2-106)/19 \geq \alpha$$
$$\mu_{2,max}(x) = (7x_1+10x_2-259)/42 \geq \alpha$$
$$\mu_{3,min}(x) = (2x_1+6x_2-148)/4 \geq \alpha$$
$$\mu_{3,max}(x) = (5x_1+9x_2-236)/8 \geq \alpha$$
$$\mu_{4,min}(x) = (3.2x_1+6.5x_2-168.2)/3.7 \geq \alpha$$
$$\mu_{4,max}(x) = (4.5x_1+7.5x_2-204)/3 \geq \alpha$$
$$(x_1, x_2) \in X \text{ and } \alpha \in [0, 1].$$

The solution of this problem is $\alpha^* = 0.50$ and $x^* = (9, 22)$. This leads to the average objective value:

$$z^* = (1/8) \sum_{i=1}^{4} [z_{i,min}(x^*) + z_{i,max}(x^*)] = 192.45.$$

The corresponding realized degrees of satisfaction are:

	$z_{1,min}$	$z_{1,max}$	$z_{2,min}$	$z_{2,max}$	$z_{3,min}$	$z_{3,max}$	$z_{4,min}$	$z_{4,max}$
$\mu_{i,k}(x^*)$	.96	.75	.68	.79	.50	.87	.97	.50

4.1.2.3 Delgado, Verdegay and Vila's Approach

Delgado, Verdegay and Vila [52] considered the problem of Equation (4.37) and assume that the imprecise objective coefficients $\tilde{c}_j$, $j = 1, 2, \ldots, n$, have the following possibility distributions (as shown in Figure 4.14):

$$\pi_j(c_j) = \begin{cases} 0 & \text{if } c_j \leq r_j \text{ or } c_j \geq R_j \\ h_j(c_j) & \text{if } r_j \leq c_j \leq c_j^L \\ 1 & \text{if } c_j^L \leq c_j \leq c_j^U \\ g_j(c_j) & \text{if } c_j^U \leq c_j \leq R_j. \end{cases} \quad (4.42)$$

Based on the following definition provided by Verdegay [250], a fuzzy objective to the fuzzy mathematical programming problem is a fuzzy set of $F(R^n)$, i.e., a function, $\Phi: F(R^n) \rightarrow [0, 1]$, where $F(R^n) = \{f \mid f:R^n \rightarrow R\}$. Let $\Phi \in P[F(R^n)]$ be a fuzzy objective. We define the following classical relation for any $\beta \in [0, 1]$:

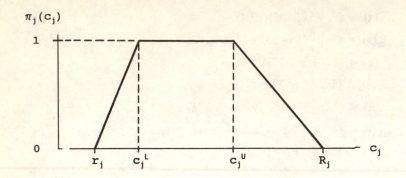

Figure 4.14   The possibility distributions of c_j

$$\forall x,y \in R^n, \; x >_\beta y \; <==> \; f(x) \geq f(y)$$

where $\forall f \in F(R^n)$ with $\Phi(f) \geq 1 - \beta$ and where $>_\beta$ must be read as "more preferred or indifferent, with β grade, than." It is clear that $\Phi(f) \geq 1 - \beta$ is a preorder (reflexive and transitive). Delgado et al. then defined:

$$c = (c_1, c_2, ..., c_n) \text{ and } \pi(c) = \inf_j \pi_j(c_j)$$

and obtained the following auxiliary problem of Equations (4.37) and (4.42):

$$\max \quad c_1 x_1 + c_2 x_2 + ... + c_n x_n$$

$$\text{s.t.} \quad \pi(c) \geq 1 - \beta \text{ and } x \in X. \tag{4.43}$$

Taking into account that:

$$\pi(c) \geq 1-\beta \; <==> \; \inf_j \pi_j(c_j) \geq 1-\beta \; <==> \; \pi_j(c_j) \geq 1-\beta$$

then, from Equation (4.42) it is obtained that:

$$\pi_j(c_j) \geq 1-\beta \; <==> \; h_j^{-1}(1-\beta) \leq c_j \leq g_j^{-1}(1-\beta).$$

Thus our problem becomes:

$$\max \; \{cx \mid x \in X, \, c \in \Gamma(1-\beta), \, \beta \in [0, 1]\} \tag{4.44}$$

where $\Gamma(1-\beta)$ denotes the set of vectors c whose components c_j are in the interval $[h_j^{-1}(1-\beta), g_j^{-1}(1-$

β)], $\forall j$. Then with fixed $\beta \in [0, 1]$ the solution of Equation (4.44) is given by $x^* \in X$ such that:

$$V c \in \Gamma(1-\beta), \forall x \in X ==> cx^* \geq x.$$

As the unity of that $x^* \in X$ is not guaranteed, let $S_{1-\beta}$ denote the set of those points being solutions of Equation (4.44). Thus, in accordance with the Decomposition Theorem for fuzzy sets, it can be defined $S = U_\beta \, \beta S_{1-\beta}$ which is a fuzzy set giving the fuzzy solution to the former problem.

Clearly, for any $\beta \in [0, 1]$, $\Gamma(1-\beta)$ is a convex set with extreme points defined by those $c \in \Gamma(1-\beta)$ such that its components are either in the lower $h_j^{-1}(1-\beta)$ or upper bound $g_j^{-1}(1-\beta)$. If $V(c) = \{x \mid cx \geq 0, c \in \Gamma(1-\beta)\}$, it is clear that $V(c)$ is a cone. Likewise, $V|\Gamma(1-\beta)| = U_c \, V(c)$, where $c \in \Gamma(1-\beta)$, is a cone [the preferential cone of the solutions of Equation (4.44)].

Now, let $E(1-\beta)$ be the subset (of $\Gamma(1-\beta)$) constituted by vectors whose jth component is equal to either the upper or the lower bound of c_j ($h_j^{-1}(1-\beta)$ or $g_j^{-1}(1-\beta)$), $\forall j$. That is:

$$E(1-\beta) = \{c_j \mid c_j = (h_j^{-1}(1-\beta) \text{ or } g_j^{-1}(1-\beta)), \forall j\}. \tag{4.45}$$

Then it is evident that the maximum number of elements in $E(1-\beta)$ is 2^n. On the other hand, it is verified that $V[\Gamma(1-\beta)] = U_c \, V(c)$ where $c \in E(1-\beta)$. Therefore, we may use the upper or the lower bound to solve Equation (4.44), rather than use the complete (infinite) family involved. That is to solve:

$$\max \ \{cx \mid x \in X, c \in E(1-\beta), \beta \in [0, 1]\}$$

or

$$\max \ (c^1 x, c^2 x, ..., c^s x)$$
$$\text{s.t.} \quad x \in X, c^k \in E(1-\beta), \ k = 1, 2, ..., 2^n$$
$$\beta \in [0, 1] \tag{4.46}$$

where $s = 2^n$, which is a conventional MOLP problem. Thus, we can solve Equation (4.45) by using any MOLP technique.

Delgado et al. proposed two approaches. The first approach is the weighing objective approach which is:

$$\max \quad \Sigma_k w_k c^k x$$
$$\text{s.t.} \quad x \in X, \ c^k \in E(1-\beta), \ k = 1, 2, \ldots, 2^n$$
$$\beta \in [0, 1], \ w_k \geq 0, \ \forall k, \text{ and } \Sigma_k w_k = 1. \tag{4.47}$$

The second approach (restricting objectives) is to use the concept of Rommelfanger et al. [216] as previously discussed. For an interval-valued objective coefficient problem (fixed β) we have:

$$\max \quad \alpha$$
$$\text{s.t.} \quad \mu_{min}(x) = (z_{min}(x)-z_{min}')/(z_{min}^*-z_{min}') \geq \alpha$$
$$\mu_{max}(x) = (z_{max}(x)-z_{max}')/(z_{max}^*-z_{max}') \geq \alpha$$
$$x \in X \text{ and } \alpha \in [0, 1] \tag{4.48}$$

whose optimal solution can be obtained from:

$$\max \quad (z_{max}(x)-z_{max}')/(z_{max}^*-z_{max}')$$
$$\text{s.t.} \quad \mu_{min}(x) = (z_{min}(x)-z_{min}')/(z_{min}^*-z_{min}') \geq \alpha$$
$$x \in X \text{ and } \alpha \in [0, 1] \tag{4.48'}$$

where $z_{max}(x) = (c_\beta^U)x$ (the upper bound) and $z_{min}(x) = (c_\beta^L)x$ (the lower bound), and other notations are correspondingly defined. But the objective function of Equation (4.48') is increasing on $(c_\beta^U)x$ and thus this problem is equivalent to:

$$\max \quad (c_\beta^U)x$$
$$\text{s.t.} \quad (c_\beta^L)x \geq (z_{min}^*-z_{min}')\alpha + z_{min}'$$
$$x \in X. \tag{4.49}$$

Similarly, to determine the compromise solution of Equation (4.46), the following problem is solved:

$$\max \quad \alpha$$
$$\text{s.t.} \quad (c^k)x \geq \alpha$$
$$c^k \in E(1-\beta), \ k = 1, 2, \ldots, 2^n$$
$$x \in X, \ \alpha \in R \text{ and } \beta \in [0, 1].$$

With fixed β (taking $1-(1-\beta)$), we then have the following equivalent problem:

$$\max \quad \alpha$$

$$\text{s.t.} \quad (c_\beta{}^k)x \geq \alpha, \, k = 1, 2, ..., 2^n$$

$$x \in X, \, \alpha \in R \text{ and } \beta \in [0, 1]. \tag{4.50}$$

However, $\forall x \in X \Rightarrow (c_\beta{}^L)x \leq (c_\beta{}^k)x \leq (c_\beta{}^U)x$ for $k = 2, 3, ..., 2^{n-1}$. Thus the first constraint of Equation (4.50) is equivalent to $(c_\beta{}^L)x \geq \alpha$. Our problem is then:

$$\max \quad (c_\beta{}^U)x$$

$$\text{s.t.} \quad (c_\beta{}^L)x \geq \alpha$$

$$x \in X, \, \alpha \in R \text{ and } \beta \in [0, 1] \tag{4.51}$$

where α is considered as a parameter.

Note that if $x^*(\alpha)$ is the optimal solution to Equation (4.51), then the optimal solution to Equation (4.50) will be the particular solution of $x^*(\alpha)$, say $x^*(\alpha^*)$, obtained for that α^* such that:

$$\alpha^* = \sup \{\alpha | \, x^*(\alpha) \text{ is the solution of Equation (4.51)}\}.$$

Moreover, the optimal solution to Equation (4.50) for the particular value $\alpha_0 = (z_{min}{}^* - z_{min}{}')\alpha + z_{min}{}'$, will coincide with the corresponding optimal solution of Equation (4.49). Equations (4.49) and (4.51) are considered equivalent in this way.

Finally, Delgado stated that if solution x^* is the optimal one, then:

$$\mu_{min}((c_\beta{}^L)x^*) = \mu_{max}((c_\beta{}^U)x^*) = \alpha \tag{4.52}$$

and because of the parametric nature of the optimal solution of Equation (4.51), say $x(\alpha)$, for each fixed $\beta \in [0, 1]$, there exists α_0 such that:

$$\mu_{min}[(c_\beta{}^L)x(\alpha_0)] = \mu_{max}[(c_\beta{}^U)x(\alpha_0)]. \tag{4.53}$$

From Equation (4.53) we can find the solution of Equation (4.48) without solving that problem.

Example. Consider the following problem:

$$\max \quad z = \tilde{c}_1 x_1 + 2x_2$$

$$\text{s.t.} \quad x \in X = \{x \, | \, 2x_1 + 3x_2 \leq 13, \, x_1 + 4x_2 \leq 16, \, 5x_1 + 4x_2 \leq 16,$$

$$x_1 \text{ and } x_2 \geq 0\}$$

where $\tilde{c}_1$ is imprecise cost and has the following possibility distribution:

$$\pi(c_1) = \begin{cases} 0 & \text{if } c_1 > 4 \text{ or } c_1 < 1/2 \\ (2c_1 - 1)/5 & \text{if } 1/2 \le c_1 \le 3 \\ 4 - c_1 & \text{if } 3 \le c_1 \le 4. \end{cases}$$

The interval representation is:

$$c_{1,0} = [1/2, 4], \quad c_{1,0.25} = [9/8, 30/8], \quad c_{1,0.5} = [7/4, 14/4], \quad c_{1,0.75} = [19/8, 26/8].$$

In this situation, Equation (4.46) is written as:

$$\begin{aligned} \max \quad & \{[(6-5\beta)/2]x_1 + 2x_2, (3+\beta)x_1 + 2x_2\} \\ \text{s.t.} \quad & x \in X \text{ and } \beta \in [0, 1]. \end{aligned} \tag{4.54}$$

Suppose $\beta = 0.75$, we can then solve Equation (4.54) by weighing objectives as in Equation (4.47). That is:

$$\max_{x \in X} \quad w_1[(19/8)x_1 + 2x_2] + w_2[(26/8)x_1 + 2x_2].$$

By taking $w_1 = w_2 = 1/2$, the optimal solution is $(3, 0)$ with $z^* = 8.43$.

On the other hand, if Equation (4.54) is solved by restricting objectives with $\beta = 0.75$, we will write Equation (4.51) as:

$$\begin{aligned} \max \quad & 3.25x_1 + 2x_2 \\ \text{s.t.} \quad & 2.375x_1 + 2x_2 \ge \alpha, \ x \in X \text{ and } \alpha \in R \end{aligned}$$

whose solutions are: $(60-8\alpha, -71.25+10\beta)$ with $z^* = 52.5 - 6\alpha$ for $\alpha \in [7.125, 7.5]$; $(3, 0)$ with $z^* = 9.75$ for $\alpha \in [0, 7.125]$. These solutions, for some particular values of α, reproduce the corresponding solutions to Equation (4.48). In fact, $z_{min}^* = 7.5$, $z_{min}' = 7.125$, $z_{max}^* = 9.75$, $z_{max}' = 7.5$. When solving

$$\mu_{min}[(c_{0.75}^L) \ x(\alpha_0)] = \mu_{max}[(c_{0.75}^U) \ x(\alpha_0)]$$

it is found that $\alpha_0 = 7.3125$ and consequently, the solution is $(1.5, 1.875)$ and α (of Equation (4.48)) is 0.5.

4.1.3 Linear Programming with Imprecise Objective and Technological Coefficients

By recalling the investment problem of Equation (4.35) in the previous section, we may consider a situation where the objective and technological coefficients are fuzzy and have triangular possibility distributions with most possible and least possible values (Lai and Hwang [138]). This is quite common, especially in the finance world. The interest rate is always not fixed in current economic systems where the float interest rate policy is adapted. However, the decision makers may know a possible interval, according to their experience. For example, the interest rate will be around 8% with no less than 6% and no larger than 11% in the next 2 years. Therefore, it is important to study the case where the interest rate has a possibility distribution.

Consider the following problem:

$$\max \quad \tilde{c}x$$

$$\text{s.t.} \quad x \epsilon X = \{x \mid \tilde{A}x \leq b \text{ and } x \geq 0\} \tag{4.55}$$

where b is precise, but $\tilde{c}$ and $\tilde{A}$ are imprecise with triangular possibilistic distributions of (c^m, c^p, c^o) and (A^m, A^p, A^o), respectively. To solve this imprecise technological coefficient, we may apply the concept of the most likely value which is $(4A^m_\beta + A^p_\beta + A^o_\beta)/6$ where β is a minimal acceptable possibility (not β-level set). Finally, we obtain [138]:

$$\min \quad z_1 = (c^m - c^p)x$$
$$\max \quad z_2 = c^m x$$
$$\max \quad z_3 = (c^o - c^m)x$$
$$\text{s.t.} \quad \{(4A^m_\beta + A^p_\beta + A^o_\beta)/6\}x \leq b$$
$$x \geq 0. \tag{4.56}$$

The weights between A^m_β, A^p_β and A^o_β can be changed subjectively. The reason for using most likely values is that A^p_β is too pessimistic and A^o_β, too optimistic. Of courses these two boundary values provide boundary solutions.

Equation (4.56) is obviously a nonlinear programming problem. However, if β is initially given by the decision maker, it will be a linear programming problem. Thus, we can provide the decision maker a solution table with $\beta = 0.0, 0.1, ..., 0.9, 1.0$.

With the consideration of the floating interest rate, the coefficients of the objective and the 3rd through 6th constraints of Equation (4.35) should be considered as imprecise or fuzzy. Interest rates of the beginning of the planning periods in the first and the second constraints are quite sure. However, interest rates will fluctuate in future years. Let us assume the following problem:

$$\max\ Z = (5.5, 5.1, 6.2)F + (-1.0, -1.2, -0.85)M + (6.0, 5.0, 6.5)D$$
$$+ (-1.065, -1.08, -1.058)B_6 + (1.046,\ 1.040,\ 1.06)L_6$$

s.t. $\quad 3F + 2M + 2D - B_1 + L_1 \leq 2$

$1F + 0.5M + 2D + 1.06B_1 - 1.04L_1 - B_2 + L_2 \leq 0.5$

$1.8F - 1.5M + 1.8D + (1.06, 1.055, 1.065)B_2 - (1.04, 1.035, 1.045)L_2 - B_3 + L_3 \leq 0.4$

$-0.4F - 1.5M - 1D + (1.06, 1.055, 1.07)B_3 - (1.04, 1.035, 1.05)L_3 - B_4 + L_4 \leq 0.38$

$-1.8F - 1.5M - 1D + (1.065, 1.06, 1.07)B_4 - (1.044, 1.038, 1.05)L_4 - B_5 + L_5 \leq 0.36$

$-1.8F - 0.2M - 1D + (1.065, 1.058, 1.075)B_5 - (1.046, 1.042, 1.055)L_5 - B_6 + L_6 \leq 0.34$

$B_i \in [0, 2],\ i = 1, 2, 3, 4, 5, 6,$

F, M and $D \in [0, 1].$ (4.57)

where (1.06, 1.055, 1.065) in the third constraint means: the most possible borrowing interest rate = 6%; the most pessimistic rate = 6.5%; the most optimistic rate = 5.5%. Similarly, for (1.04, 1.035, 1.045) the most possible lending interest rate = 4%, the most pessimistic rate = 3.5% and the most optimistic rate = 4.5%. Also, the possibility distributions of these interest rates are changed as time goes on.

The auxiliary problem of Equation (4.57) at $\beta = 0.5$ is then:

$\min\quad 0.4F - 0.2M + D - 0.015B_6 + 0.006L_6$

$\max\quad 5.5F - M + 6D - 1.065B_6 + 1.046L_6$

$\max\quad 0.7F - 0.15M + 0.5D - 0.007B_6 + 0.014L_6$

s.t. $\quad 3F + 2M + 2D - B_1 + L_1 \leq 2$

$1F + 0.5M + 2D + 1.06B_1 - 1.04L_1 - B_2 + L_2 \leq 0.5$

$1.8F - 1.5M + 1.8D + 1.06B_2 - 1.04L_2 - B_3 + L_3 \leq 0.4$

$-0.4F - 1.5M - 1D + 1.0604B_3 - 1.0408L_3 - B_4 + L_4 \leq 0.38$

$-1.8F - 1.5M - 1D + 1.065B_4 - 1.044L_4 - B_5 + L_5 \leq 0.36$

$-1.8F - 0.2M - 1D + 1.0653B_5 - 1.0464L_5 - B_6 + L_6 \leq 0.34$

$B_i \in [0, 2],\ i = 1, 2, 3, 4, 5, 6,$

F, M and $D \in [0, 1].$

The PIS is (0.23, 9.09, 0.63) and NIS (0.85, 3.13, 0.16). The final solution at $\beta = 0.5$ is shown in Table 4.3 and the possibilistic distribution of the imprecise profit obtained is given in Figure 4.12.

4.1.4 Linear Programming with Imprecise Coefficients

The income stream from previous investments may not be precise as time progresses. For example, the interest incomes of money market accounts are changed with the market. Thus, we may have the following possibilistic linear programming:

$$\max \quad \tilde{c}x$$
$$\text{s.t.} \quad \hat{A}x \le \tilde{b}, \text{ and } x \ge 0 \tag{4.58}$$

where $\hat{A}$, $\tilde{b}$ and $\tilde{c}$ are imprecise and possibility distributions. To solve Equation (4.58), we present four approaches of Lai and Hwang [138], Buckley [24], Negi [175] and Fuller [97].

4.1.4.1 Lai and Hwang's Approach

In this section, we combine the fuzzy ranking concepts with our strategy for the imprecise objective function, when the imprecise coefficients have triangular possibility distributions. We obtain the following auxiliary model [138]:

$$\min \quad z_1 = (c^m - c^p)x$$
$$\max \quad z_2 = c^m x$$
$$\max \quad z_3 = (c^o - c^m)x$$
$$\text{s.t.} \quad A^m{}_\beta x \le b^m{}_\beta, \ A^p{}_\beta x \le b^p{}_\beta, \ A^o{}_\beta x \le b^o{}_\beta$$
$$x \ge 0 \tag{4.59}$$

where the auxiliary inequality constraints have been proposed by Ramik and Rimanek as shown in Section 4.1.1. If the minimal acceptable possibility, β, is given, Equation (4.59) will be a MOLP problem.

Now, let us return to our investment problem with the modification of "income stream from previous investment" (in millions): $0.5, ($0.4, $0.35, $0.5), ($0.38, $0.35, $0.4), ($0.36, $0.34, $0.45), ($0.34, $0.3, $0.42) ($0.3, $0.28, $0.35). Other assumptions are the same as previous ones. The problem becomes:

max $z = (5.5,5.1,6.2)F+(-1.0,-1.2,-0.85)M+(6.0,5.0,6.5)D$
$+(-1.065,-1.08,-1.058)B_6+(1.046,1.04,1.06)L_6$

s.t. $3F+2M+2D-B_1+L_1 \leq 2$

$1F+0.5M+2D+1.06B_1-1.04L_1-B_2+L_2 \leq 0.5$

$1.8F-1.5M+1.8D+(1.06,1.055,1.065)B_2-(1.04,1.035,1.045)L_2-B_3+L_3 \leq(.4,.35,.5)$

$-0.4F-1.5M-1D+(1.06,1.055,1.07)B_3-(1.04,1.035,1.05)L_3-B_4+L_4 \leq (.38,.35,.4)$

$-1.8F-1.5M-1D+(1.065,1.06,1.07)B_4-(1.044,1.038,1.05)L_4-B_5+L_5 \leq(.36,.34,.45)$

$-1.8F-0.2M-1D+(1.065,1.058,1.075)B_5-(1.046,1.042,1.055)L_5-B_6+L_6 \leq(.34,.3,.42)$

$B_i \epsilon [0, 2], i = 1, ..., 6$

F, M and $D \epsilon [0, 1]$.   (4.60)

Assume $\beta = 0.5$. We obtain:

min $0.4F-0.2M+D-0.015B_6+0.006L_6$

max $5.5F-M+6D-1.065B_6+1.046L_6$

max $0.7F-0.15M+0.5D-0.007B_6+0.014L_{16}$

s.t. $3F+2M+2D-B_1+L_1 \leq 2$

$1F+0.5M+2D+1.06B_1-1.04L_1-B_2+L_2 \leq 0.5$

$1.8F-1.5M+1.8D+1.06B_2-1.04L_2-B_3+L_3 \leq 0.4$

$1.8F-1.5M+1.8D+1.0575B_2-1.0375L_2-B_3+L_3 \leq 0.375$

$1.8F-1.5M+1.8D+1.0625B_2-1.0425L_2-B_3+L_3 \leq 0.45$

$-0.4F-1.5M-D+1.06B_3-1.04L_3-B_4+L_4 \leq 0.38$

$-0.4F-1.5M-D+1.0575B_3-1.0375L_3-B_4+L_4 \leq 0.365$

$-0.4F-1.5M-D+1.065B_3-1.045L_3-B_4+L_4 \leq 0.39$

$-1.8F-1.5M-1D+1.065B_4-1.044L_4-B_5+L_5 \leq 0.36$

$-1.8F-1.5M-1D+1.0625B_4-1.041L_4-B_5+L_5 \leq 0.35$

$-1.8F-1.5M-1D+1.0675B_4-1.047L_4-B_5+L_5 \leq 0.405$

$-1.8F-0.2M-1D+1.065B_5-1.046L_5-B_6+L_6 \leq 0.34$

$-1.8F-0.2M-1D+1.0615B_5-1.044L_5-B_6+L_6 \leq 0.32$

$-1.8F-0.2M-1D+1.07B_5-1.0505L_5-B_6+L_6 \leq 0.38$

$B_i \epsilon [0, 2], i = 1, 2, 3, 4, 5, 6$

F, M and D ϵ [0, 1]. (4.61)

The PIS is (0.23, 7.19, 0.64) and NIS (0.84, 3.13, 0.16). The final solution of Equation (4.61) is shown in Table 4.3 and the possibilistic distribution of the imprecise profit obtained is shown in Figure 4.12.

4.1.4.2 Buckley's Approach

Buckley [24, 25] considered the problem of Equation (4.58) as:

$$\text{max} \quad Z = \tilde{c}x$$
$$\text{s.t.} \quad \tilde{A}x \leq \tilde{b} \text{ and } x \geq 0 \tag{4.62}$$

where $\tilde{c}$, $\tilde{b}$ and $\tilde{A}$ are all imprecise coefficient (fuzzy number) vectors and matrix. Assume that the imprecise nature of the coefficients can be modelled by trapezoidal possibility distributions, $\pi(.)$, (as shown in Figure 4.15). That denotes: $\tilde{c} = (\tilde{c}_1, ..., \tilde{c}_n)$ and $\tilde{c}_j = (c_{j,1} | c_{j,2}, c_{j,3} | c_{j,4})$; $\tilde{b} = (\tilde{b}_1, ..., \tilde{b}_m)$ and $\tilde{b}_i = (b_{i,1} | b_{i,2}, b_{i,3} | b_{i,4})$; and $\tilde{A} = [\tilde{a}_{ij}]$ and $\tilde{a}_{ij} = (a_{ij,1} | a_{ij,2}, a_{ij,3} | a_{ij,4})$.

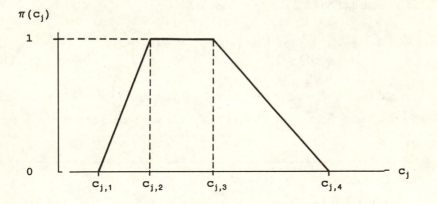

Figure 4.15 The trapezoidal possibility distribution of $\tilde{c}_j$

Buckley stated that the fuzzy numbers are the possibility distributions associated with the fuzzy variables and hence place a restriction on the possible values the variables may assume. He defined $\pi(Z = z)$ as being the possibility distribution of the objective function Z. Assume that all the fuzzy variables $\tilde{a}_{ij}$, $\tilde{b}_i$ and $\tilde{c}_j$ are non-interactive. Then, the possibility that x satisfies the ith constraint is specified as:

$$\pi(x \in F_i) = \sup_{a_i,b_i} \{\pi_{joint}(a_i,b_i) | a_i x \leq b_i\} \tag{4.63}$$

where

$$\pi_{joint}(a_i,b_i) = \min [\pi(a_{i1}), \pi(a_{i2}), ..., \pi(a_{in}), \pi(b_i)]$$

and $a_i = (a_{i1}, ..., a_{in})$ and $\pi_{joint}(a_i,b_i)$ is the joint distribution of a_{ij}, $\forall j$, and b_i. Therefore, for x ≥ 0, we have the following possibility that x is feasible or satisfies all the constraints:

$$\pi(x \in F) = \min_{1 \leq i \leq m} [\pi(x \in F_i)]. \tag{4.64}$$

Next, we construct the conditional possibility that $Z = z$ under the given x as follows:

$$\pi(Z=z| x) = \sup_c \{\pi_{joint}(c) | cx = z) \tag{4.65}$$

where $c = (c_1, ..., c_n)$ and

$$\pi_{joint}(c) = \min [\pi(c_1), \pi(c_2), ..., \pi(c_n)].$$

Thus, the possibility distribution of Z can be defined as:

$$\pi_1(Z=z) = \pi_1(Z) = \sup_{x \geq 0} \{\min [\pi(Z=z| x), \pi(x \in F)]\}. \tag{4.66}$$

Buckley also provided another alternative definition of the possibility of Z which is to construct the possibility distribution of the maximum values of Z. For given crisp values of c, b and A, solve:

$$\max \quad Z = cx$$
$$\text{s.t.} \quad Ax \leq b \text{ and } x \geq 0 \tag{4.67}$$

and let the maximum value of Z be z, and let $\pi_{joint}(c, b, A)$ be the joint possibility distribution of all the c_j, b_j and a_{ij}. Then:

$$\pi_2(Z) = \sup_{c,b,A} \{\pi_{joint}(c, b, A)| Z = z\} \tag{4.68}$$

For Equation (4.62), the decision maker may like to obtain a compromise solution x^*

which has the following features: (1) the decision maker determines a value of $\alpha = \alpha^* \epsilon [0, 1]$ and a maximum value of $Z = z(\alpha^*)$, both acceptable to him, so that $\pi[Z \geq z(\alpha^*)] = \alpha^*$; and (2) the decision maker uses this α^* to construct a crisp linear programming problem as Equation (4.67), whose optimal solution for Z is $z(\alpha^*)$, and an optimal value of x is a compromise solution x^* to the possibilistic linear programming of Equation (4.62). This value of x solves Equation (4.67) with max $Z = z(\alpha^*)$ and it will turn out that $\pi[Z \geq z(\alpha^*)] = \pi_1[Z \geq z(\alpha^*)] = \pi_2[Z \geq z(\alpha^*)] = \alpha^*$.

Buckley then employed α-cuts to develop his solution algorithm. In fact, the α-cuts of normalized and convex fuzzy numbers are bounded, closed intervals. Thus, let us assume that:

$$a_{ij}{}^\alpha = [a_{ij}{}^{\alpha L}, a_{ij}{}^{\alpha U}], \ b_i{}^\alpha = [b_i{}^{\alpha L}, b_i{}^{\alpha U}], \text{ and } c_j{}^\alpha = [c_j{}^{\alpha L}, c_j{}^{\alpha U}], \ \forall i,j.$$

Then, solve:

$$\text{max} \quad Z = \Sigma_j c_j{}^{\alpha U} x_j$$
$$\text{s.t.} \quad \Sigma_j a_{ij}{}^{\alpha L} x_j \leq b_i{}^{\alpha U}, \ \forall i$$
$$x_j \geq 0, \ \forall j, \text{ and } \alpha \epsilon [0, 1]. \tag{4.69}$$

Let the solution $z(\alpha)$ and $\Omega(\alpha)$ denote the feasible set to the problem of Equation (4.69). And, define the possibility distribution of Z as:

$$\pi(z) = \begin{cases} 1 & \text{if } z^L \leq z \leq z(1) \\ \alpha & \text{if } z = z(\alpha) \text{ and } 0 < \alpha < 1 \\ 0 & \text{if } z \geq z(0) \end{cases} \tag{4.70}$$

where $z^L = \text{min } \{\Sigma_j c_{j,2} x_j \mid x \epsilon \Omega(1)\}$ and it is the lower bound of $z(1)$. Obviously, $z(0)$ is the minimum value of the upper bound of the optimum value of Equation (4.69). Buckley proved that if $\Omega(1) \neq \emptyset$, then $\pi_1(Z=z) = \pi(z)$. Thus, the parametric linear programming of Equation (4.69) gives the solution to $\pi_1(Z)$ because: 1) $\Omega(1)$ is the largest set of $x \geq 0$ so that $\pi(x \epsilon F) = 1$; 2), $\Omega(\alpha)$ steadily increases as α decreases from one to zero; and 3) $\Omega(0)$ is the smallest set of $x \geq 0$ so that $\pi(x \epsilon F) = 0$ whenever x is not in $\Omega(0)$. Furthermore, Buckley proved that if $\Omega(1) \neq \emptyset$, $\pi_1(Z) = \pi_2(Z)$ for $z \geq z(1)$.

Similarly, for a minimization problem:

$$\text{min} \quad \tilde{c}x$$
$$\text{s.t.} \quad \tilde{A}x \geq \tilde{b} \text{ and } x \geq 0$$

we can obtain its compromise solution by solving the following auxiliary problem:

$$\min \quad Z = \Sigma_j \, c_j{}^{\alpha L} x_j$$
$$\text{s.t.} \quad \Sigma_j \, a_{ij}{}^{\alpha U} x_j \leq b_i{}^{\alpha L}, \; \forall i$$
$$x_j \geq 0, \; \forall j, \text{ and } \alpha \in [0, 1]. \tag{4.71}$$

Let the solution be $z(\alpha)$ and $\Omega(\alpha)$ denote the feasible set to the problem of Equation (4.71). And, define the possibility distribution of Z as:

$$\pi(z) = \begin{cases} 1 & \text{if } z(1) \leq z \leq z^U \\ \alpha & \text{if } z = z(\alpha) \text{ and } 0 < \alpha < 1 \\ 0 & \text{if } z \leq z(0) \end{cases} \tag{4.72}$$

where $z^U = \min \{\Sigma_j \, c_{j,3} x_j \mid x \in \Omega(1)\}$ and it is the upper bound of $z(1)$. Obviously, $z(0)$ is the maximum value of the lower bound of the optimum value of Equation (4.71).

4.1.4.2a Example: A Feed Mix (Diet) Problem

Consider the classical feed mix (diet) problem with two decision variables and the information given in Table 4.4. The LLB Poultry Farm has 2000 chickens to feed and wishes to plan their daily feeding schedule for the next few weeks. The feed consumption is around one pound of feed per chicken per day. However, in order to fatten up the chickens for market, the feed should meet certain nutritional needs. Also, these needs can be met by mixing different feeds. This problem is then formulated as:

$$\min \quad Z = 0.15x_1 + 0.40x_2$$
$$\text{s.t.} \quad 0.01x_1 + 0.025x_2 \geq 0.012(x_1 + x_2)$$
$$0.09x_1 + 0.50x_2 \leq 0.22(x_1 + x_2)$$
$$x_1 + x_2 \geq 1000, \text{ and } x_1, x_2 \geq 0$$

whose solution is $x = (5200/3, 800/3)$ and $Z^* = \$366.7$.

Now, suppose that $\tilde{c}_j$, $\tilde{b}_i$, $\tilde{a}_{ij}$, $\forall i,j$, are fuzzy variables. $\tilde{a}_{11} = (.005 \mid .01, .01 \mid .02)$, $\tilde{a}_{12} = (.01 \mid .02, .03 \mid .04)$, $\tilde{a}_{21} = (.06 \mid .08, .1 \mid .12)$, $\tilde{a}_{22} = (.4 \mid .46, .55 \mid .6)$, $\tilde{b}_1 = (.009 \mid .012, .012 \mid .015)$, $\tilde{b}_2 = (.18 \mid .2, .24 \mid .26)$, $\tilde{c}_1 = (.1 \mid .13, .17 \mid .2)$, $\tilde{c}_2 (.3 \mid .4, .4 \mid .5)$ and $\tilde{N} = (1000 \mid 2000, 2000 \mid 3000)$. The possibilistic linear programming is:

Table 4.4 Data for feed mix problem

Nutritional needs	Feeds (per lb) Corn	Soybean	Constraint
Calcium (lbs)	$a_{11}=.01$	$a_{12}=.025$	at least 1.2% calcium per day $= b_1$
Protein (lbs)	$a_{21}=.09$	$a_{22}= .50$	at most 22% protein per day $= b_2$
Cost	$c_1=\$.15$	$c_2=\$.40$	
Decision	x_1 lbs	x_2 lbs	at least 1000 lbs per day $= N$

$$\min \quad Z = \tilde{c}_1 x_1 + \tilde{c}_2 x_2$$
$$\text{s.t.} \quad \tilde{a}_{11} x_1 + \tilde{a}_{12} x_2 \geq \tilde{b}_1 (x_1 + x_2)$$
$$\tilde{a}_{21} x_1 + \tilde{a}_{22} x_2 \leq \tilde{b}_2 (x_1 + x_2)$$
$$x_1 + x_2 \geq \tilde{N} \text{ and } x_1 \text{ and } x_2 \geq 0.$$

To solve this problem, we first consider the following auxiliary:

$$\min \quad Z = c_1{}^{\alpha L} x_1 + c_2{}^{\alpha L} x_2$$
$$\text{s.t.} \quad a_{11}{}^{\alpha U} x_1 + a_{12}{}^{\alpha U} x_2 \geq b_1{}^{\alpha L}(x_1 + x_2)$$
$$a_{21}{}^{\alpha L} x_1 + a_{22}{}^{\alpha L} x_2 \leq b_2{}^{\alpha U}(x_1 + x_2)$$
$$x_1 + x_2 \geq N^{\alpha L} \text{ and } x_1 \text{ and } x_2 \geq 0. \qquad (4.73)$$

Let the solution be $x(\alpha)$ with $Z = z(\alpha)$ and feasible set $\Omega(\alpha)$, for $\alpha \in [0, 1]$. $\Omega(1)$ is shown in Figure 4.16. The first, second and third constraints of Equation (4.73) are labeled as A, B and C, respectively. As α decreases from one to zero, we can see that: 1) the line A, always passing through the origin, revolves clockwise to the x_1-axis ($\alpha = 11/13$) and then below the x_1-axis; 2) the line B, also always passing through the origin, revolves counter-clockwise towards the x_2-axis; 3) the line C moves southwestward towards the origin; and 4) the solution $x(\alpha)$ first hits x_1-axis when $\alpha = 11/13$, and then moves to the left along the x_1-axis as α approaches zero. It is not difficult to see that as α decreases from one to zero, the point $x(\alpha)$ will give the correct z value for $\pi_1(Z)$. Therefore,

$$\pi_1(Z=z) = \begin{cases} 1 & \text{if } z \geq z(1) = 314 \\ \alpha & \text{if } z = z(\alpha) \\ 0 & \text{if } z \leq z(0) = 100. \end{cases}$$

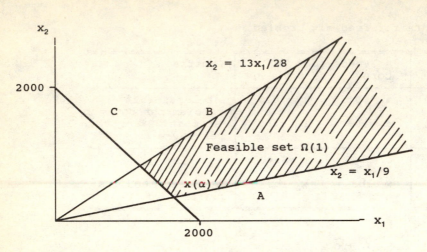

Figure 4.16 The feasible set $\Omega(1)$

For a compromise solution, suppose that the decision maker selects the satisfying solution at $\alpha^* = 0.8$. Then, the solution of Equation (4.73) is (1800, 0) with $z(0.8) = \$223.2$ and $\pi(Z \leq 223.2) = 0.8$.

4.1.4.3 Negi's Approach

Negi [175] used the concept of exceedance and strict exceedance possibilities, proposed by Dubois and Prade [BM8] , to solve Equation (4.62) in the previous section.

The exceedance possibility [denoted by $\pi(B^L \geq R^L)$] is the possibility that $\tilde{B} \geq \tilde{R}$. By referring to Figure 4.17 where $E = (x_1, \delta^1)$, we can easily calculate this possibility as follows:

$$(x_1-r_1)/(r_2-r_1) = (b_4-x_1)/(b_4-b_3) \; => \; x_1 = (b_4 r_2 - b_3 r_1)/[(b_4-b_3)+(r_2-r_1)]$$

and then $\delta^1 = (x_1-r_1)/(r_2-r_1) = (b_4-r_1)/[(b_4-b_3)+(r_2-r_1)]$. Thus we have:

$$\pi(B^U \geq R^L) = \begin{cases} 1 & \text{if } b_3 > r_2 \\ \delta^1 & \text{if } b_3 \leq r_2 \text{ and } r_1 \leq b_4 \\ 0 & \text{if } r_1 > b_4. \end{cases}$$

Similarly, we compute the strict exceedance possibility $(\pi(B^U \geq R^U))$, point S, as follows:

$$(x_2-r_3)/(r_4-r_3) = (b_4-x_2)/(b_4-b_3) \; ==> \; x_2 = (b_4 r_4 - b_3 r_3)/[(b_4-b_3)+(r_4-r_3)]$$

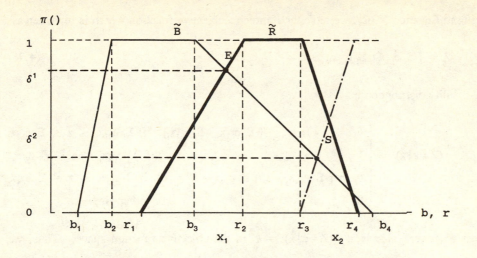

Figure 4.17 Comparison of trapezoidal fuzzy number

and then $\delta^2 = (x_2\text{-}r_3)/(r_4\text{-}r_3) = (b_4\text{-}r_3)/[(b_4\text{-}b_3)+(r_4\text{-}r_3)]$. Thus we have:

$$\pi(B^U \geq R^L) = \begin{cases} 1 & \text{if } b_3 > r_4 \\ \delta^2 & \text{if } b_3 \leq r_4 \text{ and } r_3 \leq b_4 \\ 0 & \text{if } r_3 > b_4. \end{cases}$$

When fuzzy addition is used, LHS and RHS of the ith constraint in Equation (4.62) can be considered as $\tilde{R}$ and $\tilde{B}$. If we use the exceedance possibility, we will have:

$$\pi(x\epsilon F_i) = \begin{cases} 1 & \text{if } b_{i,3} > r_{i,2} \\ \delta_i & \text{if } b_{i,3} \leq r_{i,2} \text{ and } r_{i,1} \leq b_{i,4} \\ 0 & \text{if } r_{i,1} > b_{i,4} \end{cases} \qquad (4.74)$$

where $\delta_i = (b_{i,4}\text{-}r_{i,1})/[(b_{i,4}\text{-}b_{i,3})+(r_{i,2}\text{-}r_{i,1})]$, $r_{i,1} = \Sigma_j\, a_{ij,1}x_j$, $r_{i,2} = \Sigma_j\, a_{ij,2}x_j$ and $\pi(x\epsilon F_i)$ is defined in the previous section. Then, we obtain:

$$\pi(x\epsilon F) = \min\, \{\pi(x\epsilon F_1)\, \pi(x\epsilon F_2) \ldots \pi(x\epsilon F_m)\} = \min\, \{\delta_1,\, \delta_2,\, \ldots \delta_m\} \qquad (4.75)$$

with the decision space

$$b_{i,3} \leq \Sigma_j\, a_{ij,2}x_j \text{ and } b_{i,4} \geq \Sigma_j\, a_{ij,1}x_j,\, \forall i.$$

By referring to Equation (4.62), we have the imprecise objective function which is rewritten as:

$$\tilde{Z} = [\, \Sigma_j\, c_{j,1}x_j \,|\, \Sigma_j\, c_{j,2}x_j,\, \Sigma_j\, c_{j,3}x_j \,|\, \Sigma_j\, c_{j,4}x_j] \tag{4.76}$$

and its possibility distribution is:

$$\pi(Z=z\,|\,x) = \begin{cases} \theta_1 = [z - \Sigma_j\, c_{j,1}x_j]/[\,\Sigma_j\, c_{j,2}x_j - \Sigma_j\, c_{j,1}x_j] & \text{if } \Sigma_j\, c_{j,1}x_j \leq z \leq \Sigma_j\, c_{j,2}x_j \\ 1 & \text{if } \Sigma_j\, c_{j,2}x_j < z < \Sigma_j\, c_{j,3}x_j \\ \theta_2 = [\,\Sigma_j\, c_{j,4}x_j - z]/[\,\Sigma_j\, c_{j,4}x_j - \Sigma_j\, c_{j,3}x_j] & \text{if } \Sigma_j\, c_{j,3}x_j \leq z \leq \Sigma_j\, c_{j,4}x_j \\ 0 & \text{otherwise.} \end{cases}$$

The largest objective value is at $\pi(Z=z\,|\,x) = \theta_2$ in the objective function space. Thus, we obtain:

$$\pi(Z=z) = \min\{\, \min[\delta_1, \delta_2, \dots, \delta_m],\, \theta_2\} = \min\{\delta_1, \delta_2, \dots, \delta_m,\, \theta_2\} \tag{4.77}$$

which satisfies:

$$b_{i,3} \leq \Sigma_j\, a_{ij,2}x_j,\ b_{i,4} \geq \Sigma_j\, a_{ij,1}x_j,\ \forall i,\ \text{and}\ \Sigma_j\, c_{j,3}x_j \leq z \leq \Sigma_j\, c_{j,4}x_j.$$

Clearly, the equations of δ_i and θ_2 are non-linear functions. However, if the decision maker has a lower cut-off value (α), Equation (4.77) will be equivalent to:

$$
\begin{aligned}
\max\quad & z \\
\text{s.t.}\quad & \delta_1, \delta_2, \dots, \delta_m,\, \theta_2 \geq \alpha \\
& b_{i,3} \leq \Sigma_j\, a_{ij,2}x_j,\ b_{i,4} \geq \Sigma_j\, a_{ij,1}x_j,\ \forall i. \\
& \Sigma_j\, c_{j,3}x_j \leq z \leq \Sigma_j\, c_{j,4}x_j \\
& z,\, x_j,\, \forall j,\, \geq 0
\end{aligned}
\tag{4.78}
$$

which is obviously a crisp linear programming problem.

Example. Consider the following possibilistic linear programming problem:

$$
\begin{aligned}
\max\quad & (1|2,\, 3|4)x_1 + (2|3,\, 4|6)x_2 \\
\text{s.t.}\quad & (0|1,\, 2|3)x_1 + (1|2,\, 3|5)x_2 \leq (3|5,\, 6|8) \\
& (1|2,\, 3|6)x_1 + (0|1,\, 2|3)x_2 \leq (2|4,\, 6|7) \\
& x_1 \text{ and } x_2 \geq 0.
\end{aligned}
$$

From Equation (4.74), we obtain:

$$\pi(x \epsilon F_1) = \begin{cases} 1 & \text{if } x_1 + 2x_2 \leq 6 \\ \delta_1 = (8-x_2)/(2+x_1+x_2) & \text{if } x_1 + 2x_2 \geq 6, \; x_2 \leq 8 \\ 0 & \text{if } x_2 \geq 8 \end{cases}$$

$$\pi(x \epsilon F_2) = \begin{cases} 1 & \text{if } 2x_1 + x_2 \leq 6 \\ \delta_2 = (7-x_1)/(1+x_1+x_2) & \text{if } 2x_1 + x_2 \geq 6, \; x_1 \leq 7 \\ 0 & \text{if } x_1 \geq 7 \end{cases}$$

$$\pi(Z=z \mid x) = \begin{cases} \theta_1 = (z-x_1-2x_2)/(x_1+x_2) & \text{if } x_1+2x_2 \leq z \leq 2x_1+3x_2 \\ 1 & \text{if } 2x_1+3x_2 \leq z \leq 3x_1+4x_2 \\ \theta_2 = (4x_1+6x_2-z)/(x_1+2x_2) & \text{if } 3x_1+4x_2 \leq z \leq 4x_1+6x_2 \\ 0 & \text{otherwise.} \end{cases}$$

According to Equation (4.78), we obtain:

max   z

s.t. $\delta_1 \geq \alpha$ [or $8-x_2 \geq \alpha(2+x_1+x_2)$]

$\delta_2 \geq \alpha$ [or $7-x_1 \geq \alpha(1+x_1+x_2)$]

$\theta_2 \geq \alpha$ [or $4x_1+6x_2-z \geq \alpha(x_1+2x)$]

$x_1+2x_2 \geq 6, \; x_2 \leq 8, \; 2x_1+x_2 \geq 6, \; x_1 \leq 7$

$3x_1+4x_2 \leq z \leq 4x_1+6x_2.$

The solution for $\alpha = 0.5$ is $x_1 = 3.125$, $x_2 = 3.625$ and $z = 29.0625$.

Similarly, if we adapt the strict exceedance possibility (δ^2), then we will have:

max   z

s.t. $\delta_1, \delta_2, \dots, \delta_m, \theta_2 \geq \alpha$

$b_{i,3} \leq \Sigma_j a_{ij,4} x_j, \; b_{i,4} \geq \Sigma_j a_{ij,3} x_j, \; \forall i.$

$\Sigma_j c_{j,3} x_j \leq z \leq \Sigma_j c_{j,4} x_j$

$z, \; x_j, \; \forall j, \; \geq 0$ (4.79)

where δ_i, $\forall i$, are the strict exceedance possibilities for the constraints.

For a minimization problem, we can also use the (strict) exceedance possibility to solve it. However, we just have the B and R as the LHS and RHS, respectively. (This principle can also be applied to a situation which includes both kinds of constraints and consider θ_1 instead of θ_2.) The remaining derivations are the same as previous descriptions.

4.1.4.4 Fuller's Approach

To solve Equation (4.62), Fuller [97] proposed the following auxiliary crisp linear programming:

$$\max \quad \Sigma_j d_j x_j$$
$$\text{s.t.} \quad \Sigma_j f_{ij} x_j \leq g_i, \forall i, \text{ and } x_j \geq 0, \forall j \tag{4.80}$$

where

$$d_j = \int_0^1 r[c_j(r) + C_j(r)]dr, \forall j$$

$$f_{ij} = \int_0^1 r[a_{ij}(r) + A_{ij}(r)]dr, \forall i,j$$

$$g_i = \int_0^1 r[b_j(r) + B_i(r)]dr, \forall i.$$

The fuzzy numbers of c_j, b_i and a_{ij}, $\forall i,j$, can be parametrized by $\{(c_j(r), C_j(r), r) \mid 0 \leq r \leq 1\}$, $\{(b_i(r), B_i(r), r) \mid 0 \leq r \leq 1\}$ and $\{(a_{ij}(r), A_{ij}(r), r) \mid 0 \leq r \leq 1\}$, respectively.

Furthermore, Fuller proposed that if there exist real numbers u_j, v_{ij} and w_i, $\forall i,j$, such that:

$$u_j - c_j(r) = C_j(r) - u_j$$
$$v_{ij} - a_{ij}(r) = A_{ij}(r) - v_{ij}$$
$$w_i - b_i(r) = B_i(r) - w_i$$

for all $r \in [0, 1]$. Then we will have the following auxiliary problem:

$$\max \quad \Sigma_j u_j x_j$$
$$\text{s.t.} \quad \Sigma_j v_{ij} x_j \leq w_i, x_j \geq 0, \forall i \text{ and } j \tag{4.81}$$

for the Equation (4.62).

In our particular case of trapezoidal fuzzy numbers, we have:

$$c_j(r) = (c_{j,1} - c_{j,2})r + c_{j,2}$$
$$C_j(r) = (c_{j,3} - c_{j,4})r + c_{j,4}$$
$$a_{ij}(r) = (a_{ij,1} - a_{ij,2})r + a_{ij,2}$$
$$A_{ij}(r) = (a_{ij,3} - a_{ij,4})r + a_{ij,4}$$
$$b_i(r) = (b_{i,1} - b_{i,2})r + b_{i,2}$$
$$B_i(r) = (b_{i,3} - b_{i,4})r + b_{i,4} \tag{4.82}$$

for $\forall i, j$. Finally, Equation (4.80) is obtained by substituting the following crisp numbers:

$$d_j = 1/3[c_{j,1} + c_{j,3} + (1/2)(c_{j,2} + c_{j,4})]$$
$$f_{ij} = 1/3[a_{ij,1} + a_{ij,3} + (1/2)(a_{j,2} + a_{ij,4})]$$
$$g_i = 1/3[b_{i,1} + b_{i,3} + (1/2)(b_{i,2} + b_{i,4})] \tag{4.83}$$

for $\forall i, j$. Thus, we can easily solve Equation (4.62) with trapezoidal fuzzy numbers by use of Equations (4.80) and (4.83).

4.1.5 Other Problems

In Table 4.1, we have classified the possibilistic linear programming into seven cases. These cases are further grouped into four main problems. In the previous sub-sections we have discussed some approaches to solve these problems. There remains Cases 3, 6 and 7 to be discussed.

Obviously, Case 3 is a dual problem of Case 2. Thus we can solve Case 3 by transferring it into Case 2 and by using those approaches discussed in Section 4.1.2. On the other hand, Case 6 is very similar to Case 5, except the imprecise input data are in the right-hand side instead of the left-hand side. Thus, we can solve it by using the approach discussed in Section 4.1.3. Finally, we can solve Case 7, where only A is imprecise, by using the approach discussed in Section 4.1.3, too. That is to obtain a crisp auxiliary problem by transferring the imprecise technological coefficients A into crisp representative numbers by using the concept of the most possible values.

In the next section, we are going to discuss some extensions of possibilistic linear programming problems.

4.2 Some Extensions of Possibilistic Linear Programming

In the previous section, we have discussed all the typical possibilistic linear programming problems with imprecise coefficients. However, there are some other problems we call hybrid problems such as these situations:

(1) The technological coefficients, available resources and the inequalities of the constraints are imprecise. The imprecise technological coefficients and available resources are modelled by possibilistic distributions and the imprecise/fuzzy (soft) inequalities are modelled by preference based membership functions.

(2) The coefficients of the objective function are imprecise and modelled by possibilistic distributions. And, the fuzzy constraints involve fuzzy (soft) inequalities or fuzzy resources available, which can be modelled by preference-based membership functions.

(3) The imprecise nature of the input data (c, b and/or A) should be modelled by both the possibility theory and the probability theory.

First, let us discuss the linear programing problem with imprecise coefficients and inequalities.

4.2.1 Linear Programming with Imprecise Coefficients and Fuzzy Inequalities

The first hybrid problem can be mathematically formulated as:

$$\max \quad cx$$
$$\text{s.t.} \quad \Sigma_j \tilde{a}_{ij}x_j \lesssim \tilde{b}_i, \text{ and } x_j \geq 0, \forall i,j \qquad (4.84)$$

where $\tilde{a}_{ij}$ and $\tilde{b}_i$, $\forall i,j$, are triangular fuzzy numbers (fuzzy numbers with triangular possibility distributions), and $\lesssim$ is the soft inequality.

First, the soft inequality can be considered as the inequality with allowable tolerance to accomplish the constraints. That is: it permits the constraints to be satisfied as well as possible. Thus, we can model this imprecise situation by the preference-based membership function. Consider the following constraint:

$$\Sigma_j a_{ij}x_j \lesssim b_i \qquad (4.85)$$

where the tolerance of inequality is given p_i. This inequality constraint indicates that the consumption of resources ($\Sigma_j a_{ij}x_j$) is at best less than or equal to b_i, but no more than $b_i + p_i$.

Between b_i and $b_i + p_i$, there exists a non-increasing function representing the satisfaction of our willingness concerning the consumption of the resources. Thus the soft inequality can be modelled by the membership function (as discussed in Chapter 3), $\mu_i(x)$. Assume it is linear such as:

$$\mu_i(x) = \begin{cases} 1 & \text{if } \Sigma_j\, a_{ij}x_j \leq b_i \\ 1 - (\Sigma_j\, a_{ij}x_j - b_i)/p_i & \text{if } b_i \leq \Sigma_j\, a_{ij}x_j \leq b_i + p_i \\ 0 & \text{if } \Sigma_j\, a_{ij}x_j \geq b_i + p_i. \end{cases} \tag{4.86}$$

Let $\alpha = \min_i \mu_i(x)$ be the minimal acceptable level of satisfaction. That is: $\mu_i(x) \geq \alpha$, $\forall i$. Then the constraint of Equation (4.85) will be equivalent to:

$$\Sigma_j\, a_{ij}x_j \leq b_i + p_i(1 - \alpha). \tag{4.87}$$

Next, let us consider the following constraint:

$$\Sigma_j\, \tilde{a}_{ij}x_j \leq \tilde{b}_i \tag{4.88}$$

where $\tilde{a}_{ij} = (a_{ij}{}^m, a_{ij}{}^p, a_{ij}{}^o)$ and $\tilde{b}_i = (b_i{}^m, b_i{}^p, b_i{}^o)$, $\forall j$ and have triangular possibility distributions. This notation has been used in Section 4.1.2 (see Figure 4.6 also). Equation (4.88) is actually similar to the problem discussed in Section 4.1.1. Thus, all the approaches discussed in Section 4.1.1 can by applied here.

In order to solve the constraints of Equation (4.84), Delgado, Verdegay and Vila [55] proposed a relation $\ominus$ between fuzzy numbers, which preserves the ranking when fuzzy numbers are multiplied by positive scalars. $\ominus$ can be any reasonable inequality operator. However, Delgado et al. proposed the following two operations:

$$\tilde{X} \ominus_1 \tilde{Y} \iff X^m \leq Y^m, \qquad \tilde{X} \ominus_2 \tilde{Y} \iff X^o \leq Y^p \tag{4.89}$$

where $\tilde{X} = (X^m, X^p, X^o)$ and $\tilde{Y} = (Y^m, Y^p, Y^o)$. Thus, Equation (4.84) involves fuzzy ranking problems and may be equivalent to the following auxiliary problem:

$$\begin{aligned} \max \quad & cx \\ \text{s.t.} \quad & \Sigma_j\, \tilde{a}_{ij}x_j \ominus \tilde{b}_i + \tilde{p}_i(1-\alpha) \end{aligned} \tag{4.90}$$

where $\tilde{a}_{ij}$, $\tilde{b}_i$ and $\tilde{p}_i$ are fuzzy numbers (have possibility distributions).

Furthermore, for Equation (4.90) Campos [29], and Campos and Verdegay [30] provided the following equivalent crisp linear programming models:

(1) By use of Yager's first index (see [30]), we can obtain the equivalent model as:

$$\max \quad cx$$
$$\text{s.t.} \quad (a_i + a_i^l + a_i^u)x \leq (b_i + b_i^l + b_i^u) + (t_i + t_i^l + t_i^u)(1-\alpha)$$
$$x \geq 0 \text{ and } \alpha \in [0, 1], \forall i.$$

(2) By use of Yager's second index (see [30]), we can obtain the equivalent model as:

$$\max \quad cx$$
$$\text{s.t.} \quad (2a_i + a_i^l + a_i^u)x \leq (2b_i + b_i^l + b_i^u) + (2t_i + t_i^l + t_i^u)(1-\alpha)$$
$$x \geq 0 \text{ and } \alpha \in [0, 1], \forall i.$$

(3) By use of Campo's k-preference index [29], we can obtain the equivalent model as:

$$\max \quad cx$$
$$\text{s.t.} \quad (ka_i + (1-k)a_i^u)x \leq (kb_i + (1-k)b_i^u) + (kt_i + (1-k)t_i^u)(1-\alpha)$$
$$x \geq 0 \text{ and } \alpha \in [0, 1], \forall i$$

where $k \in (0, 1]$ is fixed by the decision maker a priori.

(4) By use of Dubois and Prade's k-weak feasibility concept [66], we can obtain the equivalent model as:

$$\max \quad cx$$
$$\text{s.t.} \quad (ka_i + (1-k)a_i^l)x \leq (kb_i + (1-k)b_i^u) + (kt_i + (1-k)t_i^u)(1-\alpha)$$
$$x \geq 0 \text{ and } \alpha \in [0, 1], \forall i$$

where $k \in (0, 1]$ is fixed by the decision maker a priori.

(5) By use of Dubois and Prade's k-hard feasibility concept [66], we can obtain the equivalent model as:

$$\text{max} \quad cx$$

$$\text{s.t.} \quad (ka_i + (1-k)a_i^l)x \leq (kb_i + (1-k)b_i^l) + (kt_i + (1-k)t_i^l)(1-\alpha)$$

$$x \geq 0 \text{ and } \alpha \in [0, 1], \forall i$$

where $k \in (0, 1]$ is fixed by the decision maker a priori.

(6) By use of Tanaka, Ichihashi and Asai's [240] inequality relation, we can obtain the equivalent model as:

$$\text{max} \quad cx$$

$$\text{s.t.} \quad (ka_i + (1-k)a_i^l)x \leq (kb_i + (1-k)b_i^l) + (kt_i + (1-k)t_i^l)(1-\alpha)$$

$$(ka_i + (1-k)a_i^u)x \leq (kb_i + (1-k)b_i^u) + (kt_i + (1-k)t_i^u)(1-\alpha)$$

$$x \geq 0 \text{ and } \alpha \in [0, 1], \forall i$$

where $k \in (0, 1]$ is fixed by the decision maker a priori.

The above six possible auxiliary models of Equation (4.84) are only crisp linear programming models and can be solved by any LP package. Of course, we can also use some other ranking methods discussed in Chapter 3 to obtain crisp auxiliary models. However, some of them will cause computational difficulties, such as nonlinearity, in solving auxiliary crisp models.

4.2.1a Example: A Fuzzy Matrix Game Problem

Consider a fuzzy matrix game [29] defined by the following matrix $[a_{ij}]$:

$$
\begin{array}{c}
\\
\begin{array}{cccc} y_1 & y_2 & \cdots & y_n \end{array} \\
\begin{array}{c} x_1 \\ x_2 \\ \vdots \\ x_m \end{array}
\left[
\begin{array}{cccc}
a_{11} & a_{12} & \cdots & a_{1n} \\
a_{21} & a_{22} & \cdots & a_{2n} \\
\vdots & & & \\
a_{m1} & a_{m2} & \cdots & a_{mn}
\end{array}
\right]
\end{array}
\qquad (4.91)
$$

As is well known, the optimal strategies x^* and y^* of a classical game will be the solutions of the following linear programming problems:

$$\text{max} \quad v$$

$$\text{s.t.} \quad \Sigma_i \, a_{ij}x_i \geq v, \, j = 1, 2, \ldots n$$

$$x_i \geq 0 \text{ and } \Sigma_i \, x_i = 1, \, i = 1, \ldots, m$$

and

$$\min \quad w$$

$$\text{s.t.} \quad \Sigma_j \, a_{ij} y_j \leq w, \, i = 1, 2, \ldots, m$$

$$y_j \geq 0 \text{ and } \Sigma_j \, y_j = 1, \, j = 1, \ldots, n.$$

If we let $u_i = x_i/v$ and $s_i = y_i/w$, $\forall i,j$, we will have:

$$\min \quad \Sigma_i \, u_i$$

$$\text{s.t.} \quad \Sigma_i \, a_{ij} u_i \geq 1, \, \forall j$$

$$x_i \geq 0, \, \forall i \tag{4.92}$$

and

$$\max \quad \Sigma_j \, s_j$$

$$\text{s.t.} \quad \Sigma_j \, a_{ij} s_j \leq 1, \, \forall i$$

$$y_j \geq 0, \, \forall j. \tag{4.93}$$

Now, if $\tilde{a}_{ij}$, $\forall i,j$, are fuzzy numbers and the inequalities are soft, then Equations (4.92) and (4.93) might be considered as:

$$\min \quad \Sigma_i \, u_i$$

$$\text{s.t.} \quad \Sigma_i \, \tilde{a}_{ij} u_i \geq 1 - \tilde{p}_j(1-\alpha), \, \forall j$$

$$x_i \geq 0, \, \forall i \tag{4.94}$$

and

$$\max \quad \Sigma_j \, s_j$$

$$\text{s.t.} \quad \Sigma_j \, \tilde{a}_{ij} s_j \leq 1 + \tilde{q}_i(1-\alpha), \, \forall i$$

$$y_j \geq 0, \, \forall j \tag{4.95}$$

where $\tilde{p}_j$ and $\tilde{q}_i$, $\forall i,j$, are imprecise tolerances for the soft inequalities. Equations (4.94) and (4.95) can then be solved by any of the previous approaches.

Consider the following fuzzy game:

$$\begin{pmatrix} \tilde{180} & \tilde{156} \\ \tilde{90} & \tilde{180} \end{pmatrix}$$

where $\tilde{180} = (180, 175, 190)$, $\tilde{156} = (156, 150, 158)$, $\tilde{90} = (90, 80, 100)$. Assume that players I and II have their imprecise margins (tolerances): $\tilde{p}_1 = \tilde{p}_2 = (0.10, 0.08, 0.11)$ and $\tilde{q}_1 = \tilde{q}_2 = (0.15, 0.14, 0.17)$, respectively.

By use of Yager's second Index (2), we obtain the following crisp parametric linear programming problems:

$$\min \quad u_1 + u_2$$
$$\text{s.t.} \quad 725u_1 + 360u_2 \geq 4 - 0.39(1-\alpha)$$
$$620u_1 + 725u_2 \geq 4 - 0.39(1-\alpha)$$
$$u_1, u_2 \geq 0 \text{ and } \alpha \in [0,1]$$

and

$$\max \quad s_1 + s_2$$
$$\text{s.t.} \quad 725s_1 + 620s_2 \leq 4 + 0.61(1-\alpha)$$
$$360s_1 + 725s_2 \leq 4 + 0.61(1-\alpha)$$
$$s_1, s_2 \geq 0 \text{ and } \alpha \in [0, 1].$$

The solution is: $x^* = (0.77, 0.23)$ with $v(\alpha) = 643.45/[4-0.39(1-\alpha)]$, and $y^* = (0.23, 0.77)$ with $w(\alpha) = 643.45/[4+0.61(1-\alpha)]$, where $\alpha = (0. 1]$.

4.2.2 Linear Programming with Imprecise Objective Coefficients and Fuzzy Resources

In practice, the unit costs/profits of new products or new projects, the lending and borrowing interest rates, and cash flows are always imprecise. This imprecise nature has long been studied with the help of the probability theory. However, the probability theory might not give us the correct meaning to solve some practical decision-making problems. In addition, applying the probability theory to some optimization problems has a negative effect on the computational efficiency. In these cases, the possibility theory might be helpful.

In this subsection, let us first discuss the following particular possibilistic linear programming problem:

$$\max \quad \Sigma_j x_j/(1+\tilde{r})^j$$
$$\text{s.t.} \quad x \in X = \{x \mid Ax \leq b \text{ and } x \geq 0\} \tag{4.96}$$

where $\tilde{r} = (r^m, r^p, r^o)$ which indicates the triangular possibilistic distribution, $\pi(.)$ as in Figure 4.7. r^m, r^p and r^o are the most possible value, the most pessimistic value and the most optimistic value, respectively. If $\pi(\cdot)$ is normalized, $\pi(r^m) = 1$ and $\pi(r^p) = \pi(r^o) = 0$.

In order to solve Equation (4.96), we first represent the triangular possibilities of r by the confidence interval of its β-cut (Kaufmann [131, 132]). That is: r can be represented by $[r^L(\beta),$

$r^U(\beta)]$ or $[r^p+(r^m-r^p)\beta,\ r^o-(r^o-r^m)\beta]$ where $\beta \in [0, 1]$. Thus, the problem becomes:

$$\max_{x \in X} \ \Sigma_j\ x_j/(1+[r^L(\beta),r^U(\beta)])^j$$

or

$$\max_{x \in X} \ \Sigma_j\ x_j[1/(1+r^U(\beta))^j,\ 1/(1+r^L(\beta))^j]. \tag{4.97}$$

Since the summation of interval values will be an interval value, the objective function of Equation (4.97) is obviously a crisp interval value when β is given. Thus we have the following equivalent problem:

$$\max_{x \in X} \ [z^L(x, \beta),\ z^U(x, \beta)] \tag{4.98}$$

where $z^L(x, \beta) = \Sigma_j\ x_j/(1+r^U(\beta))^j$, $z^U(x, \beta) = \Sigma_j\ x_j/(1+r^L(\beta))^j$. To solve Equation (4.98), the simplest way is to maximize $wz^L(x,\beta) + (1-w)z^U(x,\beta)$ where $w \in [0, 1]$ indicates the possible weight. On the other hand, the objective of maximizing an interval value can also be solved by simultaneously maximizing the lower and upper bounds of the interval value. Thus we [140] propose the following bi-objective auxiliary problem:

$$\max \quad z^L(x,\beta) = \Sigma_j\ x_j/(1+r^U(\beta))^j$$
$$\max \quad z^U(x,\beta) = \Sigma_j\ x_j/(1+r^L(\beta))^j$$
$$\text{s.t.} \quad x \in X \tag{4.99}$$

where $r^L(\beta) = r^p+(r^m-r^p)\beta$, $r^U(\beta) = r^o-(r^o-r^m)\beta$ and $\beta \in [0, 1]$. If β is first given, then $r^L(\beta)$ and $r^U(\beta)$ will be constants and the problem of Equation (4.99) will be a crisp bi-objective linear programming problem which can be solved by any MODM approach. However, in this study we will propose an augmented max-min approach which is an extended version of Zimmermann's approach for solving a crisp multiple objective programming problem.

The problem of Equation (4.99) may be solved by the following two steps. The first step is to solve the following max-min problem (Zimmermann [298]):

$$\max \quad \Gamma$$
$$\text{s.t.} \quad \mu_1(z^L(x, \beta)) \geq \Gamma$$
$$\mu_2(z^U(x, \beta)) \geq \Gamma$$
$$x \in X \tag{4.100}$$

whose solution is Γ^0 and x^0. Note that in the above equation Γ is min $[\mu_1(z^L(x,\beta)), \mu_2(z^U(x, \beta))]$.

There may be multiple decisions (x) which will end up with the same Γ^0. In this case, the solution of Equation (4.100) might not be unique nor efficient, because Equation (4.100) randomly selected a solution which might be dominated by another solution with the same satisfaction level Γ^0. Moreover, we do not know whether or not multiple solutions exist. To resolve this problem, we proposed the second step which is:

$$\begin{aligned} \max \quad & w_1\mu_1(z^L(x, \beta)) + w_2\mu_2(z^U(x, \beta)) \\ \text{s.t.} \quad & \mu_1(z^L(x, \beta)) \geq \Gamma^0 \\ & \mu_2(z^U(x, \beta)) \geq \Gamma^0 \\ & x \in X \end{aligned}$$

(4.101)

where w_1 and w_2 are the weights (relative importance) among the corresponding objective functions, and $w_1 + w_2 = 1$. This second step is essentially trying to use the weighted averaging operator to allow for possible compensation among objectives, which is similar to that proposed by Li [151]. Equation (4.101) also guarantees that the overall satisfactory degree of the compromise of the objectives is at least at Γ^0.

It is noted that for a bi-objective programming problem, the solution of Equation (4.101) may be equal to the solution of Equation (4.100). For a problem with a large number objective functions, Equation (4.101) should be formed and solved for obtaining efficient solutions.

Furthermore, by observing the solution procedure of the simplex method, the second step will automatically be solved after solving the first step, if we unify both objectives of Equations (4.100) and (4.101) as follows:

$$\begin{aligned} \max \quad & \Gamma + \delta \sum_k w_k\mu_k(f_k(x)) \\ \text{s.t.} \quad & \mu_1(z^L(x, \beta)) \geq \Gamma \\ & \mu_2(z^U(x, \beta)) \geq \Gamma \\ & x \in X \end{aligned}$$

(4.102)

where δ is a sufficiently small positive number. We considered Equation (4.102) an augmented max-min approach.

With equal weights among objectives, the augmented max-min approach is, in some points of view, a dual model of the augmented minimax programming model proposed by Sakawa and

Yano [220]. However, if there are not equal weights among objectives, the proposed augmented max-min approach will not be a dual problem of augmented minimax approach at all.

From the aspect of the fuzzy sets theory, it should be noted that the proposed max-min approach allows for compensation among objectives, which cannot be reached by using the max-min operator.

Besides, the max-min model (when equal weight) is a dual model of the Tchebycheff distance criterion which has the largest measurement unit, and the average model is basically the city block distance criterion which has the smallest measurement unit among Minkowski's L_p metric. Thus, the augmented max-min approach is first to reach compromise solutions at a large unit, and then reevaluates these compromise solutions and obtains an efficient compromise solution at a smaller unit (a smaller measurement unit will provide more precise measures). Through this strategy, we will obtain an efficient compromise solution which provides more uniform achieved rates (realized degrees of satisfaction) for each objective than the solution directly from the average model. The individually realized satisfaction levels of the objectives are always more consistent from the max-min model than from the average model.

It is worth noting that the membership functions ($\mu_1()$ and $\mu_2()$), which are elicited on the basis of preference concept, can be any non-decreasing functions such as linear, piecewise linear, exponential, hyperbolic, inverse hyperbolic or some specific power functions as discussed in Chapter 2. In this study, linear membership functions are assumed, because the other types of membership functions can be transferred into equivalent linear forms by use of the variable transformation concept. That is (β is given):

$$\mu_1(x) = \begin{cases} 1 & \text{if } z^L(x) > z^{L^*} \\ [z^L(x)-z^L]/[z^{L^*}-z^L] & \text{if } z^L \leq z^L(x) \leq z^{L^*} \\ 0 & \text{if } z^L(x) < z^L \end{cases}$$

where z^{L^*} and z^L are the positive and negative ideal solutions (individual maximum and minimum), respectively. Similarly, we can establish $\mu_2(x)$. By substituting these membership functions into Equation (4.102), we will obtain a crisp linear programming problem under given β and δ.

When the interest rates (r_j) of different time periods have different possibilistic distributions, we will have the following problem:

$$\max \quad x_1/(1+r_1^U(\beta)) + \dots + x_K/\{ \prod_{k=1}^{K} [1+r_k^U(\beta)]\}$$

$$\max \quad x_1/(1+r_1^L(\beta)) + \dots + x_K/\{ \prod_{k=1}^{K} [1+r_1^L(\beta)]\}$$

s.t. $\quad x \in X$

where $r^L(\beta) = r^p + (r^m - r^p)\beta$, $r^U(\beta) = r^o - (r^o - r^m)\beta$, Vi, and $\beta \in [0, 1]$. Similarly, The above equation can be solved by the augmented max-min approach.

For the imprecise resources available, we have the following constraints of $(Ax)_i \le b_i$, Vi, where the preference-based membership functions of b_i, Vi, are assumed to be the same as Equation (4.86). Thus we have the crisp auxiliary constraints of $(Ax)_i \le b_i + p_i(1 - \alpha)$, Vi, as shown in the previous section.

In order to demonstrate the approach, let us consider the following bank hedging decision problem.

4.2.2a Example: A Bank Hedging Decision Problem

How to manage the interest rate problem in bank balance sheet management has been widely discussed by Booth and Koveos [17]. The probability theory is always a unique way to handle the uncertain interest rate problem in traditional financial analysis. However, in this study, we will assume that the interest rate is imprecise and has a possibilistic distribution, instead of a probabilistic distribution. The considered basic bank hedging decision problem is based on Booth and Koveos's model. We further modify this model by considering fuzzy/imprecise interest rates, price of futures contract, loan demand, deposit supply and ratio of desired loan to deposit. In this problem, the fuzziness of interest rates and price of futures should be modelled by (triangular) possibilistic distributions. On the other hand, the fuzziness of loan demand, deposit supply and ratio of desired loan to deposit should be modelled by preference-based membership functions because they involve the nature of the subjective satisfaction.

This fuzzy bank hedging decision model can then be modelled as:

$$\max \quad E \qquad\qquad (4.103\text{a})$$

$$\text{s.t.} \quad R \geq aD + mS_F \qquad\qquad (4.103\text{b})$$

$$S_{STS} \leq (1 - b_{STS})STS^* \qquad\qquad (4.103\text{c})$$

$$S_{LTS} \leq (1 - b_{LTS})LTS^* \qquad\qquad (4.103\text{d})$$

$$S_F \leq (1 - b_{LTS})LTS^* \qquad\qquad (4.103\text{e})$$

$$P_L \leq \tilde{\text{ld}} \qquad\qquad (4.103\text{f})$$

$$D \leq \tilde{d} \qquad\qquad (4.103\text{g})$$

$$L \leq \tilde{c}D \qquad\qquad (4.103\text{h})$$

$$STS = (1 - b_{STS})STS^* + P_{STS} - S_{STS} \qquad\qquad (4.103\text{i})$$

$$LTS = (1 - b_{LTS})LTS^* + P_{LTS} - S_{LTS} \qquad\qquad (4.103\text{j})$$

$$L = (1 - b_L)L^* + P_L \qquad\qquad (4.103\text{k})$$

$$R + STS + LTS + L = D + E \qquad\qquad (4.103\text{l})$$

where

$$E = E^* + \underset{(1)}{} + \underset{(2\text{a})}{r_{STS}^*(1-b_{STS})STS^*} + \underset{(2\text{b})}{r_{LTS}^*(1-b_{LTS})LTS^*} + \underset{(2\text{c})}{r_L^*(1-b_L)L^*} + \underset{(3\text{a})}{r_{STS}P_{STS}} + \underset{(3\text{b})}{r_{LTS}P_{LTS}} + \underset{(3\text{c})}{r_L P_L}$$

$$- \underset{(4\text{a})\ (5\text{a})}{(r_{STS}^* + k_{STS})S_{STS}} - \underset{(4\text{b})\ (5\text{b})}{(r_{LTS}^* + k_{LTS})S_{LTS}} + \underset{(6)}{[1 - (\tilde{p}_{PF}/p_{SF})]S_F} - \underset{(7)}{\tilde{r}_D D}$$

and

$$k_{STS} = 1 - \left\{ r_{STS}^* \left[\sum_{t=1}^{(.5/b_{STS})+.5} (1 + \tilde{r}_{STS})^{-t} \right] + (1 + \tilde{r}_{STS})^{-(.5/bSTS)+.5} \right\}$$

$$k_{LTS} = 1 - \left\{ r_{LTS}^* \left[\sum_{t=1}^{(.5/b_{LTS})+.5} (1 + \tilde{r}_{LTS})^{-t} \right] + (1 + \tilde{r}_{LTS})^{-(.5/bLTS)+.5} \right\}$$

and where "*" indicates historical data, and where $\tilde{p}_{PF}$, $\tilde{r}_D$, $\tilde{r}_{STS}$ and $\tilde{r}_{LTS}$ are fuzzy and have triangular possibilistic distributions, and $\tilde{\text{ld}}$, $\tilde{d}$ and $\tilde{c}$ are fuzzy and have preference-based membership functions. The specific variables and parameters of Equation (4.103) are shown in Table 4.5.

Table 4.5 Variables and parameters

Balance sheet variables:

    R   Reserves

    STS Short-term securities
    LTS Long-term securities

    L   Loans
    D   Deposits
    E   Equity

Decision variables:

    First stage:

        P_{STS} Purchase short-term securities (STS)
        P_{LTS} Purchase long-term securities (LTS)
        P_L Purchase Loans (L)
        S_F Sell futures

    Second stage:

        S_{STS} Sell short-term securities (STS)
        S_{LTS} Sell long-term securities (LTS)

Parameters and other variables:

    a   Portion of deposits (D) associated with reserves
    m   Margin per dollar of futures sold (S_F)

    b_{STS} Portion of short-term securities (STS) maturing
    b_{LTS} Portion of long-term securities (LTS) maturing
    b_L Portion of loans (L) maturing

    ld   Maximum new loan demand
    d   Maximum new deposit supply
    c   Maximum desired loan (L) to deposits (D) ratio

    r_{STS} Short-term security (STS) interest rate
    r_{LTS} Long-term security (LTS) interest rate
    r_L Loan (L) interest rate

    p_{SF} Price of futures contract sold
    p_{PF} Price of futures contract purchased

    k_{STS} Capital gains/losses of short-term security sales
    k_{LTS} Capital gains/losses of long-term security sales

The objective of Equation (4.103a) is to maximize the expected ending balance sheet value of stockholders' equity or to maximize expected profit. Stockholders' equity at the end of the period E equals: (1) beginning equity (E^*); plus the net revenues associated with (2) non-maturing and (3) newly acquired securities and loans; minus (4) the net revenues corresponding to securities sold; plus (or minus) (5) capital gains or losses (book value minus anticipated ending price); plus (or minus) (6) the gain (or loss) on the hedging transaction; minus (7) the cost of deposits.

It should be noted that the futures purchase transaction is assumed to exactly offset the sell position. Thus, per dollar profit of selling futures is $[1-(\tilde{p}_{PF}/p_{SF})]$ which reflects the observation that futures and spot prices are not mechanically related and permits the bank to be subject to basis risk.

Constraint (4.103b) requires that at the period's end the bank's cash and reserve holdings are at least a certain portion of deposits plus a margin requirement associated with the sale of futures contracts, according to mandated and managerial reserve requirements. Constraint (4.103c) indicates that the bank can sell short-term, non-maturing securities but cannot sell securities purchased. Constraint (4.103i) specifies the ending balance sheet value for short-term securities to equal short-term securities that have not yet matured plus securities purchased during the period minus securities sold. Similarly, we have constraints (4.103d) and (4.103j) that pertain for long-term securities. Constraint (4.103e) is constructed so that only the interest rate risk associated with long-term securities is hedged and the magnitude of this hedge is determined by the maximum dollar volume of long-term securities that could be sold. Constraint (4.103f) specifies that the bank cannot make loans in excess of that supplied. Constraint (4.103k) indicates that the ending balance sheet value for loans equals those loans not maturing during the planning period plus loans made. Constraint (4.103g) indicates that a bank cannot obtain more deposits than are supplied. Constraint (4.103h) brings together constraints (4.103g) and (4.103k) by defining a managerially determined relationship between loans and deposits where the loan to deposit ratio indicates the flexibility for managing liquidity. Finally, constraint (4.103l) requires that total assets equal deposits plus equity.

Now, let us assume the numerical data as in Table 4.6 where the interest rates and price of futures contract purchased are imprecise and have triangular possibilistic distributions such as "about 8%" = (0.08, 0.075, 0.085), "about 9%" = (0.09, 0.085, 0.10), "about 4%" = (0.04, 0.035, 0.05) and "about $92" = $(92, 91, 94). In the above equation, the maximum new loan demand (ld), the maximum new deposit supply (d) and the maximum desired loan to deposit ratio

(c) are subjectively determined. Thus they should be modelled by fuzzy sets. Let us assume that the membership functions of the fuzzy sets $\widetilde{ld}$, $\widetilde{d}$ and $\widetilde{c}$ are linearly decreasing functions in [400, 480], [950, 1070] and [0.8, 0.9], respectively. More precisely, we define the membership function of $\widetilde{ld}$ as:

$$\mu(ld) = \begin{cases} 1 & \text{if } ld < 400 \\ 1 - (ld\text{-}400)/80 & \text{if } 400 \leq ld \leq 480 \\ 0 & \text{if } ld > 480. \end{cases}$$

Similarly, we can also define the membership functions of $\widetilde{d}$ and $\widetilde{c}$ as the above function.

Table 4.6 Numerical data of the bank hedging decision problem

Variable/ Parameter	Historical value	Current value	Future value
Balance Sheet			
R	$200		
STS	100		
LTS	300		
L	400		
D	900		
E	100		
Interest Rate (annual)			
r_{STS}	.07	.08	"about .08"
r_{LTS}	.08	.09	"about .09"
r_L	.11	.14	
r_D			"about .04"
Futures			
p_{SF}		$92	
p_{PF}			"about $92"
m		.025	
Market Size			
ld		$400 (80)	
d			$950 (120)
Others			
a		.20	
b_{STS}		.50	
b_{LTS}		.125	
b_L		.125	
c		.8 (.1)	

Note: (1) interest rates and futures prices are expressed in annual terms, and other data are quarterly;
(2) the numbers inside the parenthesis indicate the corresponding tolerance values.

Assume $\alpha = \min(\mu(\text{ld}), \mu(d), \mu(c))$. Then we have $\mu(\text{ld}) \geq \alpha$, $\mu(d) \geq \alpha$ and $\mu(c) \geq \alpha$. That is: ld $\leq 400 + (1-\alpha)80$, d $\leq 950 + (1-\alpha)120$ and c $\leq [0.8 + (1-\alpha)0.1]$, where $\alpha \in [0, 1]$. Here, α is the minimum satisfaction level of the fuzzy constraints or the degree of feasibility.

Finally, it is noted that the annual data should be transferred into quarterly units before entering Equation (4.103). The numerical fuzzy bank hedging decision is then shown in follows:

$$\max \quad E$$

$$\text{s.t.} \quad R \geq 0.2D + 0.25S_F$$

$$S_{STS} \leq 50, \quad S_{LTS} \leq 262.5, \quad S_F \leq 262.5$$

$$P_L \leq 400 + (1-\alpha)80$$

$$D \leq 950 + (1-\alpha)120$$

$$L \leq [0.8 + (1-\alpha)0.1]D$$

$$STS = (50 + P_{STS} - S_{STS})$$

$$LTS = (262.5 + P_{LTS} - S_{LTS})$$

$$L = 350 + P_L$$

$$R + STS + LTS + L = D + E$$

where

$$E = 115.75 + 0.02P_{STS} + 0.0225P_{LTS} + 0.035P_L - (0.0175 + k_{STS})S_{STS}$$
$$- (0.02 + k_{LTS})S_{LTS} + (0, -.022, .011)S_F - (0.01, 0.00875, 0.0125)D$$

$$k_{STS} = 1 - [0.0175/(1.02, 1.01875, 1.02125) + 1.0175/(1.02, 1.01875, 1.02125)^{1.5}]$$

$$k_{LTS} = 1 - [0.02/(1.0225, 1.02125, 1.025) + 0.02/(1.0225, 1.02125, 1.025)^2 +$$
$$0.02/(1.0225, 1.02125, 1.025)^3 + 0.02/(1.0225, 1.02125, 1.025)^4 +$$
$$1.02/(1.0225, 1.02125, 1.025)^{4.5}.$$

To solve this problem, let us first assume that the acceptable degree of the feasibility of the constraints is 0.7 (or $\alpha = 0.7$) and the acceptable degree of the possibilistic occurrence of the fuzzy objective is 0.5 (or $\beta = 0.5$).

Next, we should obtain the PIS and the NIS (or individual maximum and minimum) solutions for both objectives, $z^L(x, 0.5)$ and $z^U(x, 0.5)$. The solution is then $z^{L*} = 133.6698$, $z^{U*} = 136.6401$, $z^{L-} = 94.0643$ and $z^{U-} = 100.9776$.

With the PIS and NIS, linear membership functions of both objectives of Equation (4.98) can be obtained. And the remaining problem is to solve Equation (4.101) which is (assume $\alpha = 0.7$, $\beta = 0.5$, $\delta = 0.01$ and $w_1 = w_2 = 0.5$):

$$\max \ \Gamma + 0.000126E^L + 0.000140E^U$$

$$\text{s.t.} \quad E^L - 39.6055\Gamma \geq 94.0643$$

$$E^U - 35.6625\Gamma \geq 100.9776$$

$$R - 0.2D - 0.25S_F \geq 0$$

$$S_{STS} \leq 50, \ S_{LTS} \leq 262.5, \ S_F \leq 262.5$$

$$P_L \leq 424, \ D \leq 986, \ L - 0.83D \leq 0$$

$$STS - P_{STS} + S_{STS} = 50$$

$$LTS - P_{LTS} + S_{LTS} = 50$$

$$L - P_L = 350$$

$$R + STS + LTS - D - E = 0$$

$$E^L \leq E \leq E^U$$

$$E^L - 0.02P_{STS} - 0.0225P_{LTS} - 0.035P_L + 0.01354S_{STS} + 0.026776S_{LTS} +$$
$$0.11S_F + 0.01125D = 115.75$$

$$E^U - 0.02P_{STS} - 0.0225P_{LTS} - 0.035P_L + 0.011704S_{STS} + 0.018832S_{LTS} -$$
$$0.0055S_F + 0.009375D = 115.75$$

$$\Gamma \in [0, 1]$$

where all the decision variables are non-negative. The numerical solution is provided in Table 4.7. Table 4.8 presents the solution of the balance sheet and the corresponding decisions. Furthermore, we also solve this problem with various possibility degrees. Their solutions are provided in Table 4.7 and Table 4.8. The possibility distribution of the optimal equity is presented in Figure 4.18. Obviously, it is a non-linear, bell-shaped distribution.

Table 4.7 Solutions of the bank hedging decision problem ($\alpha = 0.7$)

β	PIS z^{L*}	z^{U*}	NIS z^{L-}	z^{U-}	Compromise solution z^L	z^U	Γ
1.0					135.2494	135.2494	1.0000
0.9	134.7159	135.5271	98.5064	99.8898	134.7159	135.5240	0.9999
0.8	134.4540	135.8050	97.3944	100.1614	134.2779	135.6356	0.9952
0.7	134.1916	136.0824	96.2821	100.4325	133.8897	135.7985	0.9920
0.6	133.9312	136.3600	95.1745	100.7039	133.5161	135.9781	0.9893
0.5	133.6698	136.6401	94.0643	100.9776	133.1435	136.1662	0.9867
0.4	133.4085	137.1803	92.9546	101.2498	132.6211	136.4809	0.9805
0.3	133.1466	137.7533	91.8455	101.5211	132.0675	136.8067	0.9739
0.2	132.8852	138.3268	90.7369	101.7930	131.5050	137.1304	0.9673
0.1	132.6236	138.9025	89.6280	102.0670	130.9343	137.4550	0.9607
0.0	132.3619	139.4760	88.5194	102.3389	130.3572	137.7779	0.9543

Table 4.8 The optimal balance sheet and decision ($\alpha = 0.7$)

β	1.0	0.9	0.8	0.7	0.6	0.5
Balance sheet						
R	197.2000	197.2000	197.2000	197.2000	197.0000	197.2001
STS	0.0000	0.0000	0.0000	0.0000	0.0000	0.0000
LTS	924.0494	924.1552	924.4357	924.5985	924.7780	924.9662
L	774.0000	774.0000	774.0000	774.0000	774.0000	774.0000
D	986.0000	986.0000	986.0000	986.0000	986.0000	985.9999
E	135.2494	135.5240	135.6356	135.7985	135.9781	136.1662
Decision						
1st stage						
P_{STS}	0.0000	0.0000	0.0000	0.0000	0.0000	0.0000
P_{LTS}	924.0493	924.1552	844.5420	817.8739	805.5091	798.6815
P_L	424.0000	424.0000	424.0000	424.400	424.0000	424.0000
S_F	0.0000	0.0000	0.0000	0.0000	0.0000	0.0000
2nd stage						
S_{STS}	50.0000	50.0000	50.0000	50.0000	50.0000	50.0000
S_{LTS}	262.5000	262.5000	182.6064	155.7754	143.2311	136.2154

(Continuous)

β	0.4	0.3	0.2	0.1	0.0
Balance sheet					
R	197.2001	197.1999	197.2001	197.2000	197.2000
STS	0.0000	0.0000	0.0000	0.0000	0.0000
LTS	925.2808	925.6066	925.9304	926.2550	926.2578
L	774.0000	774.0000	774.0000	774.0000	774.0000
D	986.0000	986.0000	986.0000	986.0000	986.0000
E	136.4809	136.8067	137.1304	137.4550	137.7779
Decision					
1st stage					
P_{STS}	0.0000	0.0000	0.0000	0.0000	0.0000
P_{LTS}	823.4657	845.1398	862.0480	875.7864	887.2448
P_L	424.0000	424.0000	424.0000	424.0000	424.0000
S_F	0.0000	0.0000	0.0000	0.0000	0.0000
2nd stage					
S_{STS}	50.0000	50.0000	50.0000	50.0000	50.0000
S_{LTS}	160.6861	182.0331	198.6176	212.0314	223.1670

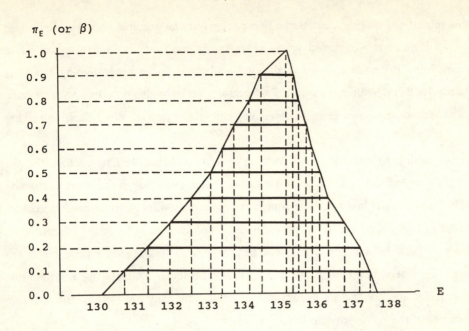

Figure 4.18 The possibility distribution of the optimal equity

From the above solution tables, the bank should obtain the funds through the maturation of its existing short-term security, long-term security and loan balance, and accept all the deposits supplied to it. Reserve balances should be reduced to the lowest level possible as indicated by the reserve and potential margin requirements. Thus, the reserve balance is reduced to 197.2 from 200. All of the non-maturing short-term securities should be sold. The amount of selling long-term securities is significantly varied with the acceptable possibility degree. Therefore, controlling the long-term securities is one of the most important decisions in order to maximize the equity. The bank should, at the same time, use some of these funds to purchase long-term securities and loans. Again, purchase of the long-term securities is very sensitive to the degree of possibility, and thus the bank should pay more attention to controlling long-term securities. On the other hand, the bank should always purchase all possible loans which will greatly increase the equity.

In this particular case, the bank should not purchase short-term securities and futures which will not increase its equity. Although we have considered the possible hedging strategy, no hedging is required. However, if the purchase price of futures is reduced, the potential hedging profit will increase; thereby we will purchase futures to hedge the interest-rate risk.

Similarly, we can obtain another distribution of the optimal equity and the corresponding balance sheet for a different α value (degree of allowable feasibility).

In this subsection, we have discussed a linear programming problem with a maximization objective of an investment discounted value where the interest rate is imprecise and has a triangular possibility distribution. An auxiliary bi-objective linear programming model is proposed to resolve this possibilistic nature. Furthermore, we have also proposed an augmented max-min approach to solve this crisp bi-objective (multiple objective, also) decision-making problem.

Through solving the numerical bank hedging decision problem, we have realized that the distribution of the objective function is a non-linear bell shape, even though the interest rates and the price of futures contract purchased have simple triangular possibility distributions. Moreover, the solution may be very sensitive to the allowable degree of the possibility of the interest rates. Therefore, a rather detailed solution table should be presented to the decision maker, in order to explain the structure of such a complex problem. In our particular example, we also find that long-term securities are key factors in maximizing equity, for they are very sensitive to the possibility degree of the corresponding interest rate.

In practice, the possibility distributions of the interest rates should be different as time goes on. To solve this dynamic system, the solution procedure discussed here can be applied also, except we need to build an interactive system to allow for the dynamic change.

Finally, we should emphasize that the desired possibilities (β) for the interest rates should not be confused with the desired degrees (α) of the feasibility. Because, the membership functions of the feasibility are elicited on the basis of the preference concept. Thus, the desired degree of feasibility may be strongly dependent upon the psychological value system. On the other hand, the possibility distributions are built on the basis of the possible occurrence of an event which is analogous to the probability concept. The possibility distribution is then determined and influenced by the (considered) physical environment. Of course, both can be subjectively evaluated by the decision makers.

4.2.3 Stochastic Possibilistic Linear Programming

In the decision-making process we might face a hybrid environment where both linguistic and frequent imprecise nature exists. For instance, the interest rate for the coming year might have a 60% chance of being "around 8%" and a 40% chance of being "around 9.5%," when a particular new government policy might either be executed (60% chance) or aborted (40% chance). The problem of frequent imprecision has been solved by using the probability theory. On the other hand, the problem of linguistic imprecision should be handled by using the fuzzy sets theory. Thus, to solve this hybrid problem, it is necessary to use both theories simultaneously. (Lai and Hwang [142])

The hybrid problems have been discussed in the literature since Zadeh in 1968 [275]. The most detailed discussions of vague statistical problems were provided by Kruse and Meyer [BM36], Negoita and Ralescu [BM41], Jones, Kaufmann and Zimmermann [BM23] and Kacprzyk and Fedrizzi [BM27]. For a linear programming problem with hybrid environment, Luhandjula [156] used the concepts of fuzzy and non-fuzzy probabilities of a fuzzy event proposed by Zadeh [275], and the concepts of the expected values and min-operator to handle the hybrid imprecision on the constraint coefficients. Wierzchon [264a] considered the problem that the right-hand side of the constraint inequalities are independently distributed normal random variables with fuzzy mean values and fuzzy standard deviations.

Here, we will discuss the problem of maximizing investment discounted value. That is:

$$\max \quad \Sigma_j \ x_j/(1+\tilde{\tilde{r}})^j$$
$$\text{s.t.} \quad x \in X \tag{4.104}$$

where $\tilde{\tilde{r}} = \{\tilde{r}_s \mid \tilde{r}_s = (r_s{}^m, r_s{}^p, r_s{}^o)$ with probability p_s and s is the state index$\}$ (see Figure 4.19 when s = 1 and 2 only). As shown in Figure 4.19, $r_s{}^m$, $r_s{}^p$ and $r_s{}^o$ are the most possible, pessimistic and optimistic values, respectively.

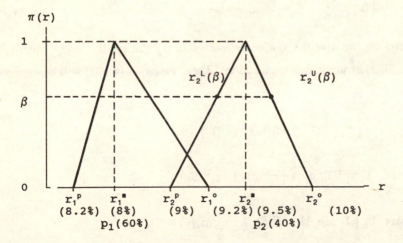

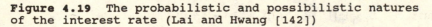

Figure 4.19 The probabilistic and possibilistic natures of the interest rate (Lai and Hwang [142])

To solve Equation (4.104), we must first resolve the stochastic and possibilistic natures. For computational efficiency, we may use the concept of expected value to resolve the probabilistic nature. That is to maximize the expected value of the stochastic and possibilistic objective. Of course, we may also minimize the variance and/or maximize the skewness of this objective function. In this study, we first consider the case of maximizing the expected value. Thus, with the assumption of the independence property we obtain the following possibilistic linear programming:

$$\max \; \Sigma_s \left\{ \Sigma_j \; x_j/(1+\tilde{r}_s)^j \right\} p_s$$
$$\text{s.t.} \quad x \in X$$

or

$$\max \; \Sigma_j \; x_j \left\{ \Sigma_s \; p_s/(1+\tilde{r}_s)^j \right\}$$
$$\text{s.t.} \quad x \in X \tag{4.105}$$

where $\tilde{r}_s = (r_s^m, r_s^P, r_s^o)$.

Obviously, our next step is to solve the possibilistic nature of Equation (4.105). To do that, let us rewrite the possibilistic distribution of $\tilde{r}_s = (r_s^m, r_s^P, r_s^o)$, $\forall s$, as the following function form (see Figure 4.19):

$$\pi_s(r_s) = \begin{cases} (r_s-r_s^P)/(r_s^m-r_s^P) & \text{if } r_s^P \leq r_s \leq r_s^m \\ (r_s^o-r_s)/(r_s^o-r_s^m) & \text{if } r_s^m \leq r_s \leq r_s^o \\ 0 & \text{otherwise.} \end{cases} \tag{4.106}$$

Then for each β-level cut, we have the confidence interval of $[r_s^L(\beta), r_s^U(\beta)] = [r_s^P+(r_s^m-r_s^P)\beta, r_s^o-(r_s^o-r_s^m)\beta]$, $\beta \in [0, 1]$ as in the previous section. Then, Equation (4.105) will be equivalent to:

$$\max_{x \in X} \; \Sigma_j \; x_j \left\{ \Sigma_s \; p_s/(1+[r_s^L(\beta), r_s^U(\beta)])^j \right\}$$

$$= \max_{x \in X} \; \Sigma_j \; x_j \left\{ \Sigma_s \; p_s/[1+r_s^L(\beta), 1+r_s^U(\beta)]^j \right\}$$

$$= \max_{x \in X} \; \Sigma_s \left\{ \Sigma_j \; x_j p_s/[1+r_s^L(\beta), 1+r_s^U(\beta)]^j \right\}$$

$$= \max_{x \in X} \; \Sigma_s \left\{ \Sigma_j \; x_j p_s([1/(1+r_s^U(\beta)), 1/(1+r_s^L(\beta))]^j \right\}$$

$$= \max_{x \epsilon X} \Sigma_s \Sigma_j [(x_j p_s)^{1/j}/(1+r_s^U(\beta)), (x_j p_s)^{1/j}/(1+r_s^L(\beta))]^j$$

$$= \max_{x \epsilon X} [z^L(x,\beta), z^U(x,\beta)]. \tag{4.107}$$

The objective function of Equation (4.107) is obviously a crisp interval value when β is given, since the summation of interval values will be an interval value. Therefore, we may use the weighted average concept. That is to maximize $wz^L(x,\beta) + (1-w)z^U(x,\beta)$, where $w \epsilon [0, 1]$ indicates the relative weight (importance). However, to maximize an interval value can also be obtained by simultaneously maximizing the lower and upper bound of the interval value. Thus we proposed the following bi-objective auxiliary problem for Equation (4.104):

$$\begin{aligned} \max \quad & z^L(x,\beta) \\ \max \quad & z^U(x,\beta) \\ \text{s.t.} \quad & x \epsilon X \end{aligned} \tag{4.108}$$

where $\beta \epsilon [0, 1]$. If β is first given, then $r^L(\beta)$ and $r^U(\beta)$ will be constants and the problem of Equation (4.108) will be a crisp bi-objective linear programming problem which can be solved by the previous augmented max-min approach . That is:

$$\begin{aligned} \max \quad & \Gamma + \delta \Sigma_k w_k \mu_k(f_k(x)) \\ \text{s.t.} \quad & \mu_1(z^L(x, \beta)) \geq \Gamma \\ & \mu_2(z^U(x, \beta)) \geq \Gamma \\ & x \epsilon X \end{aligned} \tag{4.109}$$

where δ is a sufficiently small positive number. With the membership functions $\mu_1(.)$ and $\mu_2(.)$ as discussed in the previous section, Equation (4.109) will be a crisp linear programming problem under given β and δ.

When the interest rates (r_j) of different time periods have different possibilistic distributions, we will have the following problem:

$$\max \quad x_1/(1+r_1) + ... + x_n/[\prod_{j=1}^{n}(1+\tilde{\tilde{r}}_j)]$$

$$\text{s.t.} \quad x \epsilon X \tag{4.110}$$

where $\bar{\bar{r}}_j = \{\tilde{r}_{js} \mid \tilde{r}_{js} = (r_{js}{}^m, r_{js}{}^P, r_{js}{}^o)$ with probability p_s and s is the state index$\}$ for $j = 1, ...,$ n. Similarly, Equation (4.110) can be solved by the augmented max-min approach.

In order to demonstrate the proposed approach, a real-world problem should be considered and solved.

4.2.3a Example: A Bank Hedging Decision Problem

Consider the bank hedging decision problem as discussed in Section 4.2.2a. Here we assumed the following environment (see Table 4.9 also) (Lai and Hwang [142]):

(1) Some interest rates are assumed, for instance, to have a 40% chance of being "around 8%" and a 60% chance of being "around 9.5%." That is: the decision maker can only provide linguistic values, which are then modelled by possibilistic distributions, in each (probability) state. Similarly, the price of the futures contract also has the same nature.

(2) The maximum desired values of loan demand and ratio of desired loan to deposit are often imprecise/fuzzy and have preference-based membership functions.

(3) Deposit supply has the nature of both randomness and fuzziness. The randomness is modelled by multiple states as in (1). The fuzziness is modelled by preference-based membership function as in (2).

This stochastic possibilistic bank hedging decision model is then:

$$\max \quad \Sigma_s \, p_s E_s$$
$$\text{s.t.} \quad R_s \geq aD_s + mS_F, \; \forall s$$
$$S_{s,STS} \leq (1 - b_{STS})STS^*, \; \forall s$$
$$S_{s,LTS} \leq (1 - b_{LTS})LTS^*, \; \forall s$$
$$S_F \leq (1 - b_{LTS})LTS^*$$
$$P_L \leq \tilde{ld}$$
$$D_s \leq \tilde{d}_s, \; \forall s$$
$$L \leq \tilde{c}D_s, \; \forall s$$
$$STS_s = (1 - b_{STS})STS^* + P_{STS} - S_{s,STS}, \; \forall s$$
$$LTS_s = (1 - b_{LTS})LTS^* + P_{LTS} - S_{s,LTS}, \; \forall s$$
$$L = (1 - b_L)L^* + P_L$$
$$R_s + STS_s + LTS_s + L = D_s + E_s, \; \forall s \qquad (4.111)$$

where

$$E_s = E^* + r_{STS}^*(1-b_{STS})STS^* + r_{LTS}^*(1-b_{LTS})LTS^* + r_L^*(1-b_L)L^* + r_{STS}P_{STS} + r_{LTS}P_{LTS}$$
$$+ r_L P_L - (r_{STS}^* + k_{s,STS})S_{s,STS} + (r_{LTS}^* + k_{s,LTS})S_{s,LTS} + [1-(\tilde{p}_{s,PF}/P_{s,SF})]S_F - \tilde{r}_{s,D}D_s$$

$$k_{s,STS} = 1 - \{r_{STS}^*[\sum_{t=1}^{(.5/b_{STS})+.5}(1+\tilde{r}_{s,STS})^{-t}]+[1+\tilde{r}_{s,STS}]^{(.5/bSTS)+.5}\}$$

$$k_{s,LTS} = 1 - \{r_{LTS}^*[\sum_{t=1}^{(.5/b_{LTS})+.5}(1+\tilde{r}_{s,LTS})^{-t}]+[1+\tilde{r}_{s,LTS}]^{(.5/bLTS)+.5}\}$$

where "*" indicates historical data; $\tilde{p}_{s,PF}$, $\tilde{r}_{s,D}$, $\tilde{r}_{s,STS}$ and $\tilde{r}_{s,LTS}$, $\forall s$, are fuzzy and have triangular possibilistic distributions; and $\tilde{ld}$, $\tilde{d}_s$, $\forall s$, and $\tilde{c}$ are fuzzy and have preference-based membership functions. The variables and parameters of the above equation are defined in Table 4.5. The explanation of the above equation has been given in Section 4.2.2a. However, it should be noted that the constraints in the above equation involve every state at the same time. Thus, it is not equivalent to solve the problem in Section 4.2.2a for each individual state and then to find their expected values.

Now, assume that the input data are either precisely numerical or linguistic (see Table 4.9). For simplicity, two stages ($s = 1, 2$) are considered and have equal probabilities for each stage. In Table 4.9, while other data are quarterly, interest rates and futures prices are expressed in annual terms and must be transferred into quarterly values before substituting them into Equation (4.111).

The triangular possibilistic distributions of interest rates and price of futures purchased have been discussed in the previous section. As to the fuzzy sets $\tilde{ld}$, $\tilde{d}_1$, $\tilde{d}_2$ and $\tilde{c}$, their membership functions are assumed to be linear decreasing functions within the considered ranges. For example, the membership function of the fuzzy set $\tilde{ld}$ is defined as:

$$\mu(ld) = \begin{cases} 1 & \text{if } ld < 400 \\ 1 - (ld-400)/80 & \text{if } 400 \leq ld \leq 480 \\ 0 & \text{if } ld > 480. \end{cases}$$

Similarly, we can also define the membership functions of $\tilde{d}_1$, $\tilde{d}_2$ and $\tilde{c}$ as the above function.

Now, let $\alpha = \min (\mu(ld), \mu(d_1), \mu(d_2), \mu(c))$. We have $\mu(ld) \geq \alpha$, $\mu(d_1) \geq \alpha$, $\mu(d_2) \geq \alpha$ and $\mu(c) \geq \alpha$. That is:

$$P_L \le 400 + (1-\alpha_1)80,$$
$$D_1 \le 950 + (1-\alpha_2)200, \quad D_2 \le 1050 + (1-\alpha_3)300$$
$$L \le [0.8+0.1(1-\alpha_4)]D_1, \quad L \le [0.8+0.1(1-\alpha_5)]D_2.$$

Here, α is the minimum satisfaction level of the fuzzy constraints or the degree of feasibility.

Table 4.9 Numerical data of bank balance sheet problem [142]

Variable/ Parameter	Historical value	Present value	-- Future values -- State 1 (Prob.=.5)	State 2 (Prob.=.5)
Balance Sheet				
R	$200			
STS	100			
LTS	300			
L	400			
D	900			
E	100			
Interest Rate (annual)				
r_{STS}	.07	.08	"about .08"	"about .09"
r_{LTS}	.08	.09	"about .09"	"about .10"
r_L	.11	.14		
r_D			"about .04"	"about .05"
Futures				
p_{SF}		$92		
p_{PF}			$91	$90
m		.025		
Market Size				
ld		$400 (80)		
d			$950 (200)	$1050 (300)
Others				
a		.20		
b_{STS}		.50		
b_{LTS}		.125		
b_L		.125		
c		.8 (.1)		

Note: (1) interest rates and futures prices are expressed in annual terms, and other data are quarterly;
(2) the numbers inside the parenthesis indicate the corresponding tolerances;
(3) "about .08" = (.08, .075, .085); "about .09" = (.09, .085, .10); "about .10" = (.10, .09, .12); "about .04" = (.04, .035, .05); "about .05" = (.05, .04, .06).

The numerical model of the bank hedging decision under fuzziness and randomness is then:

$$\max \quad E = .5E_1 + .5E_2$$

$$\text{s.t.} \quad R_1 \geq .2D_1 + .025S_F, \quad R_2 \geq .2D_2 + .025S_F$$

$$S_{1STS} \leq 50, \quad S_{2STS} \leq 50$$

$$S_{1LTS} \leq 262.5, \quad S_{2LTS} \leq 262.5$$

$$S_F \leq 262.5$$

$$P_L \leq 400 + (1-\alpha_1)80$$

$$D_1 \leq 950 + (1-\alpha_2)200, \quad D_2 \leq 1050 + (1-\alpha_3)300$$

$$L \leq [.8+(1-\alpha_4).1]D_1, \quad L \leq [.8+(1-\alpha_5).1]D_2$$

$$STS_1 = 50 + P_{STS} - S_{1STS}, \quad STS_2 = 50 + P_{STS} - S_{2STS}$$

$$LTS_1 = 262.5 + P_{LTS} - S_{1LTS}, \quad LTS_2 = 262.5 + P_{LTS} - S_{2LTS}$$

$$L = 350 + P_L$$

$$R_1 + STS_1 + LTS_1 + L = D_1 + E_1, \quad R_2 + STS_2 + LTS_2 + L = D_2 + E_2$$

where

$$E_1 = 115.75 + .02P_{STS} + .0225P_{LTS} + .035P_L - (.0175 + k_{1STS})S_{1STS}$$
$$-(.02 + k_{1LTS})S_{1LTS} + .01087S_F - (.01, .00875, .0125)D_{1S}$$

$$E_2 = 115.75 + .02P_{STS} + .0225P_{LTS} + .035P_L - (.0175 + k_{2STS})S_{2STS}$$
$$-(.02 + k_{2LTS})S_{2LTS} + .02174S_F - (.0125, .01, .015)D_{2S}$$

$$k_{1STS} = 1 - [.0175/(1.02, 1.01875, 1.02125) + 1.0175/(1.02, 1.01875, 1.02125)^{1.5}]$$

$$k_{2STS} = 1 - [.0175/(1.0225, 1.02125, 1.025) + 1.0175/(1.0225, 1.02125, 1.025)^{1.5}]$$

$$k_{1LTS} = 1 - \{ \Sigma_{j=1->4}.02/(1.0225, 1.02125, 1.025)^j + 1.02/((1.0225, 1.02125, 1.025)^{4.5}$$

$$k_{2LTS} = 1 - \{ \Sigma_{j=1->4}.02/(1.025, 1.0225, 1.03)^j + 1.02/((1.025, 1.0225, 1.03)^{4.5}.$$

To solve the problem presented in Table 4.9, we first assume that the acceptable degree of feasibility for the constraint is 0.7 ($\alpha = 0.7$). Also, we must elicit the membership functions of the upper and lower bounds of the expected equity from the decision maker. Here, assume these membership functions are:

$$\mu_k(E^k) = \begin{cases} 1 & \text{if } E^k > E^{k*} \\ (E^k - E^{k-})/(E^{k*} - E^{k-}) & \text{if } E^{k-} \leq E^k \leq E^{k*} \\ 0 & \text{if } E^k < E^{k-} \end{cases}$$

where E^{k*} (E^{k-}) is the individual maximum (minimum) for each objective, and k = U and L.

With the assumption of δ = 0.001 and $w_1 = w_2$ = 0.5, we can obtain E^{k*} (= $0.5E_1^{k*}$ + $0.5E_2^{k*}$), and E^{k-} (= $0.5E_1^{k-}$ + $0.5E_2^{k-}$), for k = L and U, and the corresponding compromise solutions, which are presented in Table 4.10, for β = 1.0, 0.9, ..., 0.1, 0.0.

Table 4.10 Solutions of the bank hedging decision problem
(α = 0.7)

β	PIS		NIS		Compromise solution				Γ
	E^{L*}	E^{U*}	E^{L-}	E^{U-}	$E_1^{\,L}$	$E_1^{\,U}$	$E_2^{\,L}$	$E_2^{\,U}$	
1.0					136.683	136.683	137.406	137.406	1.000
0.9	136.704	137.290	95.196	96.311	136.322	136.874	137.112	137.705	0.998
0.8	136.364	137.535	94.502	96.732	135.963	137.065	136.818	138.003	0.999
0.7	136.024	137.779	93.809	97.153	135.603	137.257	136.524	138.301	1.000
0.6	135.684	138.024	93.117	97.575	135.242	137.449	136.230	138.560	0.999
0.5	135.344	138.269	92.425	97.997	134.882	137.640	135.936	138.898	0.999
0.4	135.047	148.595	91.735	98.418	134.477	137.869	135.453	139.168	0.998
0.3	134.776	138.975	91.046	98.841	134.081	138.086	135.038	139.466	0.995
0.2	134.056	139.355	90.357	99.263	133.683	138.303	134.632	139.775	0.992
0.1	134.235	139.735	89.670	99.685	133.283	138.521	134.228	140.091	0.989
0.0	133.964	140.115	88.983	100.11	132.882	138.738	133.829	140.412	0.986

For simplicity, we only show the case of β = 0.5 in the following. First, for β = 0.5 we obtain E^{L*} = 135.3438, E^{U*} = 138.2693, E^{L-} = 92.4254 and E^{U-} = 97.9966. Then the above linear membership functions of both objectives can be easily obtained. Finally, we solve the following linear programming problem of Equation (4.109):

$$\max \quad \Gamma + 0.0000117E^L + 0.0000124E^U$$

$$\text{s.t.} \quad E^L - 42.9184\Gamma \geq 92.4254$$

$$E^U - 40.2727\Gamma \geq 97.9966$$

$$R_1 - 0.2D_1 - 0.25S_F \geq 0$$

$$R_2 - 0.2D_2 - 0.25S_F \geq 0$$

$$S_{1STS} \leq 50$$

$$S_{2STS} \leq 50$$

$$S_{1LTS} \leq 262.5$$

$$S_{2LTS} \leq 262.5$$

$$S_F \leq 262.5$$

$$P_L \leq 424$$

$$D \leq 1010$$

$$L - 0.83D_1 \leq 0$$

$$L - 0.83D_2 \leq 0$$

$$STS_1 - P_{STS} + S_{1STS} = 50$$

$$STS_2 - P_{STS} + S_{2STS} = 50$$

$$LTS_1 - P_{LTS} + S_{1LTS} = 262.5$$

$$LTS_2 - P_{LTS} + S_{2LTS} = 262.5$$

$$L - P_L = 350$$

$$R_1 + STS_1 + LTS_1 - D_1 - E_1 = 0$$

$$R_2 + STS_2 + LTS_2 - D_2 - E_2 = 0$$

$$E_1^L \leq E_1 \leq E_1^U, \; E_2^L \leq E_2 \leq E_2^U$$

$$E_1^L - 0.02P_{STS} - 0.0225P_{LTS} - 0.035P_L + 0.01354S_{1STS} + 0.026776S_{1LTS} + 0.01087S_F + 0.01125D_1 = 115.75$$

$$E_2^L - 0.02P_{STS} - 0.0225P_{LTS} - 0.035P_L + 0.018107S_{2STS} + 0.042431S_{2LTS} + 0.02174S_F + 0.1375D_2 = 115.75$$

$$E_1^U - 0.02P_{STS} - 0.0225P_{LTS} - 0.035P_L + 0.011703S_{1STS} + 0.018831S_{1LTS} - 0.01087S_F + 0.009375D_1 = 115.75$$

$$E_2^U - 0.02P_{STS} - 0.0225P_{LTS} - 0.035P_L + 0.015371S_{2STS} + 0.026776S_{2LTS} - 0.02174S_F + 0.01125D_2 = 115.75$$

$$E^L = 0.5E_1^L + 0.5E_2^L$$

$$E^U = 0.5E_1^U + E_2^U$$

$$\Gamma \in [0, 1]$$

where all the decision variables are non-negative. The numerical solution is provided in Table 4.11. Furthermore, the possibility distributions of the optimal equities for each state are presented in Figure 4.20. Obviously, the possibility distributions of expected equity and equities of both states are non-linearly bell shaped. The optimal balance-sheet and two-stage decisions at $\beta = 0.5$ are provided in Table 4.11.

If the acceptable possibility level is 0.5, then the bank should obtain the funds through the maturation of its existing short-term security, long-term security and loan balance, and accept all the deposits supplied to it (see Table 4.11).

Reserve balances should be reduced to the lowest level possible as indicated by the reserve and potential margin requirements. Thus, reserve balance is increased from 200 to 267.625 and 291.625. All of the non-maturing short-term securities should be sold. Some of long-term securities (97.258) are sold in State 1. On the other hand, no long-term security should be sold in State 2. The bank should, at the same time, use some of these funds to purchase long-term securities and loans. However, the bank should not purchase short-term securities at all. Selling futures should proceed (as much as possible) for hedging the interest rate risk. By taking the above actions at $\beta = 0.5$, the expected equity will be between \$135.40 and \$138.269.

For simplicity, we will not present all the lengthy results in all cases of $\beta = 1.0, 0.9, \ldots,$ 0.1, 0.0. A brief conclusion is that the amount of purchasing and selling long-term securities varies with the change of the possibility (β values), in both states. Therefore, the bank should pay more attention to controlling long-term securities. Other decisions are stable with the change of the possibility.

Similarly, we can also obtain another distribution of the optimal equity and the corresponding balance sheet for a different α value (degree of desirable feasibility).

Table 4.11 The optimal balance sheet and decisions ($\alpha = 0.7$ and $\beta = 0.5$)

	Initial Values	Ending Values State 1	Ending Values State 2
Balance sheet			
Reserves (R)	200	267.625	291.625
Short-term securities (STS)	100	0.000	0.000
Long-term securities (LTS)	300	880.015	977.273
Loan (L)	400	774.000	774.000
Deposits (D)	900	1010.000	1130.000
Equity (E)	100	137.640	138.898
Decision			
First stage			
Purchase short-term securities (P_{STS})		0.000	0.000
Purchase long-term securities (P_{LTS})		714.773	714.773
Purchase loans (P_L)		424.000	424.000
Sell futures (S_F)		262.500	262.500
Second stage			
Sell short-term securities (S_{STS})		50.000	50.000
Sell long-term securities (S_{LTS})		97.258	0.000

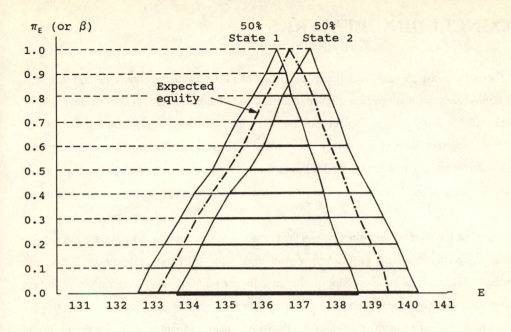

Figure 4.20 The possibility distributions of the optimal equity

5 CONCLUDING REMARKS

During the last 25 years, the fuzzy set theory has been applied in many disciplines such as operations research, management science, artificial intelligence/expert system, control theory, statistics, etc. In this study, we concentrate on fuzzy mathematical programming which is one of the most important fields in operations research and management science. This study provides readers and researchers with a capsule look into the existing methods, their characteristics and their applicability to analysis of single-objective mathematical programming under fuzziness/imprecision.

Based on the membership function and the possibilistic distribution, we have systematically classified fuzzy mathematical programming problems into two major categories: fuzzy and possibilistic mathematical programming problems. In our point of view, it is rather important to distinguish preference-based membership functions from possibility distributions. Preference-based membership concerns the degree of the decision maker's satisfaction, while possibility is the degree of the occurrence of an event. Eleven approaches of fuzzy linear programming and fourteen methods of possibilistic programming have been systematically and thoroughly classified to solve all possible existing problems. Fifteen practical problems have also been discussed in this study. Most of these methods and applications have been proposed by various researchers in the last decade, but here for the first time they are presented all together.

It is important to compare fuzzy and possibilistic approaches with stochastic or probabilistic ones for solving fuzzy/imprecise/uncertainty nature. In Section 5.1, we will summarize some essential reasons why fuzzy set theory is preferred in this study. Furthermore, we concisely present the advantages of fuzzy and possibilistic programming compared to stochastic programming in Section 5.2. Section 5.3 will present some directions of future studies. We finally present a short introduction of our following volume of *Fuzzy Multiple Objective Decision Making: Methods and Applications*, and the previous volume of *Fuzzy Multiple Attribute Decision Making: Methods and Applications* in Sections 5.4 and 5.5, respectively.

5.1 Probability Theory versus Fuzzy Set Theory

Since Fortuna, the fickle, wheel-toting goddess of chance, ruled a large and growing empire in the sciences, the law of the realm has been probability theory and statistics [BM12].

From the beginning of the mid-seventeenth century, probability theory spread in the eighteenth century from gambling problems to jurisprudence, data analysis, inductive inference (induction may be traced back at least as far as Bacon's Novum Organum in 1620s [BM5]), and insurance; and from there to sociology, to physics, to biology, and to psychology in the nineteenth century; and on to agronomy, polling, medical testing, baseball, and innumerable other practical (and not so practical) matters in the twentieth [BM12].

Thus, the concept and techniques of the probability theory are often used to deal quantitatively with imprecise goals, constraints and consequences of possible actions. Imprecision - whatever its nature - is tacitly accepted to be equated with randomness [12]. However, the imprecise nature may be associated with classes in which there is no sharp transition from membership to nonmembership. From the view of psychology, computer engineering and artificial intelligence and risk analysis, a rough but fairly functional classification of objections (of the probability theory) includes: (1) phenomenological, philosophical and semantic arguments for the existence of non-probabilistic uncertainty; (2) meta-rationalist claim that the probability theory is untenable [BM47].

For the first objection, we may have certainty about a fuzzy event (such as "is it raining") or a graded attribute (such as "John is tall"). It seems counter-intuitive to describe the vagueness inherent in the above events or attributes by subjective probabilities [288]. In Zadeh's terms, this objection is not to the axioms of the probability theory itself, but instead to its limited capacities for expression. Thus, Zadeh claimed that the probability theory may be unable to represent the meanings of propositions involving imprecision/fuzziness of predicates (small, young, ...), quantifiers (some, often, ...), probabilities (likely, long, odds, ...), truth values (very true, ...) and modifiers (very, somewhat, ...).

For the second objection, meta-rationality arguments against probability refer primarily to its computational complexity and the amount of information required, especially in the context of applying Bayes's Rule. Despite computational complexity, we must frequently add or delete individual pieces of knowledge in actually constructing a knowledge base. This cannot be done without revising the dependency information on all relevant subgroups of evidence. As to the required sufficient information, even sophisticated records cannot provide that kind of associative information. This information unavailability criticism includes (1) the absence of such information and (2) the changeable, unreliable or irreducibly contextual nature of information.

To distinguish fuzziness from frequentist's probability seems obvious and has largely been conceded by many probabilists. However, subjectivists consider probability as a degree of belief which is close to, but not equal to the fuzzy set theory. In this aspect, the fuzzy set theory does provide the attribution of a non-probabilistic meaning to the concept of graded sets. For example, we may be entirely certain that a particular object is "dark." In this case, there is no probabilistic uncertainty associated with this proposition, and the only source of uncertainty is the graded nature of the concept of darkness. Another argument is based on the concept of the possibility theory. Dubois and Prade [BM7] pointed out that a fuzzy concept like "tall" conveys a possibility distribution over the set of all heights, so that the possibility grade reflects the degree to which a given object might possibly belong to the set of tall people given his/her height. They also argued that possibility is a looser kind of uncertainty than probability, since anything that is impossible must be improbable but whatever is possible need not be probable.

As pointed out by Smithson [BM47], the majority of arguments for and against fuzziness rest primarily on appeals to axiomatics, meaningfulness, ease of expression and use, semantics, and/or philosophical distinctions. As a result, probabilists argue that we need not invent a new theory to handle uncertainty about probability, instead we may use meta-probabilities (higher-order probability)]; or that any self-consistent system for quantifying degree of belief with a single number must be probabilistic. Even such apparently non-probabilistic concepts as possibility may also be represented by probability. For example, a possibility distribution is merely subjective probabilities associated with possibilistic statements. Thus we only induce a second-order probability distribution over the first-order subjective probabilities. However, some evidence shows that high-order imprecision has little practical importance, or that human judges do not usually resort to higher-order. Furthermore, if the new frameworks deliver the goods in applications, then they will continue to be used regardless of what the probabilists are saying; unless the probabilists can show similar capabilities for their frameworks. In conclusion, the fuzzy set theory abandons the laws of the excluded middle (or additivity requirement) and contradiction, and thus abandons the standard probability calculus altogether, since it destroys the de Morgan relations. Studies on fuzzy set theory are not going to block the search for reality nor restrict the development of probability, but are going to impel humanity to develop its full power.

5.2 Stochastic versus Possibilistic Programming

Some major arguments about the differences between probability and fuzzy set theories have been discussed in Section 5.1. Here, we would like to discuss the advantages of possibilistic programming compared to stochastic programming.

First, stochastic programming deals with situations where input data (c, b and/or A) of Equation (5.1) are imprecise and described by random variables. The basic idea is to convert the random/probabilistic nature of the problem into an equivalent deterministic situation. For more than three decades, there were numerous stochastic programming models in the literature. Interested readers may refer to Wets [260] and Stancu-Minasian [BM48] for details.

Among various stochastic models, we like to introduce the chance-constrained programming developed by Charnes, Cooper and Symonds [39], and Charnes and Cooper [40].

A chance-constrained programming model can be defined as:

$$\max \quad z = \Sigma_j \bar{c}_j x_j$$
$$\text{s.t.} \quad P\{ \Sigma_j \bar{a}_{ij} x_j \le \bar{b}_i \} \ge 1 - \alpha_i, \forall i$$
$$x_j \ge 0, \forall j \tag{5.1}$$

where $\bar{c}_j$, $\bar{b}_i$ and $\bar{a}_{ij}$ are all random variables and $P\{\}$ is the degree of probability. The name "chance-constrained" follows from each constraint $\Sigma_j \bar{a}_{ij} x_j \le \bar{b}_i$ being realized with a minimum probability of $1 - \alpha_i$, $0 < \alpha_i < 1$.

For illustrative purposes, let us assume $\bar{b}_i$ and $\bar{a}_{ij}$ are normally distributed with known means and variances. If $\bar{c}_j$ is a random variable, it could always be replaced by its expected value. Thus we have the following chance-constrained programming problem:

$$\max \quad z = \Sigma_j c_j x_j$$
$$\text{s.t.} \quad P\{ \Sigma_j \bar{a}_{ij} x_j - \bar{b}_i \le 0 \} \ge 1 - \alpha_i, \forall i$$
$$x_j \ge 0, \forall j \tag{5.2}$$

where $\bar{b}_i$ and $\bar{a}_{ij}$ are normally distributed as shown in Figure 5.1.

To solve Equation (5.2), let $\bar{g}_i = \Sigma_j \bar{a}_{ij} x_j - \bar{b}_i$. Then g_i is also normally distributed with:

$$E\{g_i\} = \Sigma_j E\{a_{ij}\} x_j - E\{b_i\} \text{ and } Var\{g_i\} = X^T D_i X$$

where $X^T = (x_1, x_2, ..., x_n, 1)$ and D_i = the ith covariance matrix =

$$\begin{bmatrix} Var\{a_{i1}\} & \cdots & Cov\{a_{i1},a_{in}\} & Cov\{a_{i1},b_i\} \\ \cdot & & & \\ Cov\{a_{in},a_{i1}\} & \cdots & Var\{a_{in}\} & Cov\{a_{in},b_i\} \\ Cov\{b_i,a_{i1}\} & \cdots & Cov\{b_i,a_{in}\} & Var\{b_i\} \end{bmatrix}$$

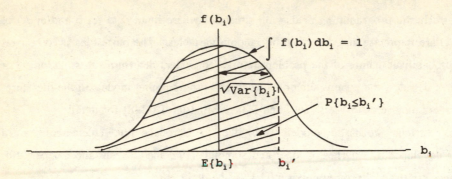

Figure 5.1 Probability density function $f(b_i)$ of b_i where
$f(b_i) = (2\pi Var\{b_i\})^{-1/2} exp[-(b_i-E\{b_i\})/2Var\{b_i\}]$ for $-\infty < b_i < \infty$

Now, the constraints of Equation (5.2) may be treated as:

$$P\{g_i \leq 0\} = P\{\frac{g_i - E\{g_i\}}{Var\{g_i\}^{1/2}} \leq \frac{0 - E\{g_i\}}{Var\{g_i\}]^{1/2}}\} \geq 1 - \alpha_i$$

or

$$P\{g_i \leq 0\} = \Phi(\frac{0 - E\{g_i\}}{Var\{g_i\}^{1/2}}) \geq 1 - \alpha_i$$

where Φ represents the CDF of the standard normal distribution.

Let $S_{\alpha i}$ be the standard normal value such that $\Phi(S_{\alpha i}) = 1 - \alpha_i$. Then $P\{g_i \leq 0\} \geq 1 - \alpha_i$ is realized, iff:

$$(0 - E\{g_i\})/(Var\{g_i\}^{1/2}) \geq S_{\alpha i}.$$

Finally, we have the following deterministic model:

$$\max \quad z = \Sigma_j c_j x_j$$
$$\text{s.t.} \quad \Sigma_j E\{a_{ij}\} x_j - E\{b_i\} + S_{\alpha i}(X^T D_i X)^{1/2}, \forall i$$
$$x_j \geq 0, \forall j \tag{5.3}$$

which can then be solved by any constrained non-linear programming algorithm.

It is noted that Equation (5.3) is actually a non-linear programming problem, but not the original linear programming problem. Obviously, it is not easy to solve a non-linear programming problem. On the other hand, possibilistic programming provides more efficient techniques to solve the imprecise nature of c, b and/or A and also preserves the original linear model. Besides, membership functions/possibility distributions provide more flexible and meaningful representation of imprecision/uncertainty.

Consider the following possibilistic programming problem:

$$\max \quad z = \Sigma_j \, c_j x_j$$
$$\text{s.t.} \quad \Sigma_j \, \tilde{a}_{ij} x_j \le \tilde{b}_i, \; \forall i$$
$$x_j \ge 0, \; \forall j \tag{5.4}$$

where $\tilde{b}_i$ and $\tilde{a}_{ij}$, $\forall i,j$, are imprecise and have their corresponding triangular possibilistic distributions $\pi_{bi} = (b_i, b_i^L, b_i^U)$ and $\pi_{aij} = (a_{ij}, a_{ij}^L, a_{ij}^U)$ as shown in Figure 5.2. $\pi.$ is the degree of possibility.

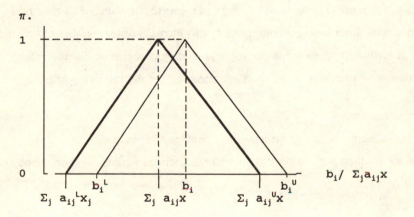

Figure 5.2 Illustration of a triangular possibilistic distribution

To solve Equation (5.4), the following auxiliary model (see Chapter 4 for details) might be used:

$$\max \quad z = \Sigma_j \, c_j x_j$$
$$\text{s.t.} \quad (\Sigma_j \, a_{ij} x_j, \; \Sigma_j \, a_{ij}^L x_j, \; \Sigma_j \, a_{ij}^U x_j) \le (b_i, b_i^L, b_i^U), \; \forall i$$
$$x_j \ge 0, \; \forall j. \tag{5.5}$$

The imprecise constraints of the above problem can be resolved by considering the following inequalities:

$$\Sigma_j \, a_{ij}x_j \le b_i, \; \Sigma_j \, a_{ij}{}^L x_j \le b_i{}^L, \; \Sigma_j \, a_{ij}{}^U x_j \le b_i{}^U \tag{5.6}$$

which preserve that the triangular possibilistic distribution of $\tilde{b}_i$ must greater than or equal to the one of $\Sigma_j \, \tilde{a}_{ij}x_j$ as indicated in Figure 5.2.

Finally, we have the following deterministic model:

$$
\begin{aligned}
\max \quad & z = \Sigma_j \, c_j x_j \\
\text{s.t.} \quad & \Sigma_j \, a_{ij}x_j \le b_i, \; \forall i \\
& \Sigma_j \, a_{ij}{}^L x_j \le b_i{}^L, \; \forall i \\
& \Sigma_j \, a_{ij}{}^U x_j \le b_i{}^U, \; \forall i \\
& x_j \ge 0, \; \forall j.
\end{aligned}
\tag{5.7}
$$

Equation (5.7) is still a linear programming problem. Of course, the number of constraints has been extended to around three times. In our opinion, this increase of the number of constraints might not cause as many trouble as a non-linear programming problem of Equation (5.3).

The question as to what is a better approach encompasses the following criteria:

(1) Execution time required.
(2) Size of the considered problem (dimensionality, number of constraints).
(3) Accuracy of the solution with respect to the optimal decision variable and/or the objective function and constraints.
(4) Simplicity of use.
(5) Simplicity of computer program to execute the algorithm.
(6) Applicability to real-world (large scale) problems.

According to these criteria, Equation (5.7) (linear) is obviously better than the Equation (5.3) (non-linear). That is, possibilistic programming is better (more efficient) than stochastic programming in the sense of methodology.

Furthermore, the assumed randomness for the input coefficients may not be good enough to represent the inherent impreciseness or uncertainty. With assumed normal distributions, for example, negative values of b_i/a_{ij} are possible, this result is clearly counter-intuitive and incorrect. Therefore, bounded distributions are better tools to represent a mathematical programming problem with imprecise input. As to the meaning of possibility and probability theories, let us consider the following example:

"Tomorrow it will *probably* rain *a lot*."

Usually, probability theory is used to model *probably* and possibility theory is used to model *a lot*. The imprecise nature of c, b and/or A may be described by the linguistic term of "around 34.5," for instance, than by the chance concept of "70% to be 34.5." Thus, possibilistic programming techniques are considered to be more efficient and meaningful than stochastic programming techniques.

5.3 Future Research

In the last decade, fuzzy set theory has provided a new research direction on both concepts and methodologies to formulate and solve mathematical programming and multiple objective decision making problems. However, a number of problems still remain to be solved in the future research. These problems are summarized in the following:

(1) In this study, we often assume that membership functions and possibility distributions are reasonably given. However, in practice, we should generate and obtain membership functions and possibility distributions from decision makers and/or historical resources. Even though there have already been some efforts on generating membership functions and possibility distributions, more research is urgently needed.

(2) To solve mathematical programming problems in various fuzzy/imprecise environments, we often use the max-min operator, although many other operators already exist. Thus, we should extend these existing operators or develop new operators to fit our specific mathematical programming problems in future studies.

(3) As indicated before, ranking approaches are very important to resolve fuzzy/imprecise constraints. However, most of the existing ranking approaches are not perfect (see Chen and Hwang [BM3]). Searching for better ranking methods is an urgent and momentous need to resolve fuzzy/imprecise constraints and other problems.

(4) In the literature, few fuzzy multiple objective decision making and possibilistic multiple objective decision making problems have been solved. At the same time, the efficiency of some existing approaches should be further improved.

(5) It frequently happens that fuzziness/imprecision and randomness simultaneously exist in a decision process. However, little research has been done on solving stochastic fuzzy/possibilistic mathematical programming and multiple objective decision-making problems.

(6) Up to date fuzzy problems are first transferred into crisp problems and then solved by traditional (crisp) computer packages. The efficiency of fuzzy methodology is reduced. Thus, research on developing fuzzy algorithm and coding existing fuzzy approaches in computer packages is urgently needed.

(7) The most important area for further research would probably involve real applications with real decision makers. Very few such applications have been reported in the literature of fuzzy mathematical programming. It should be noted once more that all concepts and techniques are developed for solving real-world problems.

5.4 Introduction of the Following Volume

Single-objective mathematical programming, however, cannot solve all decision-making problems of selecting a possible course of action from available alternatives. In most such problems, the multiplicity of criteria for judging the alternatives is pervasive. That is, for many such problems the decision maker wants to attain more than one objective or goal in selecting the course of action, while satisfying the constraints dictated by environment, processes and resources. For example, the objectives of a water quality management problem may be to raise the dissolved oxygen levels, to keep a high percentage return on industrial investment, and to retain low tax rates for cities. For a nutrition allocation problem, the objectives may be to maximize carbohydrate intake, to minimize cholesterol intake, and to minimize cost. For a bank balance sheet management problem, the objectives may be to maximize net after-tax profits, and to minimize the capital adequacy and risk assets.

Mathematically, these multiple objective mathematical programming problems (or multiple objective decision making MODM) can be formulated as:

$$\max \quad Z = [f_1(x), f_2(x), ..., f_K(x)]$$
$$\text{s.t.} \quad g_i(x) \; \{\leq, =, \geq\} \; 0, \; \forall i \text{ and } x \geq 0 \tag{5.8}$$

where f_k and $g_i(x)$ may be either linear or non-linear functions. The basic characteristic of MODM is that the objectives are apparently non-commensurable and conflict with each other. To resolve these problems, Hwang and Masud [BM19] have already provided a systematical and thorough survey. Since then, some new approaches, especially fuzzy programming techniques, have been published in the literature. These new approaches will be presented in our second volume of *Fuzzy Multiple Objective Decision Making: Methods and Applications* which is considered as a sequel of Hwang and Masud's *Multiple Objective Decision Making: Methods and Applications* [BM19].

Furthermore, the input data of the MODM problem may be fuzzy/imprecise because of incomplete or non-obtainable information. To solve these fuzzy MODM problems, we have also carried out a complete survey of all existing approaches. The classification of fuzzy MODM is similar to fuzzy mathematical programming problems as shown in Figures 1.3 and 1.4. We also distinguish fuzzy MODM from possibilistic MODM for the same reasons as discussed in this volume.

It should be noted that as with this volume, we will again make a strong effort to present practical problems. After all, all concepts and methods are developed for solving real-world problems. In presenting practical problems, we believe that the next volume can also fill the gap between researchers and practitioners (or real decision makers). Thereby, practitioners can really use the concepts and methods developed by theoretical researchers to solve real-world problems with real data.

5.5 Fuzzy Multiple Attribute Decision Making

In the study of decision making in complex environments, such terms as "multiple objectives," "multiple attributes," "multiple criteria," or "multiple dimensions" are used to describe decision situations. Often these terms are used interchangeably, and there are no universal definitions for them. Multiple criteria decision making (MCDM) has become the accepted designation for all methodologies dealing with MODM and/or MADM. In our survey, the problems of MCDM are broadly classified into two categories - MODM and MADM. The definitions to distinguish MODM from MADM are presented in Hwang and Yoon [BM20]. Basically, MADM methods are for selecting an alternative from a small, explicit list of alternatives, while MODM methods are used for an infinite set of alternatives implicitly defined by the constraints. Thus, MADM methods are useful for solving choice problems while MODM methods address design problems.

If the input data of MADM problems are fuzzy/imprecise, then we will have fuzzy MADM (FMADM) problems. Literature on FMADM methods and applications has been extensively reviewed by Chen and Hwang [BM3]. FMADM methods were systematically classified; a taxonomy of the methods is presented in Table 5.1. Detailed discussions of these methodolgies are provided in [BM3].

Table 5.1 A taxonomy of FMADM methods

	Problem size	Data type	Corresponding MADM methods	Technique involved	Approach
FMADM	n < 10 m < 10	All fuzzy	Simple Additive Weighting	α–cut	Baas and Kwakernaak's Kwakernaak's Dubois and Prade's Cheng and McInnis's
				Fuzzy arithmetic	Bonissone
			AHP	Eigenvector	Saaty's
				Weight assessing arithmetic	Laarhoven and Pedrycz's Buckley
		Crisp/fuzzy	Conjunction/ Disjunction	Possibility/ necessity	Dubois and Prade's
			MAUF	Human intuition	Efstathiou and Rajkovic's
			General MADM	Fuzzy ranking and fuzzy arithmatic	Negi's
			Outranking	Fuzzy outranking relation	Roy Siskos et al. Brans et al. Takeda
	n < 350 m: any number	All fuzzy (singleton)	Maximin	max and min operators	Bellman and Zadeh's Yager's
		Crisp/fuzzy	General MADM	Linguistic -> fuzzy set -> crisp score	Chen and Hwang

BIBLIOGRAPHY

Books, Monographs and Conference Proceedings:

[BM1] Bezdek, J.C., *Analysis of Fuzzy Information (Vol.III): Applications in Engineering and Sciences* (CRC Press, Boca Raton, 1987).

[BM2] Bouchon B., L. Saitta and R.R. Yager (eds.), *Uncertainty and Intelligent Systems* (Springer-Verlag, Heidelberg, 1988).

[BM3] Chen, S.J. and C.L. Hwang, *Fuzzy Multiple Attribute Decision Making: Methods and Applications* (Springer-Verlag, Heidelberg, 1991).

[BM4] Chvatal, V., *Linear Programming* (W.H. Freeman and Company, New York, 1983).

[BM5] Cohen, L.J., *An Introduction to the Philosophy of Induction and Probability* (Clarendon, Oxford, 1989).

[BM6] Di Nola, A., S. Sessa, W. Pedrycz and E. Sanchez, *Fuzzy Relation Equations and Their Applications to Knowledge Engineering* (Kluwer Academic, Dordrecht, 1989).

[BM7] Dubois, D. and H. Prade, *Fuzzy Sets and Systems -- Theory and Application* (Academic, New York, 1980).

[BM8] Dubois, D. and H. Prade, *Possibility Theory: An Approach to Computerized Processing of Uncertainty* (Plenum, New York, 1988).

[BM9] Elton, E.J. and M.J. Gruber, *Modern Portfolio Theory and Investment Analysis, Third Edition* (Wiley, New York, 1987).

[BM10] Evans, G.W., W. Karwowski and M.R. Wilhelm (eds.), *Applications of Fuzzy Set Methodologies in Industrial Engineering* (Elsevier Science, Amsterdam, 1989).

[BM11] Gal, T., *Postoptimal Analysis, Parametric Programming and Related Topics* (McGraw-Hill, New York, 1981).

[BM12] Gigerenzer, G., Z. Swijtink, T. Porter, L. Daston, J. Beatty and L. Kruger, *The Empire of Chance* (Cambridge University, Cambridge, 1989).

[BM13] Gould, F.J., G.D. Eppen and C. Schmidt, *Quantitative Concepts for Management: Decision Making without Algorithm, Third Edition* (Prentice-Hall, Englewood Cliffs, 1988).

[BM14] Gupta, M.M. and E. Sanchez (eds.), *Fuzzy Information and Decision Process* (North-Holland, Amsterdam, 1982).

[BM15] Gupta, M.M. and E. Sanchez (eds.), *Approximate Reasoning in Decision Analysis* (North-Holland, Amsterdam, 1982).

[BM16] Gupta, M.M. and T. Yamakawa (eds.), *Fuzzy Logic in Knowledge-Based Systems, Decision and Control* (North-Holland, Amsterdam, 1988).

[BM17] Hall, A.D. III, *Metasystems Methodology: A New Synthesis and Unification* (Pergamon, Oxford, 1989).

[BM18] Hollnagel, E., G. Mancini and D.D. Woods (eds.), *Intelligent Decision Support in Process Environments* (Springer-Verlag, Heidelberg, 1985).

[BM19] Hwang, C.L. and A.S.M. Masud, *Multiple Objective Decision Making: Methods and Applications* (Springer-Verlag, Heidelberg, 1979).

[BM20] Hwang, C.L. and K. Yoon, *Multiple Attribute Decision Making: Methods and Applications* (Springer-Verlag, Heidelberg, 1981).

[BM21] Hwang, C.L. and Lin M.-J., *Group Decision Making under Multiple Criteria: Methods and Applications* (Springer-Verlag, Heidelberg, 1987).

[BM22] Janko, W.H., M. Roubens and H.-J. Zimmermann (eds.), *Progressive in Fuzzy Sets and Systems* (Kluwer Academic, Dordrecht, 1989).

[BM23] Jones, A., A. Kaufmann and H.-J. Zimmermann (eds.), *Fuzzy Sets Theory and Applications* (D. Reidel, Dordrecht, 1985).

[BM24] Kacprzyk, J., *Multistage Decision-Making under Fuzziness: Theory and Applications* (Verlag TUV Rheinland, New York, 1981).

[BM25] Kacprzyk, J. and R.R. Yager (eds.), *Management Decision Support Systems Using Fuzzy Sets and Possibility Theory* (Verlag TUV Rheinland, New York, 1985).

[BM26] Kacprzyk, J. and S.A. Orlovski (eds.), *Optimization Models Using Fuzzy Sets and Possibility Theory* (D. Reidel, Dordrecht, 1987).

[BM27] Kacprzyk, J. and M. Fedrizzi (eds.), *Combining Fuzzy Imprecision with Probabilistic Uncertainty in Decision Making* (Springer-Verlag, Heidelberg, 1988).

[BM28] Kacprzyk, J. and M. Roubens (eds.), *Non-Conventional Preference Relations in Decision Making* (Springer-Verlag, Heidelberg, 1988).

[BM29] Kandel, A., *Fuzzy Mathematical Techniques with Applications* (Addison-Wesley, Reading, 1987).

[BM30] Karwowski, W. and A. Mital (eds.), *Applications of Fuzzy Set Theory in Human Factors* (Elsevier Science, Amsterdam, 1986).

[BM31] Kaufmann, A. and M.M. Gupta, *Introduction to Fuzzy Arithmetic: Theory and Applications* (Van Nostrand Reinhold, New York, 1985).

[BM32] Kaufmann, A. and M.M. Gupta, *Fuzzy Mathematical Models in Engineering and Management Science* (North-Holland, Ansterdam, 1988).

[BM33] Kickert, W.J.M., *Fuzzy Theories on Decision-Making* (Martinus Nijhoff, Leiden, 1978).

[BM34] Klir, G.J. (ed.), *Applied General Systems Research* (Wiley, New York, 1978).

[BM35] Klir, G.J. and T.A. Folger, *Fuzzy Sets Uncertainty, and Information* (Prentice-Hall, Englewood Cliff, 1988).

[BM36] Kruse, R. and K.D. Meyer, *Statistics with Vague Data* (D. Reidel, Dordrecht, 1987).

[BM37] Lasker, G.E. (ed.), *Applied Systems and Cybernetics Vol. VI* (Pergamon, New York, 1981).

[BM38] Leung, Y., *Spatial Analysis and Planning under Imprecision* (North-Holland, Amsterdam, 1988).

[BM39] Mitra, G., H.J. Greenberg, F.A. Lootsma, M.J. Rijckaert and H.-J. Zimmermann (eds.), *Mathematical Models for Decision Support* (Springer-Verlag, Heidelberg, 1988).

[BM40] Negoita, C.V., *Management Applications of System Theory* (Birkhauser Verlag, Bucharest, 1979).

[BM41] Negoita, C.V. and D. Ralescu, *Simulation, Knowledge-Based, Computing, and Fuzzy Statistics* (Van Nostrand Reinhold, New York, 1987).

[BM42] Nemhauser, G.L., A.H.G. Rinnooy Kan and M.J. Todd (eds.), *Handbooks in Operations Research and Management Science (Vol. 1): Optimization* (North-Holland, Amsterdam, 1989).

[BM43] Sanchez, E.(ed.), *Fuzzy Information , Knowledge Representation and Decision Analysis*, A Proceedings of the IFAC Symposium, Marseille, France (Oxford, New York, 1983).

[BM44] Schmucker, K.J., *Fuzzy Sets, Natural Language, Computations, and Risk Analysis* (Computer Science, Rockville, 1984).

[BM45] Seo, F. and M. Sakawa, *Multiple Criteria Decision Analysis in Regional Planning - Concepts, Methods and Applications* (D. Reidel, Dordrecht, 1988).

[BM46] Smithson, M., *Fuzzy Set Analysis for Behavioral and Social Sciences* (Springer-Verlag, Heidelberg, 1987).

[BM47] Smithson, M., *Ignorance and Uncertainty* (Springer-Verlag, Heidelberg, 1989).

[BM48] Stancu-Minasian, I.M., *Stochastic Programming with Multiple Objective Functions* (D. Reidel, Dordrecht, 1984).

[BM49] Trappl R., G.J. Klir and L. Ricciardi (eds.), *Progress in Cybernetics and Systems Research: General System Methodology, Fuzzy Mathematics, Fuzzy Systems, Biocybernetics and Theoretical Neurobiology, vol.III* (Hemisphere, Washington, 1978).

[BM50] Trappl R., G.J. Klir and F.R. Pichler (eds.), *Progress in Cybernetics and Systems Research: General System Methodology, Mathematical System Theory, and Fuzzy Sets vol.VIII* (Hemisphere, Washington, 1982).

[BM51] Trappl, R. (ed.), *Cybernetics and Systems Research* (North-Holland, Amsterdam, 1982).

[BM52] Trappl, R. (ed.), *Cybernetics and Systems Research* (North-Holland, Amsterdam, 1984).

[BM53] Trappl, R. (ed.), *Cybernetics and Systems* (D. Reidel, Dordrecht, 1986).

[BM54] Wagner, H.M., *Principles of Operations Research with Applications to Managerial Decisions* (Prentice-Hall, Englewood Cliffs, 1969).

[BM55] Wang, P.P. (ed.), *Advance in Fuzzy Sets, Possibility Theory and Applications* (Plenum, New York, 1983).

[BM56] Yager, R.R. (ed.), *Fuzzy Set and Possibility Theory: Recent Developments* (Pergamon, New York, 1982).

[BM57] Yager, R.R., S. Ovchinnikov, R.M. Tong and H.T. Nguyen (eds.), *Fuzzy Sets and Applications: Selected Papers by L.A. Zadeh* (Wiley, New York, 1987).

[BM58] Yu, P.L., *Multiple Criteria Decision Making: Concepts, Techniques and Extensions* (Plenum, New York, 1985).

[BM59] Yu, P.L., *Forming Winning Strategies: An Integrated Theory of Habitual Domains* (Springer-Verlag, Heidelberg, 1990).

[BM60] Zadeh, L.A., K.S. Fu, K. Tanaka and M. Shirmura (eds.), *Fuzzy Sets and Their Applications to Cognitive and Decision Processes* (Academic, New York, 1975).

[BM61] Zetenyi, T. (ed.), *Fuzzy Sets in Psychology* (North-Holland, Amsterdam, 1988).

[BM62] Zimmermann, H.-J., L.A. Zadeh and B.R. Gaines (eds.), *Fuzzy Sets and Decision Analysis* (Elsevier Science, Amsterdam, 1984).

[BM63] Zimmermann, H.-J., *Fuzzy Set Theory and Its Applications* (Kluwer-Nijhoff, Hinghum, 1985).

[BM64] Zimmermann, H.-J., *Fuzzy Sets, Decision Making and Expert Systems* (Kluwer Academic, Norwell, 1987).

Journal Articles, Technical Reports, and Theses:

[1] Ackoff, R.L., The future of operational research is past, *Journal of Operational Research Society* 30 (1979) 93-104.

[2] Ackoff, R.L., Resurrecting the future of operational research, *Journal of Operational Research Society* 30 (1979) 189-199.

[3] Ammar, S., Determining the 'best' decision in the presence of imprecise information, *Fuzzy Sets and Systems* 29 (1989) 293-302.

[4] Asai, K., H. Tanaka and T. Dkuda, On discrimination of fuzzy states in probability space, *Kybernetes* 6 (1977) 185-192.

[5] Aubin, J.P., Cooperative fuzzy games: the static and dynamic points of view, in [BM62] (1984) 407-428.

[6] Baldwin, J.F. and N.C.F. Guild, Comparison of fuzzy sets on the same decision space, *Fuzzy Sets and Systems* 2 (1979) 213-231.

[7] Bandemer, H., From fuzzy data to functional relationships, *Mathematical Modelling* 9 (1987) 419-426.

[8] Baptistella, L.F.B. and A. Ollero, Fuzzy methodologies for interactive multicriteria optimization, *IEEE Transactions on Systems, Man, and Cybernetics* 10 (1980) 355-365.

[9] Beale, E.M.L., On minimizing a convex function subject to linear inequalities, *Journal of Royal Statistist Society* B17 (1955) 173-184.

[10] Bedrosian, S.D., A role for fuzzy concepts in interactive decision-making, in [BM25] (1985) 38-47.

[11] Behringer, F.A., A simplex method based algorithm for the lexicographically extended linear maxmin problem, *European Journal of Operational Research* 7 (1981) 274-283.

[12] Bellman, R.E. and L.A. Zadeh, Decision-making in a fuzzy environment, *Management Science* 17 (1970) B141-B164.

[13] Bellman, R. and M. Giertz, On the analytic formalism of the theory of the fuzzy sets, *Information Sciences* 5 (1973) 149-156.

[14] Bensoussan, A. and J.M. Proth, Planning horizons for production planning models in the case of concave costs, in [BM43] (1983) 393-398.

[15] Bezdek, J.C. and R.J. Hathaway, Clustering with relational c-means partitions from pairwise distance data, *Mathematical Modelling* 9 (1987) 435-439.

[16] Booth, G.G. and G.H. Dash, Jr., Alternative programming structures for bank portfolios, *Journal of Banking and Finance* 3 (1979) 67-82.

[17] Booth, G.G. and P.E. Koveos, A programming model for bank hedging decisions, *The Journal of Financial Research* IX (1986) 271-279.

[18] Booth, G.G. and W. Bessler, Goal programming models for managing interest-rate risk, *OMEGA International Journal of Management Science* 17 (1989) 81-89.

[19] Booth, G.G., W. Bessler and W.G. Foote, Managing interest rate risk in banking institutions, *European Journal of Operational Research* 41 (1989) 302-313.

[20] Bortolan, G. and R. Degani, A review of some methods for ranking fuzzy subsets, *Fuzzy Sets and Systems* 15 (1985) 17-19.

[21] Bouchon, B., Preferences deduced from fuzzy questions, in [BM26] (1987) 110-120.

[22] Buckley, J.J., Possibility and necessity in optimization, *Fuzzy Sets and Systems* 25 (1988) 1-13.

[23] Buckley, J.J., Generalized and extended fuzzy sets with application, *Fuzzy Sets and Systems* 25 (1988) 159-174.

[24] Buckley, J.J., Possibilistic linear programming with triangular fuzzy numbers, *Fuzzy Sets and Systems* 26 (1988) 135-138.

[25] Buckley, J.J., Solving possibilistic linear programming problems, *Fuzzy Sets and Systems* 31 (1989) 329-341.

[26] Buckley, J.J., Fuzzy PERT, in [BM10] (1989) 103-114.

[27] Buckley, J.J., Stochastic versus possibilistic programming, *Fuzzy Sets and Systems* 34 (1990) 173-177.

[28] Buckley, J.J. and Y. Qu, Solving linear and quadratic fuzzy equations, *Fuzzy Sets and Systems* 38 (1990) 43-59.

[29] Campos, L., Fuzzy linear programming models to solve fuzzy matrix games, *Fuzzy Sets and Systems* 32 (1989) 271-289.

[30] Campos, L. and J.L. Verdegay, Linear programming problems and ranking of fuzzy numbers, *Fuzzy Sets and Systems* 32 (1989) 1-11.

[31] Carlsson, C. and P. Korhonen, A parametric approach to fuzzy linear programming, *Fuzzy Sets and Systems* 20 (1986) 17-30.

[32] Chanas, S. and J. Kamburowski, The use of fuzzy variables in PERT, *Fuzzy Sets and Systems* 5 (1981) 11-19.

[33] Chanas, S. and W. Kolodziejczyk, Maximum flow in a network with fuzzy arc capacities, *Fuzzy Sets and Systems* 8 (1982) 165-173.

[34] Chanas, S., The use of parametric programming in FLP, *Fuzzy Sets and Systems* 11 (1983) 243-251.

[35] Chanas, S. and M. Kulej, A fuzzy linear programming problem with equality constraints, *Control and Cybernetics* 13 (1984) 195-201.

[36] Chanas, S., Fuzzy optimization in networks, in [BM26] (1987) 303-327.

[37] Chanas, S., Fuzzy programming in multiobjective linear programming - a parametric approach, *Fuzzy Sets and Systems* 29 (1989) 303-313.

[38] Chang, T.C. and K.C. Crandall, An algorithm for solving expected possibility and its application in construction resource allocation, *Fuzzy Sets and Systems* 34 (1990) 157-171.

[39] Charnes, A., W.W. Cooper and G.H. Symonds, Cost horizons and certainty equivalents: an approach to stochastic programming of heating oil, *Management Science* 4 (1958) 235-263.

[40] Charnes, A. and W.W. Cooper, Chance-constrainted programming, *Management Science* 5 (1959) 73-79.

[41] Chen, G.Q., S.C. Lee and E.S.H Yu, Application of fuzzy set theory to economics, in [BM55] (1983) 277-305.

[42] Chen, S.H., Ranking fuzzy numbers with maximizing set and minimizing set, *Fuzzy Sets and Systems* 17 (1985) 113-129.

[43] Chen, Y.C., Fuzzy linear programming and fuzzy multiple objective linear programming, Master's Thesis (1984), Dept. of I.E., Kansas State University.

[44] Chien, T.Q., Medium distances of probability fuzzy-points and an application to linear programming, *Kybernetika* 25 (1989) 494-504.

[45] Cooley, J.W., and J.D. Hicks, Jr., A fuzzy set approach to aggregating internal control judgments, *Management Science* 29 (1983) 317-334.

[46] Cooley, J.W. and J.D. Hicks, Jr., The use of fuzzy sets in the accounting spectrum: the evaluation of internal control systems, in [BM25] (1985) 334-342.

[47] Dantzig, G.B., Linear programming under uncertainty, *Management Science* 1 (1955) 197-204.

[48] Darby-Dowman, K, C. Lucas, G. Mitra and J. Yadegar, Linear, integer, separable and fuzzy programming problems: a unified approach towards reformulation, *Journal of Operational Research Society* 29 (1988) 161-171.

[49] Darzentas, J., A discrete location model with fuzzy accessibility measures, *Fuzzy Sets and Systems* 23 (1987) 149-154.

[50] Darzentas, J., On fuzzy location models, in [BM26] (1987) 328-341.

[51] de Campos Ibanez, L.M., A subjective approach for ranking fuzzy numbers, *Fuzzy Sets and Systems* 29 (1989) 145-153.

[52] Delgado, M., J.L. Verdegay, and M.A. Vila, Imprecise costs in mathematical programming problems, *Control and Cybernetics* 16 (1987) 113-121.

[53] Delgado, M., J.L. Verdegay, and M.A. Vila, Fuzzy transportation problems: a general analysis, in [BM26] (1987) 342-358.

[54] Delgado, M., A procedure for ranking fuzzy number using fuzzy relations, *Fuzzy Sets and Systems* 26 (1988) 49-62.

[55] Delgado, M., J.L. Verdegay, and M.A. Vila, A general model for fuzzy linear programming, *Fuzzy Sets and Systems* 29 (1989) 21-29.

[56] Delgado, M., J.L. Verdegay, and M.A. Vila, Relating different approaches to solve linear programming problems with imprecise costs, *Fuzzy Sets and Systems* 37 (1990) 33-42.

[57] Delgado, M., J.L. Verdegay, M. A. Vila, On valuation and optimization problems in fuzzy graphs: a general approach and some particular cases, *ORSA Journal on Computing* 2 (1990) 74-83.

[58] Dias, O.P., Jr., The R & D project selection problem with fuzzy coefficients, *Fuzzy Sets and Systems* 26 (1988) 299-316.

[59] Dishkant, H.C., About membership-function estimation, *Fuzzy Sets and Systems* 5 (1981) 141-148.

[60] Dombi, J., A general class of fuzzy operators, the Demorgan class of fuzzy operators, and fuzziness measures induced by fuzzy operators, *Fuzzy Sets and Systems* 8 (1982) 149-163.

[61] Dombi, J., Membership function as an evaluation, *Fuzzy Sets and Systems* 35 (1990) 1-21.

[62] Dowlatshahi, F. and L.J. Kohout, Intentional possibilistic approach to dealing with incompleteness, vagueness and uncertainty, *Fuzzy Sets and Systems* 25 (1988) 277-295.

[63] Dubois, D., A fuzzy, heuristic, interactive approach to the optimal network problem, in [BM55] (1983) 253-275.

[64] Dubois, D. and H. Prade, System of linear fuzzy constraints, *Fuzzy Sets and Systems* 3 (1980) 37-48.

[65] Dubois, D. and H. Prade, The use of fuzzy numbers in decision making, in [BM15] (1982) 309-321.

[66] Dubois, D. and H. Prade, Ranking fuzzy numbers in the setting of possibility theory, *Information Sciences* 30 (1983) 183-224.

[67] Dubois, D. and H. Prade, Criteria aggregation and ranking of alternatives in the framework of fuzzy set theory, in [BM62] (1984) 209-240.

[68] Dubois, D. and H. Prade, The advantages of fuzzy approach in OR/MS demonstrated on two examples of resources allocation problems, in [BM52] (1984) 491-497.

[69] Dubois, D. and H. Prade, Fuzzy-theoretic difference and inclusions and their use in analysis of fuzzy equations, *Control and Cybernetics* 13 (1984) 129-145.

[70] Dubois, D. and H. Prade, A review of fuzzy set aggregation connectives, *Information Sciences* 36 (1985) 85-121.

[71] Dubois, D., An application of fuzzy arithmetic to the optimization of industrial machining processes, *Mathematical Modelling* 9 (1987) 461-475.

[72] Dubois, D., Linear programming with fuzzy data, in [BM1] (1987) 241-263.

[73] Dubois, D. and H. Prade, Properties of measurer of information in evidence and possibility theories, *Fuzzy Sets and Systems* 24 (1987) 161-182.

[74] Dubois, D. and H. Prade, Necessity measures and the resolution principle, *IEEE Transactions on Systems, Man, and Cybernetics* 17 (1987) 474-478.

[75] Dubois, D. and H. Prade, Incomplete conjunctive information, *Computers and Mathematics with Applications* 15 (1988) 797-810.

[76] Dubois, D., Fuzzy knowledge in an artificial intelligence system for job shop scheduling, in [BM10] (1989) 73-89.

[77] Dumitru, V. and F. Luban, Membership functions, some mathematical programming models and production scheduling, *Fuzzy Sets and Systems* 8 (1982) 19-33.

[78] Dumitru, V. and F. Luban, On some optimization problems under uncertainty, *Fuzzy Sets and Systems* 18 (1986) 257-272.

[79] Dyson, R.G., Maximin programming, Fuzzy linear programming and multicriteria decision making, *Journal of Operational Research Society* 31 (1981) 263-267.

[80] Esogbue, A.O. and Z.M. Ahipo, Fuzzy sets and water resources planning, in [BM56] (1982) 450-465.

[81] Evans, G.W., M.R. Wilhelm and W. Karwowski, A layout design heuristic employing the theory of fuzzy sets, *International Journal of Production Research* 25 (1987) 1431-1450.

[82] Evans, G.W., W. Karwowski and M.R. Wilhelm, An introduction to fuzzy set methodologies for industrial and systems engineering, in [BM10] (1989) 3-11.

[83] Fabian, Cs. and M. Stoica, Fuzzy integer programming, in [BM62] (1984) 123-131.

[84] Fabian, Cs., Gh. Ciobanu and M. Stoica, Interactive polyoptimization for fuzzy mathematical programming, in [BM26] (1987) 272-291.

[85] Falk, J.E., Exact solutions of inexact linear programs, *Operations Research* 22 (1974) 783-787.

[86] Fedrizzi, M., J. Kacprzyk and S. Zadrozny, An interactive multi-user decision support system for consensus reaching processes using fuzzy logic with linguistic quantifiers, *Decision Support Systems* 4 (1988) 313-327.

[87] Feng, Y.J., A method using fuzzy mathematics to solve the vector maximum problem, *Fuzzy Sets and Systems* 9 (1983) 129-136.

[88] Feng, Y.J., Fuzzy programming - a new model of optimization, in [BM26] (1987) 216-225.

[89] Feng, Y.-J., M.-Z. Lin and C. Jiang, Application of fuzzy decision-making in earthquake research, *Fuzzy Sets and Systems* 36 (1990) 15-26.

[90] Fiksel, J., Applications of fuzzy sets and possibility theory to systems management, in [BM56] (1982) 494-503.

[91] Flachs, J. and M.A. Pollatschek, Further results on fuzzy mathematical programming, *Information and Control* 38 (1978) 241-257.

[92] Freeling, A.N.S., Fuzzy sets and decision analysis, *IEEE Transactions on Systems, Man, and Cybernetics* 10 (1980) 341-354.

[93] Freeling, A.N.S., Possibilities versus fuzzy probabilities - two alternative decision aids, in [BM62] (1984) 67-81.

[94] French, S., Fuzzy decision analysis: some criticisms, in [BM62] (1984) 29-44.

[95] Friedman, M., M. Schneider and A. Kandel, The use of weighted fuzzy expected value (WFEV) in fuzzy expert systems, *Fuzzy Sets and Systems* 31 (1989) 37-45.

[96] Fu, K.S., M. Ishizuka and J.T.P. Yao, Applications of fuzzy sets in earthquake engineering, in [BM56] (1982) 504-523.

[97] Fuller, R., On a special type of fuzzy linear programming, *Colloquia Mathematica Societatis Janos Bolyai* 50 (1986) 511-519.

[98] Fuller, R., On stability in fuzzy linear programming problems - a short communication, *Fuzzy Sets and Systems* 30 (1989) 339-344.

[99] Fuller, R., On stability in possibilistic linear equality systems with Lipschitzian fuzzy numbers, *Fuzzy Sets and Systems* 34 (1990) 347-353.

[100] Furuta, H., K. Furukawa, and N. Shiraishi, Application of fuzzy optimality criterion to earthquake-resistant design, in [BM1] (1987) 17-27.

[101] Gaines, B.R. and L.J. Kohout, The fuzzy decade: a bibliography of fuzzy systems and closely related topics, *International Journal of Man-Machine Studies* 9 (1977) 1-68.

[102] Gaines, B.R., Precise past - fuzzy future, *International Journal of Man-Machine Studies* 19 (1983) 117-134.

[103] Gaines, B.R., Fundamentals of decision: probabilistic, possibilistic and other forms of uncertainty in decision analysis, in [BM62] (1984) 47-65.

[104] Gaines, B.R., New paradigms in systems engineering: from hard to soft approaches, in [BM26] (1987) 3-12.

[105] Gambarelli, G., J. Holubiec and J. Kacprzyk, Modeling and optimization of international economic cooperation via fuzzy mathematical programming and cooperative games, *Control and Cybernetics* 17 (1988) 325-335.

[106] Giles, R., The concept of grade of membership, *Fuzzy Sets and Systems* 25 (1988) 297-323.

[107] Ghassan, K., New utilization of fuzzy optimization method, in [BM15] (1982) 239-246.

[108] Goguen, J.A., L-fuzzy sets, *Journal of Mathematical Analysis and Its Applications* 18 (1967) 145-174.

[109] Grobelny, J., The fuzzy approach to facilities layout problems, *Fuzzy Sets and Systems* 23 (1987) 175-190.

[110] Grobelny, J., On one possible 'fuzzy' approach to facilities layout problems, *International Journal of Production Research* 25 (1987) 1123-1141.

[111] Grobelny, J., The linguistic pattern method for a workstation layout analysis, *International Journal of Production Research* 26 (1988) 1779-1798.

[112] Hamacher, H., H. Leberling and H.-J. Zimmermann, Sensitivity analysis in fuzzy linear programming, *Fuzzy Sets and Systems* 1 (1978) 269-281.

[113] Hammerbacher, I.M. and R.R. Yager, Predicting television revenues using fuzzy subsets, in [BM62] (1984) 469-477.

[114] Hersh, H.M. and A. Caramazzn, A fuzzy set approach to modifiers and vagueness in natural language, *Journal of Experimental Psychology: General* 105 (1976) 254-276.

[115] Hessel, M. and M. Zeleny, Optimal system design: towards new interpretation of shadow prices in linear programming, *Computers and Operations Research* 14 (1987) 265-271.

[116] Hidding, G.J., DSS conceptualizes in O.R.: should you apply AI?, in [BM39] (1988) 665-694.

[117] Hisdal, E., Are grades of membership probabilities?, *Fuzzy Sets and Systems* 25 (1988) 325-348.

[118] Hisdal, E., The philosophical issues raised by fuzzy set theory, *Fuzzy Sets and Systems* 25 (1988) 349-356.

[119] Hisdal, E., A flexible classification structure, in [BM16] (1988) 11-67.

[120] Inuiguchi, M., H. Ichihashi and Y. Kume, A solution algorithm for fuzzy linear programming with piecewise linear membership functions, *Fuzzy Sets and Systems* 34 (1990) 15-31.

[121] Inuiguchi, M. and H. Ichihashi, Relative modalities and their use in possibilistic linear programming, *Fuzzy Sets and Systems* 35 (1990) 303-323.

[122] Kabbara, G., New utilization of fuzzy optimization method, in [BM15] (1982) 239-245.

[123] Kacprzyk, J. and A. Straszak, A fuzzy approach to the stability of integrated regional development, in [BM37] (1981) 2997-3004.

[124] Kacprzyk, J., Multistage decision process in a fuzzy environment: a survey, in [BM14] (1982) 251-263.

[125] Kacprzyk, J. and A. Straszak, Determination of stable regional development trajectories via fuzzy decision-making model, in [BM56] (1982) 531-541.

[126] Kacprzyk, J. and P. Staniewski, Long-term inventory policy-making through fuzzy decision-making models, *Fuzzy Sets and Systems* 8 (1982) 117-132.

[127] Kacprzyk, J., Towards human-consistent multistage decision making and control models using fuzzy sets and fuzzy logic, *Fuzzy Sets and Systems* 18 (1986) 299-314.

[128] Kacprzyk, J. and S.A. Orlovski, Fuzzy optimization and mathematical programming: a brief introduction and survey, in [BM26] (1987) 50-72.

[129] Karwowski, W., T. Marek, C. Noworol and K. Ostaszewski, Fuzzy modeling of risk factors for industrial accident prevention: some empirical results, in [BM10] (1989) 141-153.

[130] Kaufmann, A., Fuzzy zero-base budgeting, in [BM62] (1984) 479-487.

[131] Kaufmann, A., Hybrid data - various associations between fuzzy subsets and random variables, in [BM23] (1985) 171-211.

[132] Kaufmann, A., Fuzzy subsets applications in O.R. and management, in [BM29] (1985) 257-300.

[133] Kim, J.B., A. Baartmans and N.S. Sahadin, Determinant theory for fuzzy matrices, *Fuzzy Sets and Systems* 29 (1989) 349-356.

[134] Kim, K. and F.W. Roush, Fuzzy flow on networks, *Fuzzy Sets and Systems* 8 (1982) 35-38.

[135] Kim, K. and K.S. Park, Ranking fuzzy numbers with index of optimism, *Fuzzy Sets and Systems* 35 (1990) 143-150.

[136] Kuzmin, V.B., A parametric approach to description of linguistic values of variables and hedges, *Fuzzy Sets and Systems* 6 (1981) 27-41.

[137] Lai, Y.J. and C.L. Hwang, Interactive fuzzy linear programming, *Fuzzy Sets and Systems* 45 (1992) 169-183.

[138] Lai, Y.J. and C.L. Hwang, A new approach to some possibilistic linear programming problem, *Fuzzy Sets and Systems* 49 (1992).

[139] Lai, Y.J., Fuzzy mathematical programming and fuzzy multiple objective decision making, Ph.D. Dissertation (1991), Dept. of I.E., Kansas State University.

[140] Lai, Y.J. and C.L. Hwang, Possibilistic linear programming for managing interest rate risk, *Fuzzy Sets and Systems* (forthcoming).

[141] Lai, Y.J. and C.L. Hwang, IFLP-II: a decision support system, *Fuzzy Sets and Systems* (forthcoming).

[142] Lai, Y.J. and C.L. Hwang, Stochastic possibilistic linear programming (submitted).

[143] Lai, Y.J., T.Y. Liu and C.L. Hwang, TOPSIS for MODM, *European Journal of Operational Research* (forthcoming).

[144] Lee, E.S. and R.-J. Li, Comparison of fuzzy numbers based on probability measure of fuzzy events, *Computers and Mathematics with Applications* 15 (1988) 887-896.

[145] Lee, Y.Y., Fuzzy Sets Theory Approach to Aggregate Production Planning and Inventory Control, Ph.D. Dissertation (1990), Dept. of I.E., Kansas State University.

[146] Lee, Y.Y., B.A. Kramer and C.L. Hwang, Part-period balancing with uncertainty: a fuzzy sets theory approach, *International Journal of Production Research* 28 (1990) 1771-1778.

[147] Lehtimaki, A.K., An approach for solving decision problems of master scheduling by utilizing theory of fuzzy sets, *International Journal of Production Research* 25 (1987) 1781-1793.

[148] Leung, Y., Interregional equilibrium and fuzzy linear programming - 1, *Environment and Planning A* 20 (1988) 25-40.

[149] Leung, Y., Interregional equilibrium and fuzzy linear programming - 2, *Environment and Planning A* 20 (1988) 219-230.

[150] Li Calzi, M., Towards a general setting for the fuzzy mathematics of finance, *Fuzzy Sets and Systems* 35 (1990) 265-280.

[151] Li, R.J., Multiple Objective Decision Making in A Fuzzy Environment (1990), Ph.D. Dissertation, Dept. of I.E., Kansas State University.

[152] Liu, X., M. Wang and P. Wang, A fuzzy mathematical method for earthquake disaster prediction, in [BM1] (1987) 55-65.

[153] Llena, J., On fuzzy linear programming, *European Journal of Operational Research* 22 (1985) 216-213.

[154] Lokwick, W.A., Analysis of structure in fuzzy linear programs, *Fuzzy Sets and Systems* 38 (1990) 15-26.

[155] Luhandjula, M.K., Compensatory operators in fuzzy linear programming with multiple objectives, *Fuzzy Sets and Systems* 8 (1982) 245-252.

[156] Luhandjula, M.K., Linear programming under randomness and fuzziness, *Fuzzy Sets and Systems* 10 (1983) 45-55.

[157] Luhandjula, M.K., On possibilistic linear programming, *Fuzzy Sets and Systems* 18 (1986) 15-30.

[158] Luhandjula, M.K., Satisfying solution for a possibilistic linear program, *Information Sciences* 40 (1986) 247-265.

[159] Luhandjula, M.K., Linear programming with a possibilistic objective function, *European Journal of Operational Research* 31 (1987) 110-117.

[160] Luhandjula, M.K., Multiple objective programming problems with possibilistic coefficients, *Fuzzy Sets and Systems* 21 (1987) 135-145.

[161] Luhandjula, M.K., Fuzzy optimization an appraisal, *Fuzzy Sets and Systems* 30 (1989) 257-282.

[162] Mabuchi, S., An approach to the comparison of fuzzy subsets with α-cut dependent index, *IEEE Transactions on Systems, Man, and Cybernetics* 18 (1988) 264-272.

[163] Marek, T. and C. Noworol, Fuzzy rational aggregation for mental load and fatigue assessment, in [BM10] (1989) 155-165.

[164] Mares, M., Network analysis of fuzzy technologies, in [BM10] (1989) 115-125.

[165] Martine, J.A., Some geometrical properties of trapezoidal functions for their association with linguistic labels leading to a reduced 2-D representation, in [BM2] (1988) 47-54.

[166] Mccahon, C.S., Fuzzy Set Theory Applied to Production and Inventory Control (1987), Ph.D. Dissertation, Dept. of I.E., Kansas State University.

[167] Mccahon, C.S. and E.S. Lee, Project network analysis with fuzzy activity times, *Computers and Mathematics with Applications* 15 (1988) 829-838.

[168] Mital, A. and W. Karwowski, A framework of the fuzzy linguistic approach to facilities location problem, in [BM10] (1989) 323-330.

[169] Mitra, G., Models for decision making: an overview of problems, tools and major issues, in [BM39] (1989) 17-53.

[170] Mizumoto, M., Fuzzy sets and their operators - I, *Information Sciences* 48 (1981) 30-48.

[171] Mizumoto, M., Fuzzy sets and their operators - II, *Information Sciences* 50 (1981) 160-174.

[172] Mjelde, K.M., Fuzzy resource allocation, *Fuzzy Sets and Systems* 19 (1986) 239-250.

[173] Nahmias, S., Fuzzy variables, *Fuzzy Sets and Systems* 1 (1978) 97-110.

[174] Nakamura, K., Some extensions of fuzzy linear programming, *Fuzzy Sets and Systems* 14 (1984) 211-229.

[175] Negi, D.S., Fuzzy Analysis and Optimization (1989), Ph.D. Dissertation, Dept. of I.E., Kansas State University.

[176] Negoita, C.V. and M. Sularia, On fuzzy mathematical programming and tolerances in planning, *Economic Computer and Economic Cybernetic Studies and Researches* 1 (1976) 3-15.

[177] Negoita, C.V., P. Flondor and M. Sularia, On fuzzy environment in optimization problems, *Economic Computer and Economic Cybernetic Studies and Researches* 2 (1977) 13-24.

[178] Negoita, C.V. and D.A. Ralescu, On fuzzy optimization, *Kybernetes* 6 (1977) 193-196.

[179] Negoita, C.V., The current interest in fuzzy optimization, *Fuzzy Sets and Systems* 6 (1981) 261-269.

[180] Negoita, C.V. and A.C. Stefanesca, On fuzzy optimization, in [BM15] (1982) 247-250.

[181] Negoita, C.V., Structure and logic in optimization, *Control and Cybernetics* 13 (1984) 121-127.

[182] Norwich, A.M. and I.B. Turksen, The construction of membership functions, in [BM56] (1982) 61-67.

[183] Norwich, A.M. and I.B. Turksen, Meaningfulness in fuzzy set theory, in [BM56] (1982) 68-74.

[184] Oden, G.C., Integration of fuzzy logical information, *Journal of Experimental Psychology: Human Perception and Performance* 3 (1976) 565-575.

[185] OhEigeartaigh, M., A fuzzy transportation algorithm, *Fuzzy Sets and Systems* 8 (1982) 235-243.

[186] Okuda, T., H. Tanaka and K. Asai, A formulation of fuzzy decision problems with fuzzy information using probability measures of fuzzy events, *Information and Control* 38 (1978) 135-147.

[187] Okuda, T., H. Tanaka and K. Asai, Discrimination problem with fuzzy states and fuzzy information, in [BM62] (1984) 97-106.

[188] Onisawa, T., A representation of human reliability using fuzzy concepts, *Information Sciences* 45 (1988) 153-173.

[189] Onisawa, T., M. Sugeno, Y. Nishiwak, H. Kawai and Y. Harima, Fuzzy measure analysis of public attitude towards the use of nuclear energy, *Fuzzy Sets and Systems* 20 (1988) 259-289.

[190] Onisawa, T., Fuzzy concepts in human reliability, in [BM16] (1988) 317-335.

[191] Onisawa, T., An application of fuzzy concepts to modelling of reliability analysis, *Fuzzy Sets and Systems* 37 (1990) 267-286.

[192] Orlovsky, S.A., On programming with fuzzy constraint sets, *Kybernetes* 6 (1977) 197-202.

[193] Orlovsky, S.A., Decision-making with a Fuzzy Preference Relation, *Fuzzy Sets and Systems* 1 (1978) 155-167.

[194] Orlovsky, S.A., On formalization of a general fuzzy mathematical problem, *Fuzzy Sets and Systems* 3 (1980) 311-321.

[195] Orlovsky, S.A., Effective alternatives for multiple fuzzy preference relations, in [BM51] (1982) 185-189.

[196] Orlovsky, S.A., Multiobjective programming problems with fuzzy parameters, *Control and Cybernetics* 13 (1984) 175-183.

[197] Orlovsky, S.A., Mathematical programming problems with fuzzy parameters, in [BM25] (1985) 136-145.

[198] Ostasiewicz, W., A new approach to fuzzy programming, *Fuzzy Sets and Systems* 7 (1982) 139-152.

[199] Ostermark, R., Sensitivity analysis of fuzzy linear programs: an approach to parametric interdependence, *Kybernetes* 16 (1987) 113-120.

[200] Ostermark, R., Profit apportionment in concerns with mutual ownership - an application of fuzzy inequalities, *Fuzzy Sets and Systems* 26 (1988) 283-297.

[201] Ostermark, R., Fuzzy linear constraints in the capital asset pricing model, *Fuzzy Sets and Systems* 30 (1989) 93-102.

[202] Owsinski, J.W., S. Zadrozny and J. Kacprzyk, Analysis of water use and needs in agriculture through a fuzzy programming model, in [BM26] (1987) 328-341.

[203] Pedrycz, W., Fuzzy models and relational equations, *Mathematical Modelling* 9 (1987) 427-434.

[204] Pedrycz, W., Methodological and applicational aspects of fuzzy models for systems engineering, in [BM10] (1989) 13-26.

[205] Pence, J.A. and A.L. Soyster, Relationship between fuzzy set programming and semi infinite linear programming, in [BM10] (1989) 237-252.

[206] Ponsard, C., Fuzzy sets in economics: foundations of soft decision theory, in [BM25] (1985) 25-37.

[207] Prade, H., Using fuzzy set theory in a scheduling problem: a case study, *Fuzzy Sets and Systems* 2 (1979) 153-165.

[208] Ralescu, D., Optimization in fuzzy environment, in [BM43] (1983) 301-305.

[209] Ramik, J., Extension principle and fuzzy mathematical programming, *Kybernetica* 19 (1983) 516-525.

[210] Ramik, J. and J. Rimanek, Inequality relation between fuzzy numbers and its use in fuzzy optimization, *Fuzzy Sets and Systems* 16 (1985) 123-138.

[211] Ramik, J., Extension principle in fuzzy optimization, *Fuzzy Sets and Systems* 19 (1986) 29-35.

[212] Ramik, J. and J. Rimanek, Fuzzy parameters in optimal allocation of resources, in [BM26] (1987) 359-374.

[213] Rapoport, A., T.S. Wallsten and J.A. Cox, Direct and indirect scaling of membership functions of probability phrases, *Mathematical Modelling* 9 (1987) 397-417.

[214] Rinks, D.B., A heuristic approach to aggregate production scheduling using linguistic variables: Methodology and application, in [BM56] (1982) 562-581.

[215] Rodder, W. and H.-J. Zimmermann, Duality in fuzzy linear programming, in Fiacco, A.V. and K.D. Kortanek (eds.), *External Methods and Systems Analysis* (Springer-Verlag, Heidelberg, 1980) 415-427.

[216] Rommelfanger, H., R. Hanuscheck and J. Wolf, Linear programming with fuzzy objectives, *Fuzzy Sets and Systems* 29 (1989) 31-48.

[217] Rommelfanger, H., Interactive decision making in fuzzy linear optimization problems, *European Journal of Operational Research* 41 (1989) 210-217.

[218] Roubens, M. and Ph. Vincke, Fuzzy preferences in an optimization perspective, in [BM26] (1987) 77-90.

[219] Roventa, E., On the degree of fuzziness of a fuzzy set, *Fuzzy Sets and Systems* 36 (1990) 259-264.

[220] Sakawa, M. and H. Yano, An interactive fuzzy satisfying method using augmented minimax problems and its application to environmental systems, *IEEE Transactions on Systems, Man, and Cybernetics* 5 (1985) 720-729.

[221] Shafai, B. and G. Sotirov, Uniqueness of solution in FLP under parameter perturbations, *Fuzzy Sets and Systems* 34 (1990) 179-186.

[222] Sher, A.P., Solution of the mathematical programming problem with a linear objective function and fuzzy constraints, *Automata* (1981) 997-1002.

[223] Simon, H., Dynamic programming under uncertainty with a quadratic criterion function, *Econometrica* 24 (1956) 74-81.

[224] Simon, H., Two heads are better then one: the collaboration between AI and OR, *Interfaces* 17 (1987) 8-15.

[225] Slowinski, R., Multiobjective network scheduling with efficient use of renewable and nonrenewable resources, *European Journal of Operational Research* 7 (1982) 265-273.

[226] Slowinski, R., An interactive method for multiobjective linear programming with fuzzy parameters and its application to water supply planning, in [BM26] (1987) 396-414.

[227] Slowinski, R., A multicriteria fuzzy linear programming method for water supply system development planning, *Fuzzy Sets and Systems* 19 (1986) 217-237.

[228] Slowinski, R. and J. Teghem, Jr., Fuzzy versus stochastic approaches to multicriteria linear programming under uncertainty, *Naval Research Logistics* 35 (1988) 673-695.

[229] Smets, P. and P. Magrez, The measure of the degree of truth and of the grade of membership, *Fuzzy Sets and Systems* 25 (1988) 67-72.

[230] Smithson, M., Fuzzy set theory and the social sciences: the scope for application, *Fuzzy Sets and Systems* 26 (1988) 1-21.

[231] Sommer, G. and M.A. Pollatschek, A fuzzy programming approach to an air pollution regulation problem, in [BM49] (1978) 303-313.

[232] Soyster, A.L., Convex programming with set-inclusive constraints and applications to inexact linear programming, *Operations Research* 21 (1973) 1154-1157.

[233] Soyster, A.L., A duality theory for convex programming with set-inclusive constraints, *Operations Research* 22 (1974) 892-898.

[234] Soyster, A.L., Convex programming with set inclusive constraints application to inexact programming, *Operations Research* 6 (1981) 333-348.

[235] Stohr, E.A., Automated support for formulating linear programs, in [BM39] (1988) 519-538.

[236] Tanaka, H., T. Okuda and K. Asai, On fuzzy mathematical programming, *Journal of Cybernetics* 3 (1974) 37-46.

[237] Tanaka, H., T. Okuda and K. Asai, A formulation of fuzzy decision problems and its application to an investment problem, *Kybernetes* 5 (1976) 25-30.

[238] Tanaka, H. and K. Asai, Fuzzy solution in fuzzy linear programming problems, *IEEE Transactions on Systems, Man, and Cybernetics* 14 (1984) 325-328.

[239] Tanaka, H. and K. Asai, Fuzzy linear programming problems with fuzzy numbers, *Fuzzy Sets and Systems* 13 (1984) 1-10.

[240] Tanaka, H., H. Ichihashi and K. Asai, A formulation of fuzzy linear programming problems based on comparison of fuzzy numbers, *Control and Cybernetics* 13 (1984) 186-194.

[241] Tanaka, H., H. Ichihashi and K. Asai, Fuzzy decision in linear programming problems with trapezoid fuzzy parameters, in [BM25] (1985) 146-154.

[242] Tanaka, H., H. Ichihashi and K. Asai, A value of information in FLP problem via sensitivity analysis, *Fuzzy Sets and Systems* 18 (1986) 119-129.

[243] Tanaka, H., J. Watada and K. Asai, Linear regression analysis by possibilistic models, in [BM26] (1987) 186-199.

[244] Tanaka, H., Fuzzy data analysis by possibilistic linear models, *Fuzzy Sets and Systems* 24 (1987) 363-375.

[245] Thole, U., H.-J. Zimmermann and P. Zysno, On the suitability of minimum and product operators for the intersection of fuzzy sets, *Fuzzy Sets and Systems* 2 (1979) 167-180.

[246] Thompson, S.D. and W.J. Davis, An integrated approach for modeling uncertainty in aggregate production planning, *IEEE Transactions on Systems, Man, and Cybernetics* 20 (1990) 1000-1012.

[247] Tintner, G, Stochastic linear programming with application to argricultural economics, in H.A. Antonosiewicz (ed.), *Proceedings of the Second Symposium in Linear Programming* (National Bureau Standard, Washington, 1955) 197-228.

[248] Trappey, J.F.C., C., R. Liu and T.C. Chang, Fuzzy non-linear programming: theory and application in manufacturing, *International Journal of Production Research* 26 (1988) 957-985.

[249] Triantaphyllou, E. and S.H. Mann, An evaluation of the eigenvalue approach for determining the membership values in fuzzy sets, *Fuzzy Sets and Systems* 35 (1990) 295-301.

[250] Verdegay, J.L., Fuzzy mathematical programming, in [BM15] (1982) 231-236.

[251] Verdegay, J.L., A dual approach to solve the fuzzy linear programming problem, *Fuzzy Sets and Systems* 14 (1984) 131-141.

[252] Verdegay, J.L., Application of fuzzy optimization in operational research, *Control and Cybernetics* 13 (1984) 229-239.

[253] Wagenknecht, M. and K. Hartmann, On the existence of minimal solutions for fuzzy equations with tolerances, *Fuzzy Sets and Systems* 34 (1990) 237-244.

[254] Wang, S. and Y.Z. Lu, Knowledge-based successive linear programming for linearly constraint optimization, *International Journal of Man-Machine Studies* 29 (1988) 625-636.

[255] Wang, Z. and Y. Dang, Application of fuzzy decision making and fuzzy linear programming in prediction and planning of animal husbandry system in farming region, in [BM52] (1984) 563-566.

[256] Ward, T.L., Fuzzy discounted cash flow analysis, in [BM14] (1989) 91-102.

[257] Werners, B., Interactive multiple objective programming subject to flexible constraints, *European Journal of Operational Research* 31 (1987) 342-349.

[258] Werners, B., An interactive fuzzy programming system, *Fuzzy Sets and Systems* 23 (1987) 131-147.

[259] Werners, B., Aggregation models in mathematical programming, in [BM39] (1988) 295-305.

[260] Wets, R.J.B., Stochastic programming, in G.L. Nemhauser, A.H.G. Rinnooy Kan and M.J. Todd (eds.), *Handbooks in Operations Research and Management Science: Optimization* (North-holland, Amsterdam, 1989) 573-629.

[261] Whalen, Th., Introduction to decision making under various kinds of uncertainty, in [BM26] (1987) 27-49.

[262] White, D.J., A bibliography on the applications of mathematical programming multiple-objective methods, *Journal of Operational Research Society* 41 (1990) 669-691.

[263] Wiedey, G. and H.-J. Zimmermann, Media selection and fuzzy linear programming, *Journal of Operational Research Society* 29 (1978) 1071-1084.

[264] Wierzchon, S.T., Linear programming with fuzzy sets: a general approach, *Mathematical Modelling* 9 (1987) 447-459.

[264a] Wierzchon, S.T., Randomness and fuzziness in a linear programming problem, in [BM27] (1988) 227-239.

[265] Wu, J.-Y., V. Van Brunt, W.-R. Zhang and J.C. Bezdek, Tower packing evaluation using linguistic variables, *Computers and Mathematics with Applications* 15 (1988) 863-869.

[266] Yager, R.R., Mathematical programming with fuzzy constraints and a preference on the objective, *Kybernetes* 8 (1979) 285-292.

[267] Yager, R.R., A measurement informational discussion of fuzzy union and intersection, *International Journal of Man-Machine Studies* 11 (1979) 189-200.

[268] Yager, R.R., Possibilistic decision making, *IEEE Transactions on Systems, Man, and Cybernetics* 9 (1979) 388-392.

[269] Yager, R.R., A foundation for a theory of possibility, *Journal of Cybernetics* 10 (1980) 177-204.

[270] Yager, R.R., Quantifiers in the formulation of multiple objective decision functions, *Information Sciences* 31 (1983) 107-139.

[271] Yazenin, A.V., Fuzzy and stochastic programming, *Fuzzy Sets and Systems* 22 (1987) 171-180.

[272] Yu, P.L., Dissolution of fuzziness for better decision - perspective and techniques, in [BM62] 171-207.

[273] Zadeh, L.A., From circuit theory to system theory, *Proceedings of Institute of Radio Engineering* 50 (1962) 856-865.

[274] Zadeh, L.A., Fuzzy sets, *Information and Control* 8 (1965) 338-353.

[275] Zadeh, L.A., Probability measures of fuzzy events, *Mathematical Analysis and Applications* 23 (1968) 421-427.

[276] Zadeh, L.A., Fuzzy algorithms, *Information and Control* 12 (1968) 94-102.

[277] Zadeh, L.A., Similarity relations and fuzzy orderings, *Information Sciences* 3 (1971) 177-200.

[278] Zadeh, L.A., A fuzzy-set-theoretic interpretation of linguistic hedges, *Journal of Cybernetics* 2 (1972) 4-34.

[279] Zadeh, L.A., Outline of a new approach to the analysis of complex systems and decision processes, *IEEE Transactions on Systems, Man, and Cybernetics* 3 (1973) 28-44.

[280] Zadeh, L.A., The concept of a linguistic variable and its application to approximate reasoning - part 1, *Information Sciences* 8 (1975) 199-249.

[281] Zadeh, L.A., The concept of a linguistic variable and its application to approximate reasoning - part 2, *Information Sciences* 8 (1975) 301-357.

[282] Zadeh, L.A., The concept of a linguistic variable and its application to approximate reasoning - part 3, *Information Sciences* 9 (1976) 43-80.

[283] Zadeh, L.A., A fuzzy-algorithmic approach to the definition of complex or imprecise concepts, *International Journal of Man-Machine Studies* 8 (1976) 249-291.

[284] Zadeh, L.A., Fuzzy sets as a basis for a theory of possibility, *Fuzzy Sets and Systems* 1 (1978) 3-28.

[285] Zadeh, L.A., PRUF - a meaning representation language for natural languages, *International Journal of Man-Machine Studies* 10 (1978) 395-460.

[286] Zadeh, L.A., A theory of approximate reasoning, *Machine Intelligence* 9 (1979) 149-194.

[287] Zadeh, L.A., The role of fuzzy logic in the management of uncertainty in expert systems, *Fuzzy Sets and Systems* 11 (1983) 199-227.

[288] Zadeh, L.A., A computational approach to fuzzy quantifiers in natural language, *Computers and Mathematics with Applications* 9 (1983) 149-184.

[289] Zadeh, L.A., Syllogistic reasoning in fuzzy logic and its application to usuality and reasoning with dispositions, *IEEE Transactions on Systems, Man, and Cybernetics* 15 (1985) 754-762.

[290] Zadeh, L.A., Test-score semantics as a basis for a computational approach to the representation of meaning, *Literary and Linguistic Computing* 1 (1986) 24-35.

[291] Zadeh, L.A., Outline of a theory of usuality based on fuzzy logic, in [BM34], 1988, 79-97.

[292] Zeleny, M., Multiobjective design of high-productivity systems, *Proceedings of Joint Automatic Control Conference*, Paper Appl9-4 (ASME, New York, 1976) 297-300.

[293] Zeleny, M., Optimal system design with multiple criteria: De Novo programming approach, *Engineering Cost and Production Economics* 10 (1986) 89-94.

[294] Zeleny, M., An external reconstruction approach (ERA) to linear programming, *Computers and Operations Research* 13 (1986) 95-100.

[295] Zeleny, M., Parallelism, integration, autocoordination and ambiguity in human support systems, in [BM16] (1988) 107-121.

[296] Zimmermann, H.-J., Description and optimization of fuzzy system, *International Journal of General System* 2 (1976) 209-216.

[297] Zimmermann, H.-J., Results of empirical studies in fuzzy set theory, in [BM34] (1978) 303-312.

[298] Zimmermann, H.-J., Fuzzy programming and linear programming with several objective functions, *Fuzzy Sets and Systems* 1 (1978) 45-55.

[299] Zimmermann, H.-J. and P. Zysno, Latent connectives in human decision making, *Fuzzy Sets and Systems* 4 (1980) 37-51.

[300] Zimmermann, H.-J., Fuzzy mathematical programming, *Computers and Operations Research* 10 (1983) 291-298.

[301] Zimmermann, H.-J., Using fuzzy sets in operational research, *European Journal of Operational Research* 13 (1983) 201-216.

[302] Zimmermann, H.-J. and P. Zysno, Decisions and evaluations by hierarchical aggregation of information, *Fuzzy Sets and Systems* 10 (1983) 244-260.

[303] Zimmermann, H.-J. and M.A. Pollatschek, Fuzzy 0-1 linear programming, in [BM62] (1984) 133-145.

[304] Zimmermann, H.-J., Application of fuzzy set theory to mathematical programming, *Information Sciences* 36 (1985) 29-58.

[305] Zimmermann, H.-J. and P. Zysno, Quantifying vagueness in decision models, *European Journal of Operational Research* 22 (1985) 148-158.

[306] Zimmermann, H.-J., Fuzzy sets theory and mathematical programming, in [BM23] (1985) 99-114.

[307] Zimmermann, H.-J., Address by first EURO gold metal recipient, *European Journal of Operational Research* 31 (1987) 173-176.

[308] Zimmermann, H.-J., Modeling and solving ill-structured problems in operations research, in [BM1] (1987) 217-239.

[309] Zimmermann, H.-J., Interactive decision support for semi-structured mathematical programming problems, in [BM39] (1988) 307-319.

[310] Zysno, P., The integration of concepts within judgmental and evaluative processes, in [BM50] (1982) 509-514.

Lecture Notes in Economics and Mathematical Systems

For information about Vols. 1–210
please contact your bookseller or Springer-Verlag

Vol. 211: P. van den Heuvel, The Stability of a Macroeconomic System with Quantity Constraints. VII, 169 pages. 1983.

Vol. 212: R. Sato and T. Nôno, Invariance Principles and the Structure of Technology. V, 94 pages. 1983.

Vol. 213: Aspiration Levels in Bargaining and Economic Decision Making. Proceedings, 1982. Edited by R. Tietz. VIII, 406 pages. 1983.

Vol. 214: M. Faber, H. Niemes und G. Stephan, Entropie, Umweltschutz und Rohstoffverbrauch. IX, 181 Seiten. 1983.

Vol. 215: Semi-Infinite Programming and Applications. Proceedings, 1981. Edited by A.V. Fiacco and K.O. Kortanek. XI, 322 pages. 1983.

Vol. 216: H.H. Müller, Fiscal Policies in a General Equilibrium Model with Persistent Unemployment. VI, 92 pages. 1983.

Vol. 217: Ch. Grootaert, The Relation Between Final Demand and Income Distribution. XIV, 105 pages. 1983.

Vol 218: P. van Loon, A Dynamic Theory of the Firm: Production, Finance and Investment. VII, 191 pages. 1983.

Vol. 219: E. van Damme, Refinements of the Nash Equilibrium Concept. VI. 151 pages. 1983.

Vol. 220: M. Aoki, Notes on Economic Time Series Analysis: System Theoretic Perspectives. IX, 249 pages. 1983.

Vol. 221: S. Nakamura, An Inter-Industry Translog Model of Prices and Technical Change for the West German Economy. XIV, 290 pages. 1984.

Vol. 222: P. Meier, Energy Systems Analysis for Developing Countries. VI, 344 pages. 1984.

Vol. 223: W. Trockel, Market Demand. VIII, 205 pages. 1984.

Vol. 224: M. Kiy, Ein disaggregiertes Prognosesystem fur die Bundesrepublik Deutschland. XVIII, 276 Seiten. 1984.

Vol. 225: T.R. von Ungern-Sternberg, Zur Analyse von Märkten mit unvollständiger Nachfragerinformaton. IX, 125 Seiten. 1984.

Vol. 226: Selected Topics in Operations Research and Mathematical Economics. Proceedings, 1963. Edited by G. Hammer and D. Pallaschke IX, 478 pages. 1984.

Vol. 227: Risk and Capital. Proceedings, 1983. Edited by G. Bamberg and K. Spremann VII, 306 pages. 1984.

Vol. 228: Nonlinear Models of Fluctuating Growth. Proceedings, 1983. Edited by R.M. Goodwin, M. Krüger and A. Vercelli. XVII, 277 pages. 1984.

Vol. 229: Interactive Decision Analysis. Proceedings, 1983. Edited by M. Grauer and A.P. Wierzbicki. VIII, 269 pages. 1984.

Vol. 230: Macro-Economic Planning with Conflicting Goals. Proceedings, 1982. Edited by M. Despontin, P. Nijkamp and J. Spronk. VI, 297 pages. 1984.

Vol. 231: G.F. Newell, The M/M/8 Service System with Ranked Servers in Heavy Traffic. XI, 126 pages. 1984.

Vol. 232: L. Bauwens, Bayesian Full Information Analysis of Simultaneous Equation Models Using Integration by Monte Carlo. VI, 114 pages. 1984.

Vol. 233: G. Wagenhals, The World Copper Market. XI, 190 pages. 1984.

Vol. 234: B.C. Eaves, A Course in Triangulations for Solving Equations with Deformations. III, 302 pages. 1984.

Vol. 235: Stochastic Models in Reliability Theory Proceedings, 1984. Edited by S. Osaki and Y. Hatoyama. VII, 212 pages. 1984.

Vol. 236: G. Gandolfo, P.C. Padoan, A Disequilibrium Model of Real and Financial Accumulation in an Open Economy. VI, 172 pages. 1984.

Vol. 237: Misspecification Analysis. Proceedings, 1983. Edited by T.K. Dijkstra. V, 129 pages. 1984.

Vol. 238: W. Domschke, A. Drexl, Location and Layout Planning. IV, 134 pages. 1985.

Vol. 239: Microeconomic Models of Housing Markets. Edited by K. Stahl. VII, 197 pages. 1985.

Vol. 240: Contributions to Operations Research. Proceedings, 1984. Edited by K. Neumann and D. Pallaschke. V, 190 pages. 1985.

Vol. 241: U. Wittmann, Das Konzept rationaler Preiserwartungen. XI, 310 Seiten. 1985.

Vol. 242: Decision Making with Multiple Objectives. Proceedings, 1984. Edited by Y.Y. Haimes and V. Chankong. XI, 571 pages. 1985.

Vol. 243: Integer Programming and Related Areas. A Classified Bibliography 1981–1984. Edited by R. von Randow. XX, 386 pages. 1985.

Vol. 244: Advances in Equilibrium Theory. Proceedings, 1984. Edited by C.D. Aliprantis, O. Burkinshaw and N.J. Rothman. II, 235 pages. 1985.

Vol. 245: J.E.M. Wilhelm, Arbitrage Theory. VII, 114 pages. 1985.

Vol. 246: P.W. Otter, Dynamic Feature Space Modelling, Filtering and Self-Tuning Control of Stochastic Systems. XIV, 177 pages.1985.

Vol. 247: Optimization and Discrete Choice in Urban Systems. Proceedings, 1983. Edited by B.G. Hutchinson, P. Nijkamp and M. Batty VI, 371 pages. 1985.

Vol. 248: Pural Rationality and Interactive Decision Processes. Proceedings, 1984. Edited by M. Grauer, M. Thompson and A.P. Wierzbicki. VI, 354 pages. 1985.

Vol. 249: Spatial Price Equilibrium: Advances in Theory, Computation and Application. Proceedings, 1984. Edited by P.T. Harker. VII, 277 pages. 1985.

Vol. 250: M. Roubens, Ph. Vincke, Preference Modelling. VIII, 94 pages. 1985.

Vol. 251: Input-Output Modeling. Proceedings, 1984. Edited by A. Smyshlyaev. VI, 261 pages. 1985.

Vol. 252: A. Birolini, On the Use of Stochastic Processes in Modeling Reliability Problems. VI, 105 pages. 1985.

Vol. 253: C. Withagen, Economic Theory and International Trade in Natural Exhaustible Resources. VI, 172 pages. 1985.

Vol. 254: S. Müller, Arbitrage Pricing of Contingent Claims. VIII, 151 pages. 1985.

Vol. 255: Nondifferentiable Optimization: Motivations and Applications. Proceedings, 1984. Edited by V.F. Demyanov and D. Pallaschke. VI, 350 pages. 1985.

Vol. 256: Convexity and Duality in Optimization. Proceedings, 1984. Edited by J. Ponstein. V, 142 pages. 1985.

Vol. 257: Dynamics of Macrosystems. Proceedings, 1984. Edited by J.-P. Aubin, D. Saari and K. Sigmund. VI, 280 pages. 1985.

Vol. 258: H. Funke, Eine allgemeine Theorie der Polypol- und Oligopolpreisbildung. III, 237 pages. 1985.

Vol. 259: Infinite Programming. Proceedings, 1984. Edited by E.J. Anderson and A.B. Philpott. XIV, 244 pages. 1985.

Vol. 260: H.-J. Kruse, Degeneracy Graphs and the Neighbourhood Problem. VIII, 128 pages. 1986.

Vol. 261: Th.R. Gulledge, Jr., N.K. Womer, The Economics of Made-to-Order Production. VI, 134 pages. 1986.

Vol. 262: H.U. Buhl, A Neo-Classical Theory of Distribution and Wealth. V, 146 pages. 1986.

Vol. 263: M. Schäfer, Resource Extraction and Market Struucture. XI, 154 pages. 1986.

Vol. 264: Models of Economic Dynamics. Proceedings, 1983. Edited by H.F. Sonnenschein. VII, 212 pages. 1986.

Vol. 265: Dynamic Games and Applications in Economics. Edited by T. Basar. IX, 288 pages. 1986.

Vol. 266: Multi-Stage Production Planning and Inventory Control. Edited by S. Axsäter, Ch. Schneeweiss and E. Silver. V, 264 pages.1986.

Vol. 267: R. Bemelmans, The Capacity Aspect of Inventories. IX, 165 pages. 1986.

Vol. 268: V. Firchau, Information Evaluation in Capital Markets. VII, 103 pages. 1986.

Vol. 269: A. Borglin, H. Keiding, Optimality in Infinite Horizon Economies. VI, 180 pages. 1986.

Vol. 270: Technological Change, Employment and Spatial Dynamics. Proceedings, 1985. Edited by P. Nijkamp. VII, 466 pages. 1986.

Vol. 271: C. Hildreth, The Cowles Commission in Chicago, 1939–1955. V, 176 pages. 1986.

Vol. 272: G. Clemenz, Credit Markets with Asymmetric Information. VIII,212 pages. 1986.

Vol. 273: Large-Scale Modelling and Interactive Decision Analysis. Proceedings, 1985. Edited by G. Fandel, M. Grauer, A. Kurzhanski and A.P. Wierzbicki. VII, 363 pages. 1986.

Vol. 274: W.K. Klein Haneveld, Duality in Stochastic Linear and Dynamic Programming. VII, 295 pages. 1986.

Vol. 275: Competition, Instability, and Nonlinear Cycles. Proceedings, 1985. Edited by W. Semmler. XII, 340 pages. 1986.

Vol. 276: M.R. Baye, D.A. Black, Consumer Behavior, Cost of Living Measures, and the Income Tax. VII, 119 pages. 1986.

Vol. 277: Studies in Austrian Capital Theory, Investment and Time. Edited by M. Faber. VI, 317 pages. 1986.

Vol. 278: W.E. Diewert, The Measurement of the Economic Benefits of Infrastructure Services. V, 202 pages. 1986.

Vol. 279: H.-J. Büttler, G. Frei and B. Schips, Estimation of Disequilibrium Modes. VI, 114 pages. 1986.

Vol. 280: H.T. Lau, Combinatorial Heuristic Algorithms with FORTRAN. VII, 126 pages. 1986.

Vol. 281: Ch.-L. Hwang, M.-J. Lin, Group Decision Making under Multiple Criteria. XI, 400 pages. 1987.

Vol. 282: K. Schittkowski, More Test Examples for Nonlinear Programming Codes. V, 261 pages. 1987.

Vol. 283: G. Gabisch, H.-W. Lorenz, Business Cycle Theory. VII, 229 pages. 1987.

Vol. 284: H. Lütkepohl, Forecasting Aggregated Vector ARMA Processes. X, 323 pages. 1987.

Vol. 285: Toward Interactive and Intelligent Decision Support Systems. Volume 1. Proceedings, 1986. Edited by Y. Sawaragi, K. Inoue and H. Nakayama. XII, 445 pages. 1987.

Vol. 286: Toward Interactive and Intelligent Decision Support Systems. Volume 2. Proceedings, 1986. Edited by Y. Sawaragi, K. Inoue and H. Nakayama. XII, 450 pages. 1987.

Vol. 287: Dynamical Systems. Proceedings, 1985. Edited by A.B. Kurzhanski and K. Sigmund. VI, 215 pages. 1987.

Vol. 288: G.D. Rudebusch, The Estimation of Macroeconomic Disequilibrium Models with Regime Classification Information. VII,128 pages. 1987.

Vol. 289: B.R. Meijboom, Planning in Decentralized Firms. X, 168 pages. 1987.

Vol. 290: D.A. Carlson, A. Haurie, Infinite Horizon Optimal Control. XI, 254 pages. 1987.

Vol. 291: N. Takahashi, Design of Adaptive Organizations. VI, 140 pages. 1987.

Vol. 292: I. Tchijov, L. Tomaszewicz (Eds.), Input-Output Modeling. Proceedings, 1985. VI, 195 pages. 1987.

Vol. 293: D. Batten, J. Casti, B. Johansson (Eds.), Economic Evolution and Structural Adjustment. Proceedings, 1985. VI, 382 pages.

Vol. 294: J. Jahn, W. Knabs (Eds.), Recent Advances and Historical Development of Vector Optimization. VII, 405 pages. 1987.

Vol. 295. H. Meister, The Purification Problem for Constrained Games with Incomplete Information. X, 127 pages. 1987.

Vol. 296: A. Börsch-Supan, Econometric Analysis of Discrete Choice. VIII, 211 pages. 1987.

Vol. 297: V. Fedorov, H. Läuter (Eds.), Model-Oriented Data Analysis. Proceedings, 1987. VI, 239 pages. 1988.

Vol. 298: S.H. Chew, Q. Zheng, Integral Global Optimization. VII, 179 pages. 1988.

Vol. 299: K. Marti, Descent Directions and Efficient Solutions in Discretely Distributed Stochastic Programs. XIV, 178 pages. 1988.

Vol. 300: U. Derigs, Programming in Networks and Graphs. XI, 315 pages. 1988.

Vol. 301: J. Kacprzyk, M. Roubens (Eds.), Non-Conventional Preference Relations in Decision Making. VII, 155 pages. 1988.

Vol. 302: H.A. Eiselt, G. Pederzoli (Eds.), Advances in Optimization and Control. Proceedings, 1986. VIII, 372 pages. 1988.

Vol. 303: F.X. Diebold, Empirical Modeling of Exchange Rate Dynamics. VII, 143 pages. 1988.

Vol. 304: A. Kurzhanski, K. Neumann, D. Pallaschke (Eds.), Optimization, Parallel Processing and Applications. Proceedings, 1987. VI, 292 pages. 1988.

Vol. 305: G.-J.C.Th. van Schijndel, Dynamic Firm and Investor Behaviour under Progressive Personal Taxation. X, 215 pages.1988.

Vol. 306: Ch. Klein, A Static Microeconomic Model of Pure Competition. VIII, 139 pages. 1988.

Vol. 307: T.K. Dijkstra (Ed.), On Model Uncertainty and its Statistical Implications. VII, 138 pages. 1988.

Vol. 308: J.R. Daduna, A. Wren (Eds.), Computer-Aided Transit Scheduling. VIII, 339 pages. 1988.

Vol. 309: G. Ricci, K. Velupillai (Eds.), Growth Cycles and Multisectoral Economics: the Goodwin Tradition. III, 126 pages. 1988.

Vol. 310: J. Kacprzyk, M. Fedrizzi (Eds.), Combining Fuzzy Imprecision with Probabilistic Uncertainty in Decision Making. IX, 399 pages. 1988.

Vol. 311: R. Färe, Fundamentals of Production Theory. IX, 163 pages. 1988.

Vol. 312: J. Krishnakumar, Estimation of Simultaneous Equation Models with Error Components Structure. X, 357 pages. 1988.

Vol. 313: W. Jammernegg, Sequential Binary Investment Decisions. VI, 156 pages. 1988.

Vol. 314: R. Tietz, W. Albers, R. Selten (Eds.), Bounded Rational Behavior in Experimental Games and Markets. VI, 368 pages. 1988.

Vol. 315: I. Orishimo, G.J.D. Hewings, P. Nijkamp (Eds), Information Technology: Social and Spatial Perspectives. Proceedings 1986. VI, 268 pages. 1988.

Vol. 316: R.L. Basmann, D.J. Slottje, K. Hayes, J.D. Johnson, D.J. Molina, The Generalized Fechner-Thurstone Direct Utility Function and Some of its Uses. VIII, 159 pages. 1988.

Vol. 317: L. Bianco, A. La Bella (Eds.), Freight Transport Planning and Logistics. Proceedings, 1987. X, 568 pages. 1988.

Vol. 318: T. Doup, Simplicial Algorithms on the Simplotope. VIII, 262 pages. 1988.

Vol. 319: D.T. Luc, Theory of Vector Optimization. VIII, 173 pages. 1989.

Vol. 320: D. van der Wijst, Financial Structure in Small Business. VII, 181 pages. 1989.

Vol. 321: M. Di Matteo, R.M. Goodwin, A. Vercelli (Eds.), Technological and Social Factors in Long Term Fluctuations. Proceedings. IX, 442 pages. 1989.

Vol. 322: T. Kollintzas (Ed.), The Rational Expectations Equilibrium Inventory Model. XI, 269 pages. 1989.

Vol. 323: M.B.M. de Koster, Capacity Oriented Analysis and Design of Production Systems. XII, 245 pages. 1989.

Vol. 324: I.M. Bomze, B.M. Pötscher, Game Theoretical Foundations of Evolutionary Stability. VI, 145 pages. 1989.

Vol. 325: P. Ferri, E. Greenberg, The Labor Market and Business Cycle Theories. X, 183 pages. 1989.

Vol. 326: Ch. Sauer, Alternative Theories of Output, Unemployment, and Inflation in Germany: 1960–1985. XIII, 206 pages. 1989.

Vol. 327: M. Tawada, Production Structure and International Trade. V, 132 pages. 1989.

Vol. 328: W. Güth, B. Kalkofen, Unique Solutions for Strategic Games. VII, 200 pages. 1989.

Vol. 329: G. Tillmann, Equity, Incentives, and Taxation. VI, 132 pages. 1989.

Vol. 330: P.M. Kort, Optimal Dynamic Investment Policies of a Value Maximizing Firm. VII, 185 pages. 1989.

Vol. 331: A. Lewandowski, A.P. Wierzbicki (Eds.), Aspiration Based Decision Support Systems. X, 400 pages. 1989.

Vol. 332: T.R. Gulledge, Jr., L.A. Litteral (Eds.), Cost Analysis Applications of Economics and Operations Research. Proceedings. VII, 422 pages. 1989.

Vol. 333: N. Dellaert, Production to Order. VII, 158 pages. 1989.

Vol. 334: H.-W. Lorenz, Nonlinear Dynamical Economics and Chaotic Motion. XI, 248 pages. 1989.

Vol. 335: A.G. Lockett, G. Islei (Eds.), Improving Decision Making in Organisations. Proceedings. IX, 606 pages. 1989.

Vol. 336: T. Puu, Nonlinear Economic Dynamics. VII, 119 pages. 1989.

Vol. 337: A. Lewandowski, I. Stanchev (Eds.), Methodology and Software for Interactive Decision Support. VIII, 309 pages. 1989.

Vol. 338: J.K. Ho, R.P. Sundarraj, DECOMP: an Implementation of Dantzig-Wolfe Decomposition for Linear Programming. VI, 206 pages.

Vol. 339: J. Terceiro Lomba, Estimation of Dynamic Econometric Models with Errors in Variables. VIII, 116 pages. 1990.

Vol. 340: T. Vasko, R. Ayres, L. Fontvieille (Eds.), Life Cycles and Long Waves. XIV, 293 pages. 1990.

Vol. 341: G.R. Uhlich, Descriptive Theories of Bargaining. IX, 165 pages. 1990.

Vol. 342: K. Okuguchi, F. Szidarovszky, The Theory of Oligopoly with Multi-Product Firms. V, 167 pages. 1990.

Vol. 343: C. Chiarella, The Elements of a Nonlinear Theory of Economic Dynamics. IX, 149 pages. 1990.

Vol. 344: K. Neumann, Stochastic Project Networks. XI, 237 pages. 1990.

Vol. 345: A. Cambini, E. Castagnoli, L. Martein, P Mazzoleni, S. Schaible (Eds.), Generalized Convexity and Fractional Programming with Economic Applications. Proceedings, 1988. VII, 361 pages. 1990.

Vol. 346: R. von Randow (Ed.), Integer Programming and Related Areas. A Classified Bibliography 1984–1987. XIII, 514 pages. 1990.

Vol. 347: D. Ríos Insua, Sensitivity Analysis in Multi-objective Decision Making. XI, 193 pages. 1990.

Vol. 348: H. Störmer, Binary Functions and their Applications. VIII, 151 pages. 1990.

Vol. 349: G.A. Pfann, Dynamic Modelling of Stochastic Demand for Manufacturing Employment. VI, 158 pages. 1990.

Vol. 350: W.-B. Zhang, Economic Dynamics. X, 232 pages. 1990.

Vol. 351: A. Lewandowski, V. Volkovich (Eds.), Multiobjective Problems of Mathematical Programming. Proceedings, 1988. VII, 315 pages. 1991.

Vol. 352: O. van Hilten, Optimal Firm Behaviour in the Context of Technological Progress and a Business Cycle. XII, 229 pages. 1991.

Vol. 353: G. Ricci (Ed.), Decision Processes in Economics. Proceedings, 1989. III, 209 pages 1991.

Vol. 354: M. Ivaldi, A Structural Analysis of Expectation Formation. XII, 230 pages. 1991.

Vol. 355: M. Salomon. Deterministic Lotsizing Models for Production Planning. VII, 158 pages. 1991.

Vol. 356: P. Korhonen, A. Lewandowski, J . Wallenius (Eds.), Multiple Criteria Decision Support. Proceedings, 1989. XII, 393 pages. 1991.

Vol. 357: P. Zörnig, Degeneracy Graphs and Simplex Cycling. XV, 194 pages. 1991.

Vol. 358: P. Knottnerus, Linear Models with Correlated Disturbances. VIII, 196 pages. 1991.

Vol. 359: E. de Jong, Exchange Rate Determination and Optimal Economic Policy Under Various Exchange Rate Regimes. VII, 270 pages. 1991.

Vol. 360: P. Stalder, Regime Translations, Spillovers and Buffer Stocks. VI, 193 pages . 1991.

Vol. 361: C. F. Daganzo, Logistics Systems Analysis. X, 321 pages. 1991.

Vol. 362: F. Gehrels, Essays In Macroeconomics of an Open Economy. VII, 183 pages. 1991.

Vol. 363: C. Puppe, Distorted Probabilities and Choice under Risk. VIII, 100 pages . 1991

Vol. 364: B. Horvath, Are Policy Variables Exogenous? XII, 162 pages. 1991.

Vol. 365: G. A. Heuer, U. Leopold-Wildburger. Balanced Silverman Games on General Discrete Sets. V, 140 pages. 1991.

Vol. 366: J. Gruber (Ed.), Econometric Decision Models. Proceedings, 1989. VIII, 636 pages. 1991.

Vol. 367: M. Grauer, D. B. Pressmar (Eds.), Parallel Computing and Mathematical Optimization. Proceedings. V, 208 pages. 1991.

Vol. 368: M. Fedrizzi, J. Kacprzyk, M. Roubens (Eds.), Interactive Fuzzy Optimization. VII, 216 pages. 1991.

Vol. 369: R. Koblo, The Visible Hand. VIII, 131 pages.1991.

Vol. 370: M. J. Beckmann, M. N. Gopalan, R. Subramanian (Eds.), Stochastic Processes and their Applications. Proceedings, 1990. XLI, 292 pages. 1991.

Vol. 371: A. Schmutzler, Flexibility and Adjustment to Information in Sequential Decision Problems. VIII, 198 pages. 1991.

Vol. 372: J. Esteban, The Social Viability of Money. X, 202 pages. 1991.

Vol. 373: A. Billot, Economic Theory of Fuzzy Equilibria. XIII, 164 pages. 1992.

Vol. 374: G. Pflug, U. Dieter (Eds.), Simulation and Optimization. Proceedings, 1990. X, 162 pages. 1992.

Vol. 375: S.-J. Chen, Ch.-L. Hwang, Fuzzy Multiple Attribute Decision Making. XII, 536 pages. 1992.

Vol. 376: K.-H. Jöckel, G. Rothe, W. Sendler (Eds.), Bootstrapping and Related Techniques. Proceedings, 1990. VIII, 247 pages. 1992.

Vol. 377: A. Villar, Operator Theorems with Applications to Distributive Problems and Equilibrium Models. XVI, 160 pages. 1992.

Vol. 378: W. Krabs, J. Zowe (Eds.), Modern Methods of Optimization. Proceedings, 1990. VIII, 348 pages. 1992.

Vol. 379: K. Marti (Ed.), Stochastic Optimization. Proceedings, 1990. VII, 182 pages. 1992.

Vol. 380: J. Odelstad, Invariance and Structural Dependence. XII, 245 pages. 1992.

Vol. 381: C. Giannini, Topics in Structural VAR Econometrics. XI, 131 pages. 1992.

Vol. 382: W. Oettli, D. Pallaschke (Eds.), Advances in Optimization. Proceedings, 1991. X, 527 pages. 1992.

Vol. 383: J. Vartiainen, Capital Accumulation in a Corporatist Economy. VII, 177 pages. 1992.

Vol. 384: A. Martina, Lectures on the Economic Theory of Taxation. XII, 313 pages. 1992.

Vol. 385: J. Gardeazabal, M. Regúlez, The Monetary Model of Exchange Rates and Cointegration. X, 194 pages. 1992.

Vol. 386: M. Desrochers, J.-M. Rousseau (Eds.), Computer-Aided Transit Scheduling. Proceedings, 1990. XIII, 432 pages. 1992.

Vol. 387: W. Gaertner, M. Klemisch-Ahlert, Social Choice and Bargaining Perspectives on Distributive Justice. VIII, 131 pages. 1992.

Vol. 388: D. Bartmann, M. J. Beckmann, Inventory Control. XV, 252 pages. 1992.

Vol. 389: B. Dutta, D. Mookherjee, T. Parthasarathy, T. Raghavan, D. Ray, S. Tijs (Eds.), Game Theory and Economic Applications. Proceedings, 1990. ??, ?? pages. 1992.

Vol. 390: G. Sorger, Minimum Impatience Theorem for Recursive Economic Models. X, 162 pages. 1992.

Vol. 391: C. Keser, Experimental Duopoly Markets with Demand Inertia. X, 150 pages. 1992.

Vol. 392: K. Frauendorfer, Stochastic Two-Stage Programming. VIII, 228 pages. 1992.

Vol. 393: B. Lucke, Price Stabilization on World Agricultural Markets. XI, 274 pages. 1992.

Vol. 394: Y.-J. Lai, C.-L. Hwang, Fuzzy Mathematical Programming. XIII, 301 pages. 1992.

Vol. 395: G. Haag, U. Mueller, K. G. Troitzsch (Eds.), Economic Evolution and Demographic Change. XVI, 409 pages. 1992.